AF716968

Agricultural Soil Science

SCIENCES

Agronomy and Food Science, Field Directors –
Jack Legrand and Gilles Trystram

Agronomy, Subject Heads –
Yves Coquet, Marianne Le Bail, Jean Roger-Estrade

Agricultural Soil Science

Sustainable Management of Agricultural Soils

Coordinated by
Yves Coquet
Joël Michelin

iSTE

WILEY

First published 2025 in Great Britain and the United States by ISTE Ltd and John Wiley & Sons, Inc.

Apart from any fair dealing for the purposes of research or private study, or criticism or review, as permitted under the Copyright, Designs and Patents Act 1988, this publication may only be reproduced, stored or transmitted, in any form or by any means, with the prior permission in writing of the publishers, or in the case of reprographic reproduction in accordance with the terms and licenses issued by the CLA. Enquiries concerning reproduction outside these terms should be sent to the publishers at the undermentioned address:

ISTE Ltd
27-37 St George's Road
London SW19 4EU
UK

www.iste.co.uk

John Wiley & Sons, Inc.
111 River Street
Hoboken, NJ 07030
USA

www.wiley.com

© ISTE Ltd 2025
The rights of Yves Coquet and Joël Michelin to be identified as the authors of this work have been asserted by them in accordance with the Copyright, Designs and Patents Act 1988.

Any opinions, findings, and conclusions or recommendations expressed in this material are those of the author(s), contributor(s) or editor(s) and do not necessarily reflect the views of ISTE Group.

Library of Congress Control Number: 2024950286

British Library Cataloguing-in-Publication Data
A CIP record for this book is available from the British Library
ISBN 978-1-78945-185-6

ERC code:
PE10 Earth System Science
PE10_12 Sedimentology, soil science, palaeontology, earth evolution

Contents

Introduction

Yves COQUET and Joël MICHELIN
UMR ECOSYS, AgroParisTech, INRAE, Université Paris Saclay, Palaiseau, France

Toward the multifunctional usage of agricultural soils

Historically, the study of soils has been inseparable from that of agriculture. Backed by a utilitarian approach and the scientific renewal of the 18th century, its initial focus was on understanding the determining factors of agricultural soil fertility on the basis of its two founding disciplines: soil chemistry and agricultural hydraulics. The end of the 19th century saw the emergence of a naturalistic approach to soils with the birth of pedology. These two methods of soil study continued until the end of the 20th century, particularly in the teachings of the agricultural colleges in France. During the second half of the 20th century, the emergence of several environmental challenges brought to light the fact that soils have other functions as well as ensuring sufficient plant production. Today, soil is at the heart of new challenges, such as climate regulation, pollution control, water flow regulation, waste management and the maintenance of biodiversity. The broad-reaching nature of these issues has led soil science to evolve from a disciplinary approach divided into subdisciplines (soil chemistry, soil biology, pedology, etc.) to a truly trans-disciplinary approach, examining the interrelations between the different functions of the soil and drawing on other disciplines (ecology, economics, sociology, etc.). This evolution is reflected in an increased integration and globalization of knowledge regarding soils, supported by the use of systemic approaches and increasingly effective computer software.

Agricultural Soil Science,
coordinated by Yves COQUET and Joël MICHELIN. © ISTE Ltd 2025.

In recent decades, training on soils in agricultural higher education programs have changed significantly, from an approach essentially centered on plant production capabilities (in agriculture and forestry) – heir to 19th century agricultural chemistry – to a multifunctional approach to soils. This paradigm shift owes a great deal to the work of ecologists, and in particular to the work carried out during the Millennium Ecosystem Assessment at the very beginning of the 21st century. Beyond the multifunctional approach of soils, this work has highlighted the many ecosystem services provided by soils, in addition to agricultural, forest and pasture production services. For the past 20 years, the evolution of soil-related training in higher education has followed this dynamic, particularly at AgroParisTech, France, with greater integration with other disciplines, more attention paid to functions other than ones related to agriculture and forestry production, and a more prominent role given to the assessment of ecosystem services provided by soils. The goal of this book is to illustrate this evolution, focusing on a limited number of themes.

Overview of the book

Chapter 1, addressing *soil tillage* and its impact on the structure of the soil, demonstrates the role played by plowing and the consequences of its elimination in the context of conservation agriculture. Chapter 2 focuses on the impacts of agricultural practices on soil *biodiversity* and the mechanisms for managing this biodiversity. Chapter 3 concerns the study of the *spatial variability* and the *mapping* of agricultural soils, with an update on the information available on soils in France and around the world, as well as examinations on the use of relatively recent tools such as drones. Chapter 4 is devoted to soil *erosion* in connection with agricultural activities and the techniques for remedying it through integrated watershed management. Chapter 5 presents the *legal aspects* of soil protection and agricultural land. Chapter 6 concerns the methods for *estimating the agronomic value* of soils, while Chapter 7 looks at the application of the notion of *ecosystem services* to soils. The last three chapters are devoted to specific types of soils: *Mediterranean agricultural soils* (Chapter 8), *tropical soils* (Chapter 9) and *urban agricultural soils* (Chapter 10). The chapters that comprise this work cover only a small part of the entirety of the topics relating to agricultural soil management. Rather, the goal is to demonstrate how some of these issues are now addressed within the framework of higher education as it is practiced today, and in particular at AgroParisTech and at Paris-Saclay University, where the majority of the contributors teach. We hope that these elements can be useful to all readers who wish to bring their knowledge up-to-date on the topics covered here, or simply to become familiar with them.

1

Tillage and Structure of Agricultural Soils

Yves COQUET[1] and Lionel ALLETTO[2]

[1]*UMR ECOSYS, AgroParisTech, INRAE, Université Paris Saclay, Palaiseau, France*

[2]*UMR AGIR, INRAE, Université de Toulouse, Castanet Tolosan, France*

1.1. What is tillage? Why should soils be "tilled"?

Soil tillage likely dates back to the same time as sedentarization of human societies during the Neolithic age, as the need arose to transform the natural environment to allow for agricultural activities. The first stage of this settlement process consisted of clearing or deforesting a given location, often accompanied by the burning of vegetation – a process which still occurs today in the form of "slash-and-burn" practices. Thus, it is conceivable that a major concern for Neolithic farmers was to maintain the gains derived from slash-and-burn techniques, namely, a "clean" plot – a term still used today by many farmers, referring to a plot where no vegetation grows other than what has been planted for cultivation. This brings to light one of the primary functions of tillage: to fight against weeds.

After clearing comes the sowing of the crops, and then the second function of tillage: the burying of seeds. Historically, the plow was used primarily to sow seeds (Sewell 1919; Sigaut 1977). The first tools for soil tillage were made from wood, some of which featured very simple designs, such as the use of simple sticks for burying seeds or the well-known African daba. These appeared at the same time as agriculture: the first appearance of ox-drawn plows dates back to 3000 BCE in Egypt and Mesopotamia. The first iron tools date back to the Roman Empire (around 200 BCE). Over time, the plow was gradually improved by adding elements such as

Agricultural Soil Science,
coordinated by Yves COQUET and Joël MICHELIN. © ISTE Ltd 2025.

wheels and moldboards, until the advent of the tractor around the middle of the 19th century, which increased the farmer's capacity to work the soil tenfold.

Figure 1.1. *Image showing the tillage process in ancient Egypt at the tomb of Sennedjem, 13th century BCE, Deir el Medineh, Egypt (source: The Yorck Project (2002)). For a color version of this figure, see www.iste.co.uk/coquet/agricultural.zip*

In today's highly mechanized agriculture, tillage is no longer used to sow seeds in the ground. Instead, it is used to prepare the soil to facilitate the installation of the crop, burying the seeds directly in the prepared soil using such techniques being carried out by tools specially designed for sowing (seeders). The function of burying weeds continues, but also concerns the organic materials present on the surface of the soil, whether from the previous harvest – thus referred to as "crop residues", such as grain stubble if it is not exported in the form of straw – or exogenous inputs, such as manure, slurry, composts and other organic waste products. Tillage makes it possible to incorporate these organic materials by mixing them over the full breadth of the surface layer of the soil being tilled, and thus to increase the organic matter content of this layer.

Finally, the fourth and final function of tillage is to modify the soil structure. Through their mechanical action on the soil, soil tillage tools increase and/or modify soil porosity. This porosity allows for better rooting of the plants and a better infiltration and circulation of the water in the soil. Tillage is a very effective technique for increasing the porosity of the soil by fragmenting the soil into structural elements (i.e. clods and aggregates) of various sizes. When these structural elements are too large in relation to the size of the planted seeds, the farmer may have to carry out "surface techniques", consisting of an additional tilling at a level shallower than the plowing. This creates a "finer" soil structure, consisting of smaller structural elements, of sizes suitable for the seeds that have been sown. This is referred to as a "seedbed". The objective of these tilling operations is to obtain a good soil structure ("a good tilth" as referred to by Kuipers (1963)), which will allow for the optimal development of the crop.

Figure 1.2. *Mechanized plowing, Grignon plateau, 2002. Crop residues (corn) located on the surface of the soil are buried by tillage (photo credit: C. Coutadeur)*

1.2. Soil structure

Soil structure describes the way in which the different particles that constitute the solid phase of the soil are arranged in physical space. These particles, whether of a mineral or organic nature, can vary greatly in size, ranging in scales from micrometers or smaller (as in the case of clays) to decimeters (coarse, "block"-type elements). The porosity of the soil is the space located between the soil particles. The size of the porosity can also vary to a large extent. The soil particles are connected to each other through forces of various different kinds. In particular, the bonds between mineral particles and organic particles play an important role in the determination and stability of the soil structure. Only part of the soil structure is visible to the naked eye (visual resolution is around 0.1 mm). This is the soil structure that is described by soil scientists or agronomists based on observation pits.

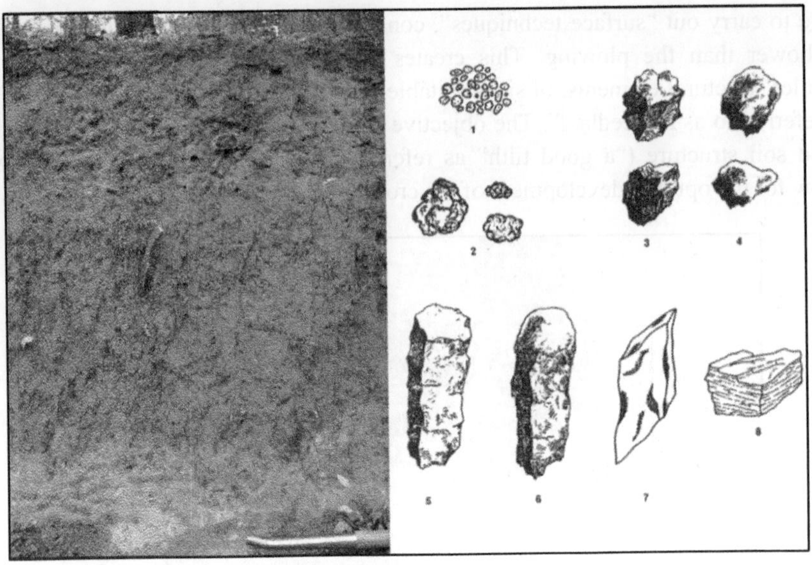

Figure 1.3. *Left: profile of forest soil (Luvic Cambisol, Grignon park – soil thickness: 1 m) (photo credit: L.M. Bresson). Right: types of soil structure (1: granular; 2: crumb; 3: angular blocky; 4: subangular blocky; 5: prismatic; 6: columnar; 7: wedgelike; 8: platy) (source: Baize and Jabiol 2011). For a color version of this figure, see www.iste.co.uk/coquet/agricultural.zip*

The structure of natural soils is the result of their pedogenesis (soil formation process), which involves mechanisms of rock weathering, in conjunction with the climate, the topography and the living organisms that are present. Through cultivation, the farmer modifies the natural structure of the soil (Or et al. 2021) and, above all, gives it a much more marked temporal dynamic. The search for an "optimal" soil structure for a given crop or a given cultivation system has led to the question of how to objectify this structure, particularly by seeking to "measure" this structure. This has been a problem for a long time, and to this day, a satisfactory answer has not been found.

1.2.1. *Porosity and bulk density*

As mentioned earlier, one of the objectives of plowing is to increase the porosity of the soil. This porosity, P (as a fraction of volume), corresponds to the volume of pores contained within a given volume of soil. When a plow is passed, it cuts and fragments the soil into clods (structural elements larger than 2 cm) and fine earth (structural elements of a size less than or equal to 2 cm). The result is a "fluffy" soil surface level, with a high roughness (Figure 1.2) and whose average surface level is slightly higher than the initial average surface level.

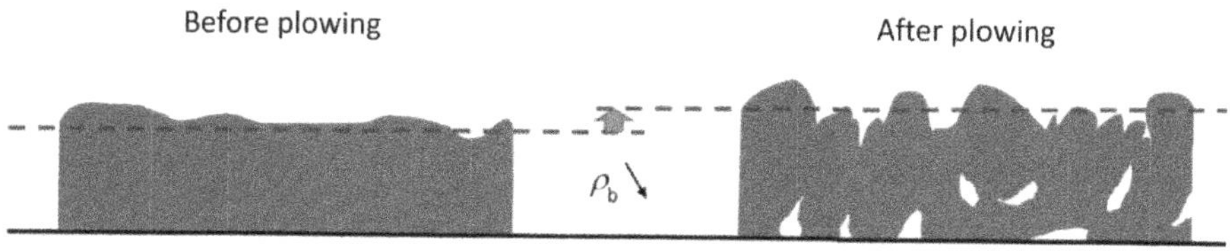

Figure 1.4. *Expansion effect of plowing on the soil surface layer. The average altitude of the soil surface (red dashed line) increases compared to the bottom of the plowed area (black solid line). The bulk density, ρ_b, of the plowed layer decreases. For a color version of this figure, see www.iste.co.uk/coquet/agricultural.zip*

The porosity of the surface layer is therefore increased by plowing. One indirect and very common way of measuring soil porosity consists of taking a known volume of soil (generally a cylinder) and measuring its dry mass (which is its mass after being dried in an oven at 105°C). The bulk density of the soil ρ_b (in kg.dm^{-3}) can therefore be obtained by:

$$\rho_b = \frac{M}{V},$$

where M is the dry mass of the soil sample and V, the volume of the sampling cylinder. If the density of the soil solid phase ρ_s is known, the porosity can then be calculated as:

$$P = 1 - \frac{\rho_b}{\rho_s}.$$

The density of the solid phase of a soil can be measured in the laboratory using a pycnometer, but for most soils, if they are not too organic, we can use the value of 2.65 kg.dm^{-3}, which is the same as that of quartz or calcite. Therefore, a soil with a bulk density of 1.5 kg.dm^{-3} will have a porosity of about 43%. When plowing a 25-cm thick soil layer with an initial bulk density of 1.5 kg.dm^{-3}, this results in a layer with a bulk density of 1.25 kg.dm^{-3}; the height of the average soil surface will increase by 5 cm and the porosity by 10 percentage points (increasing from approximately 43% to 53%). Similarly, any compaction of the soil, whether by mechanical devices or trampling by animals, may under certain conditions, cause a local lowering of the height of the soil surface and an increase in the bulk density of the soil.

In fact, bulk density has been and remains an effective measurement of soil structure. However, it is a very crude measurement that provides little information. Indeed, two soils of the same bulk density can have very different pore size distributions. The bulk density does not make it possible to determine whether the porosity consists of large or small pores (Gupta et al. 1989), nor does it offer any information about their level of connectivity, which plays a major role in the circulation of fluids in the soil.

1.2.2. *Soil structuring mechanisms*

Since natural soils have a structure, we will distinguish between natural mechanisms and anthropogenic mechanisms, all of which lead to changes in the soil structure. The natural mechanisms are linked to the action of the climate and that of living beings. These include the physical processes of freezing–thawing and shrinkage–swelling, as well as phenomena such as slaking and hardsetting. The processes that involve biological agents are aggregation and bioturbation (including root growth). Mechanisms of anthropogenic origin include tillage and settlement by agricultural machinery. Settlement can also occur due to the trampling of animals and can be therefore natural in origin. However, this phenomenon is often accentuated by human activity (by increasing the load of animals per hectare, or by the introduction of ungulates, as has occurred on the Australian continent).

1.2.2.1. *The freezing–thawing cycle*

Freeze–thaw cycles are a very effective way of changing the soil structure from large structural elements to smaller structural elements (crumbling). This is a phenomenon that is well known to farmers who, in regions where frost is significant, make use of it to refine the plowing they carry out at the beginning of winter on plots whose soil is particularly rich in clay. This allows energy to be saved for the preparation of the seedbed for spring crops.

The fragmentation that occurs in freeze–thaw cycles is due to the significant expansion of water in the soil when it freezes. This is incommensurable with the thermal expansion coefficients of the other constituents of the soil. When water shifts to a solid state, its volume increases by 9%, implying a linear expansion coefficient of around 3%, while by contrast this coefficient is around 10^{-6} m.m^{-1}.K^{-1} for the mineral constituents of the soil and 10^{-4} m.m^{-1}.K^{-1} for liquid water. Thus, the expansion of water when it freezes creates very significant stresses inside the pore space of the soil, and leads to ruptures within the structural elements, which then fragment into smaller elements.

The effectiveness of the freeze–thaw cycles decreases rapidly with depth, in parallel with changes in temperature. Their fragmentation effect hardly exceeds the depth of tillage in temperate climates. The freeze–thaw cycles play a prominent role in the fragmentation process, more than the seasonal changes in soil temperature (the depth at which water pipes are buried to prevent freezing is much greater than that of the soil fragmentation effect).

1.2.2.2. *Shrink–swell phenomena*

The other family of processes that have significant effects in modifying the soil structure are shrink–swell phenomena. When a soil dries out, its solid-phase particles may move closer to each other due to the effects of capillary forces, which become more and more intense when it is dried. This is known as soil "shrinkage". If the solid particles cannot move relative to each other (such as in the case of sandy soils), it is considered to be a "rigid" soil; otherwise, the soil is considered "deformable" (e.g. in the case of clay-rich Vertisol-type soils). Conversely, when the soil becomes moistened again, its volume may increase. This is referred to as the "swelling" of the soil.

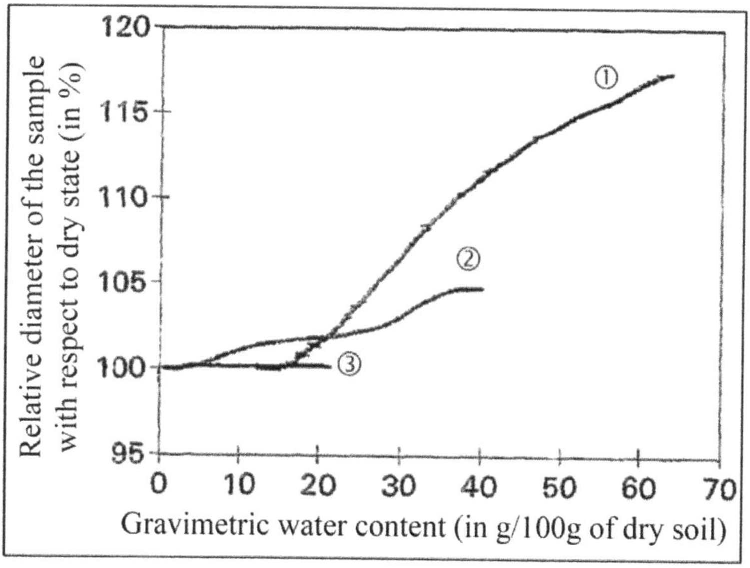

Figure 1.5. *Shrinkage curves of three soils (1: clay-rich soil – Vertisol; 2: loamy soil – Luvisol; 3: sandy soil – Ferralsol) (from Coquet 1996)*

Figure 1.5 shows the linear shrinkage curve of three types of soil: a Vertisol (from New Caledonia), a Luvisol (from Grignon, France) and a Ferralsol (from Senegal). These shrinkage curves were obtained by measuring the variation in diameter of a cylindrical soil sample while it was dried. The Vertisol, which contains swelling clays, has a linear shrinkage coefficient on the order of 17%, the Luvisol has a coefficient on the order of 5%, while the Ferralsol can be considered to be a rigid soil (linear shrinkage coefficient of <0.2%). When it dries, a Vertisol presents dramatic cracks, which close when it is re-humidified. Considerable pressure forces can appear along the faces of these closed cracks, generating slickensides that are typical of Vertisols (see item 7 in Figure 1.3). The magnitude of the shrink–swell phenomenon is much lower for loamy soils (such as Luvisols), but can nevertheless give rise to blocky or prismatic structures (3–5 in Figure 1.3). Since the changes in water content may affect great soil depths (particularly under forests), the shrink–swell changes concern the entire profile of the soil, unlike the freeze–thaw changes.

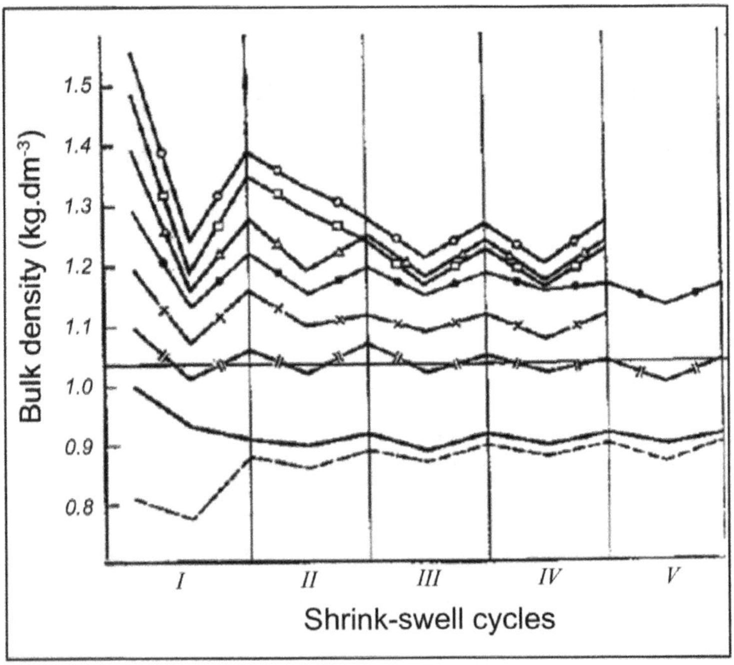

Figure 1.6. *Changes in bulk density of the tilled horizon of a remolded loamy soil during shrink–swell cycles for different initial bulk densities (as given by Kuznetsova and Danilova 1988)*

1.2.2.3. *Slaking and hardsetting*

In some particular cases, the moistening of a soil can lead not to swelling, but rather to a reduction in the soil's porosity, and therefore its apparent volume. This reduction is particularly striking in the case of the "hardsetting" of the soil. This hardsetting phenomenon is well known in the case of irrigation of salty clay soils. The abundance of sodium on the exchangeable complex of clays combined with the dilution occurring due to irrigation water leads to the dispersion of clay particles and a total loss of the soil structure. This phenomenon can also be observed to a lesser extent in the case of predominantly loamy soils (Bresson and Moran 2004) or even sandy soils (Lamotte et al. 1997) under specific conditions. The result of this hardsetting is a soil without any macrostructure, appearing as a continuous volume, which is difficult for water to penetrate.

Figure 1.7. *Cross-section of the surface of a loamy soil with a slaked crust (photo credit: C. Roth). For a color version of this figure, see www.iste.co.uk/coquet/agricultural.zip*

Slaking also causes the soil structure to disappear, but over a very limited depth of soil (up to a few millimeters), at the surface of the soil, under the effects of rain. This phenomenon is characteristic of loamy soils with a low clay content (<15%) and organic matter content. In the phenomenon of slaking, several mechanisms can play a role. These include the physical impact of raindrops when they fall on the surface of the soil, the bursting of the aggregates when the water suddenly penetrates them under the effects of capillary forces where air is unable to escape freely from the aggregates (dead-end pores), and the dispersion of clay particles.

In the case of slaking as well as in the case of hardsetting, the tillage of the soil must be accompanied by other corrective actions (such as the addition of organic matter to combat slaking or calcium, in the form of carbonates or sulfates, to avoid hardsetting) if it is desired for the soil to "recover" an adequate structure that will be relatively stable over time.

1.2.2.4. *Aggregation and bioturbation*

Living beings play a decisive role in the structuring of soils. Aggregation refers to the phenomenon in which the soil attains a structure that is visible to the naked eye, where it is possible to distinguish "aggregates", or structural elements, that vary in size (from a few tens of micrometers to a few millimeters), within which organic matter is closely bound to the mineral matter. The bonds between clay minerals and organic molecules, in particular those secreted by soil microorganisms, are particularly important in determining soil aggregation. For its part, bioturbation refers to the mixing effect of the solid phase of the soil resulting from the activities of the soil fauna, particularly earthworms.

In practice, it is difficult to distinguish aggregation and bioturbation effects, as the two phenomena frequently occur simultaneously, particularly near the surface of soils. Apart from castings from earthworms, particularly those left on the surface of the soils by anecic worms, bioturbation effects are rather visible in soil micromorphology, that is, through observing thin soil slides with a microscope, which allows for the identification of "papules" (bioturbation remains) within the soil (Sauzet et al. 2016). Vegetation also plays a very important role in the structuring of the soils, both through the mechanical actions due to the growth of the roots in the soil, as well as by stimulating the microbial processes by the contribution of organic materials by the roots (exudates or root decomposition).

1.2.2.5. *Compaction*

With the development of mechanized agriculture, the machines used for field work (tractors, combines, trailers for harvesting, etc.) have continuously improved, with the machines gaining in size and weight. This has given rise to an increase in the risk of soil compaction (or settlement), or even effects seen in the subsoil (in the agronomic sense, meaning the soil layer located under the tilled layer). Some authors consider that the increase in soil compaction may be one of the explanations for the stagnation in the yields observed in Europe since the end of the 1990s (Keller et al. 2019).

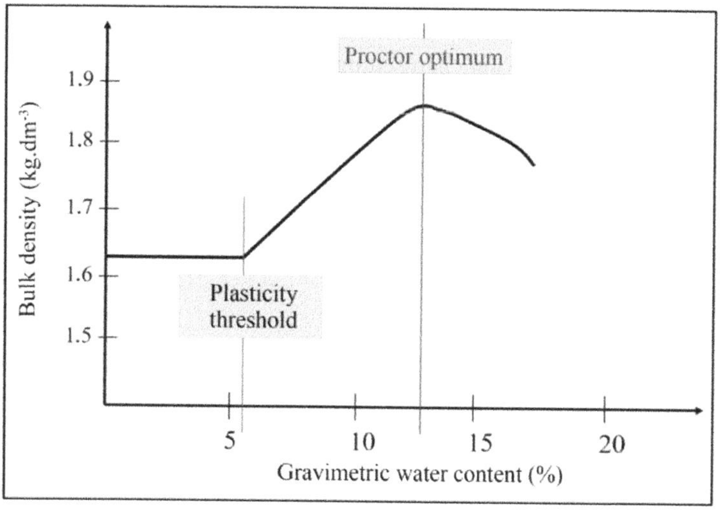

Figure 1.8. *Soil compaction curve as determined by the Proctor test (according to D. Tessier)*

Soil compaction depends on the mechanical stresses applied to it (pressure, shear), as well as on its own mechanical characteristics (shear strength and deformation capacity), which are in turn dependent on the properties and the state of the soils at the time when they are subjected to the stresses. In general, a sandy soil is much less sensitive to compaction than a loamy or clay-rich soil (O'Sullivan and Ball 1993; Unger and Kaspar 1994).

The soil water status at the time the stresses are applied is an essential factor in the compaction. A dry clay soil will be insensitive to compaction, while it will be much more sensitive if it is wet. In geotechnics, where a civil engineer seeks to obtain soils that are as stable as possible (and therefore as compacted as possible), the Proctor test makes it possible to define the optimal water content, which will enable the maximum bulk density to be achieved during compaction (Figure 1.8). This is exactly what farmers seek to avoid when they work their fields with agricultural machinery.

Soil compaction can also occur due to trampling by some of the heavier animals. Cattle and equine animals can generate stress levels of several hundred kPa under their hooves, stresses which are equivalent to (or greater than) those exerted by the heaviest machines in mechanized agriculture.

1.2.2.6. *Soil tillage*

The phenomena of slaking, hardsetting and above all compaction lead to a decrease in soil porosity and to a change in soil structure, which can pose a problem for farmers. When a slaked top layer is dry, this can form an impenetrable obstacle for sprouting plants, and thus prevent crops from emerging. For its part, soil compaction can seriously limit the rooting of the crop (Unger and Kaspar 1994). In order to recover a structural state of the soil surface layer that is favorable to the development of crops, farmers carry out tillage, which can be done in several stages using different tools.

The deepest type of tillage (with the exception of subsoiling) is plowing (Figure 1.2). This type of plowing has various objectives (see section 1.1), including that of recreating porosity inside the soil (Figure 1.4). In Europe, plowing is generally carried out using a moldboard plow (Figure 1.2), which makes it possible to cut a strip of soil and rotate it from 130° to 160°. By contrast, plowing in North America is carried out using a disc plow, which is less efficient than the moldboard plow in rotating the soil, but more efficient in mixing it. Currently, the depth of plowing in Europe is on the order of 20–25 cm, but in the past, it was possible for significantly deeper plowing to be carried out (of up to 35 cm or more). A fairly deep tillage can be carried out using tined tools, such as cultivators, which are used in particular for post-harvest stubble removal of grain straw. Rotary cultivators make it possible to obtain a more homogeneous relief of the soil surface than a cultivator with fixed tines.

Figure 1.9. *Tillage tools: (a) front cultivator; (b) rotary harrow; (c) subsoiler (photo credits: Pottinger, Quivogne). For a color version of this figure, see www.iste.co.uk/coquet/agricultural.zip*

In general, the soil structure created by plowing is not suitable for seeding a crop immediately after plowing. As indicated previously (section 1.2.2), the freeze–thaw and wet–dry cycles can contribute significantly to the fragmentation of large structural elements (clods) created by plowing and to a "finer" structure being obtained (that is to say, one consisting of smaller structural elements). An alternative (or complementary) solution to the effects of the climate consists of carrying out a "resumption" of the plowing using another tool that works at a shallower depth, which will make it possible to obtain a finer and above all more homogeneous structure. Various different tools, such as pulverisers (or "cover-crops") or harrows (fixed, alternating, or rotating), make it possible to effectively fragment the upper part of the plowed layer. This surface tillage of the soil can be completed using a roller, which will "recompact" the soil surface and thus promote better contact between the soil and the seeds that have been planted. The degree of fragmentation and settlement of the seedbed will depend on the size of the seed; for instance, a corn seedling will not need a seedbed as thin as a rapeseed seedling.

Certain tools, such as the cultivator (rotary harrow), are designed to remedy slaking when it appears. This is a superficial tillage (2–3 cm) that seeks to fragment the slaked crust and recreate the circulation of fluids at the surface of the soil. This tillage can also allow sprouts to rise if the slaked crust has formed immediately after seeding.

The last major category of tillage tools has the explicit objective of counteracting soil compaction. This category includes decompactors, which plow the soil to the tillage depth and can be used to decompact the soil after a harvest carried out in bad conditions (wet soil), which has led to soil compaction, and subsoilers, deeper tools, whose objective is to decompact the soil under the plowed layer.

As this section makes clear, there are a wide variety of tillage tools, which have only briefly been touched upon here. These tools are typical of highly capitalistic agriculture, since all such tools have a high cost and require large amounts of energy to be used. They are one of the factors that explain the tremendous productivity gains made by agriculture during the 19th and 20th centuries.

1.2.3. *A method for characterizing the structure of cultivated soils: the* "Profil cultural"

We have seen that soil structure can be quantified in a rudimentary fashion by measuring the soil's bulk density (section 1.2.1). This measurement is generally carried out on a volume of soil that barely exceeds 1 dm^3 and cannot account for the spatial and temporal variability of the structure of cultivated soils without multiplying the measurements over time and in space (Strudley et al. 2008). The original idea that the agronomist Hubert Manichon had for studying the structure of cultivated soils while preparing his thesis (Manichon 1982) was to observe the soil, using the methods of a soil scientist, but focusing on the effects of the different tools used during crop succession (Manichon 1987). The observation was carried out from a pit that was excavated perpendicular to the direction of movement of agricultural machinery on the plot, wide enough to include, at a minimum, the greatest working width of the machinery and a depth greater than the maximum depth of the soil to be tilled.

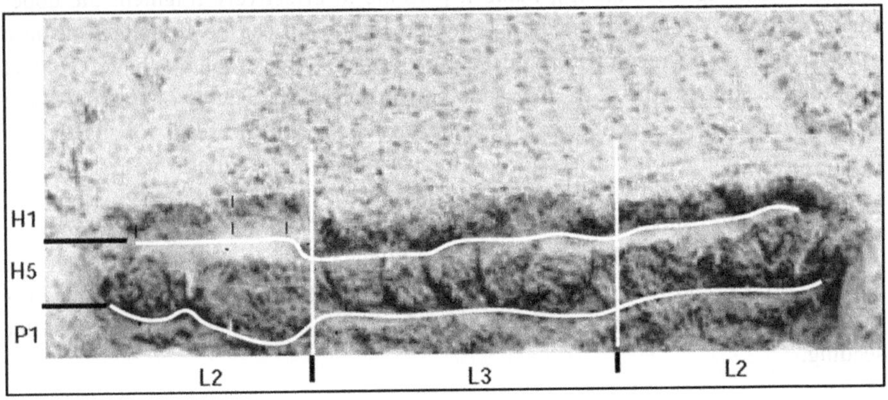

Figure 1.10. *Crop profile of a plot where corn has been sown (Grignon, France). H1: seedbed; H5: plowed layer not taken up by surface tillage; P1: soil under the tillage. L2: Path of the wheels of the tractor pulling the harrow for the creation of the seedbed; L3: area not affected by the wheel paths (as given by Desbourdes-Coutadeur 2002)*

A compartmentalization of the soil profile is first carried out, taking into consideration a vertical compartmentalization according to the depths of the different tillage tools. We may distinguish certain elements such as (given in order from the surface) the seedbed, the plowed layer not modified by superficial tillage, the plow pan, and the soil that is not tilled. Additionally, it also considers a lateral compartmentalization on the basis of the location of the paths of machines' wheels, which can generate soil compaction. The soil structure within each compartment is then described according to a purely morphological approach. The large structural elements (clods) are described geometrically and according to their "internal state". A state noted as "Γ" is distinguished, representing a soil with a macroporosity clearly visible to the naked eye, as well as a condition noted as "Δ", corresponding to a compacted soil without macroporosity, and a state as "Φ", corresponding to a state Δ transformed by the shrink–swell and freeze–thaw phenomena and with a more advanced state of cracking. Structural elements with a size of less than 2 cm are classified as "fine earth".

Figure 1.11. *The three internal structural states of the Profil cultural: (a) state Γ, (b) state Δ and (c) state Φ (as given by Desbourdes-Coutadeur 2002). For a color version of this figure, see www.iste.co.uk/coquet/agricultural.zip*

The description of the soil profile that is tilled according to the Profil cultural method makes it possible to situate the heterogeneity of the soil structure observed in relation to the various cropping actions (plowing, sowing, introduction of inputs, harvesting, etc.) and thus to carry out a diagnosis on the cultivation system and its impact on the soil structure. The proportion of state Δ has proven to be an interesting indicator for qualifying the structure of an agricultural soil, given the adverse impact of compaction on the development of crops (Roger-Estrade et al. 2000). It should be noted that the Profil cultural method, initially developed for cultivation systems that include plowing, has been adapted for systems that exclude plowing (Boizard et al. 2017). A number of methods that are "lighter" to implement have been developed from the Profil cultural method, in particular, the spade test, and more recently, the 3D profile method (Tomis et al. 2019).

1.3. Consequences of the different methods for tilling soil

In general, it is extremely difficult to demonstrate the systematic effects of different tillage practices. This can be explained by, on the one hand, the great diversity of tillage tools and the methods for using them, and, on the other hand, the diverse array of pedoclimatic situations – the same tillage tool can be used identically and have different effects depending on soil texture, its water status at the time when the tool is used, or its organic matter content. In addition, these effects can vary broadly in space and in time (Strudley et al. 2008; Alletto et al. 2010), making it difficult to clearly demonstrate their effects through measurements, which must take into account such differences. The “Profil cultural” method is an elegant possibility for giving consideration, at least partially, to the spatial component of these variations (see section 1.2.3).

For the rest of this chapter, we will narrow our focus to two main methods of tillage, which are widely studied in the literature and whose major trends can be identified: first, cultivation systems in which the tillage is achieved with a moldboard plow – characteristic of “conventional” agriculture as practiced in France; and second, “no-till” systems, which are typical of so-called “conservation” agriculture. It should be noted that conservation agriculture is characterized not merely by tillage-free agriculture, but also includes the permanent presence of a soil cover (whether in the form of crops or their residues) and a more elaborate crop rotation than in conventional agriculture (featuring in particular a diversification of the species that are cultivated).

1.3.1. *Systems that include tillage (“conventional” agriculture)*

In section 1.1, we made reference to the different functions of tillage. One of these is to increase soil porosity. This increase in porosity represents a decrease in the soil’s bulk density (section 1.2.1), but what are the functional components of this porosity? Figure 1.12 shows the water retention characteristics of the soil (a), which represents the volumetric water content of the soil in relation to the matric potential of the water, expressed in pF, and the hydraulic conductivity characteristics of the soil (b), which relates its hydraulic conductivity to the matric potential. The pF is the base-10 logarithm of the absolute value of the matric potential, expressed in hPa. The matric potential of water in the soil quantifies the energy with which water is retained by the soil. This is a negative value, with the value of zero representing free water. The lower the matric potential is (and thus the more positive the pF is), the higher the strength of the water retention by the soil.

In the case of the soil studied (silt loam soil in Grignon), the tillage increases the water retention of the soil; at a given pF value, excluding the values at 1 and 4.2, the volumetric water content of the tilled soil is higher than that of the non-tilled soil (under the tilled layer). On the other hand, there is hardly any difference between the tilled layer not affected by the surface tillage (in this case, carried out using a rotary harrow) and the seedbed. At 4.2 pF, the permanent wilting point of the plants, the water content of the tilled soil is identical to that of the non-tilled soil. On the other hand, the water content at the field capacity, which can be located between 2 and 2.5 pF, is higher in the tilled soil than in the non-tilled soil. As a result, the tillage has a slightly positive effect on the reserve of water usable by the plants, arising from the increase in porosity that results from the plowing.

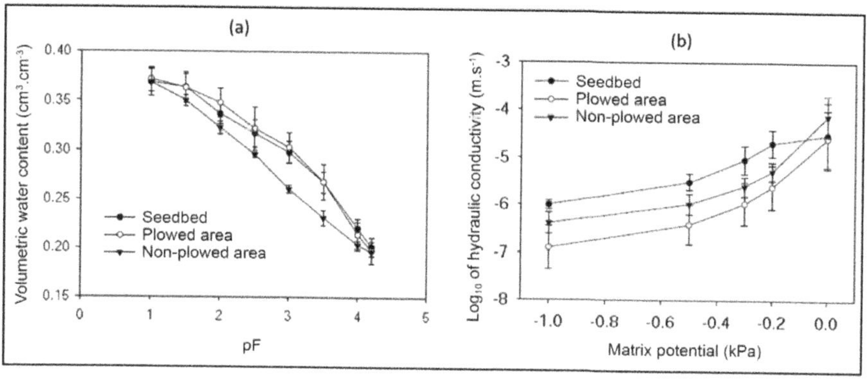

Figure 1.12. *Hydraulic properties of a silt loam soil seeded with corn (April 1999). (a): water retention characteristics; (b): hydraulic conductivity characteristics close to saturation (according to Desbourdes-Coutadeur 2002)*

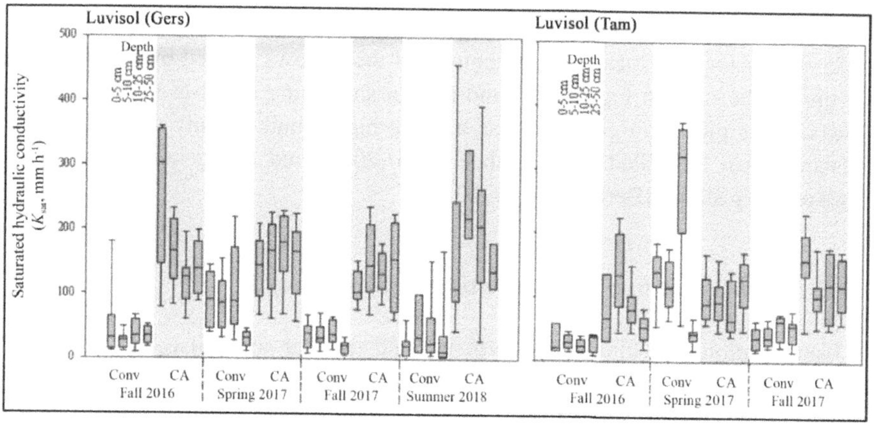

Figure 1.13. *Temporal dynamics of the saturated hydraulic conductivity at different depths for two types of cultivation systems (Conv: conventional agriculture with plowing; CA: conservation agriculture without plowing) applied to two soils (boulbènes soils, with a rich topsoil) from the Southwest of France (according to Alletto et al. 2022)*

The effect of tillage on hydraulic conductivity (Figure 1.12(b)) is different from its effect on water retention. On average, the hydraulic conductivity of the seedbed is eight times higher than that of the plowed layer and three times higher than that of the non-tilled soil, except at saturation (zero matric potential). For the plowed layer, the hydraulic conductivity is three times lower than that of the non-tilled soil. In fact, plowing tends to reduce the continuity of the macropores present in the non-tilled soil (such as holes from roots, earthworms, etc.), which strongly impacts the ability of the soil to conduct water when the soil is close to saturation (pF 1 represents a matric potential of –1 kPa). The superficial tillage carried out for the seedbed production fragments and homogenizes the tilled soil, making it possible to attain a good hydraulic conductivity close to saturation, which will facilitate the infiltration of water and exchanges with the atmosphere.

The effects of tillage on the porosity and on the hydraulic properties of soils most often have a transient effect, which can last from a few weeks to a few months, depending on the characteristics of the soil, the climate and agricultural practices. In the case of the *boulbène* soils from Southwest France (Figure 1.13), which are loamy soils whose structure is particularly unstable, the increase in hydraulic conductivity of the soil arising from plowing is clearly visible in the spring, but quickly fades from the summer and is completely gone by the fall.

Tillage, when carried out in relatively humid conditions, can lead to a compaction of the soil located just below the plowing depth, over a thickness of a few centimeters. This layer, called the "plow pan", results both from the compaction under the tractor tires located at the bottom of the hollows opened by the plow (see the tire on the right in Figure 1.2) and from a smoothing effect at the level of the plow(s). This plow pan generally has a much higher bulk density and penetration resistance than non-tilled soil (Jeřábek et al. 2017) and can generate subsurface flows on steep slopes (Bertolino et al. 2010).

1.3.2. *No-till systems ("conservation" agriculture)*

Conservation agriculture (the term implied to refer to "soil conservation") is based on the desire to protect soils from damage caused by erosion, whether from water or wind. Historically, it was initially developed in the United States and in South America. It has proven to be particularly well suited to tropical soils, which are often poor in organic matter and have low structural stability. In addition, the low content of swelling clays in these soils makes it difficult to naturally recover a favorable structure after compaction. Therefore, the compaction of these soils tends to persist and generate intense runoff, and consequently, water erosion.

Cultivation systems in conservation agriculture imply a lack of tillage (thus the name "no-till") or a very limited tillage (tillage only on the seeded rows) and/or a simplified tillage (with the use of a single tool, such as a seeder adapted for direct seeding). They also imply a permanent covering of the soil by crop residues or service plants and diversified cash crops (Scopel et al. 2013; Palm et al. 2014). The presence of a permanent ground cover makes it possible to greatly reduce the risks of erosion. This soil cover often combines a mulch, consisting of the residues of the previous crop, with the crop in place, whether it is a service crop (such as an intermediate nitrate trap crop) or a cash crop (Figure 1.14).

By creating a rough soil surface, it allows runoff water to be slowed down. In addition, no-till techniques allow for the settlement and development of macrofauna, particularly earthworms, as well as the establishment of root galleries, which will last over time (Figure 1.15). This results in an increased porosity that is biological in origin and presents a much better connectivity than in the case of plowed soils. Therefore, although the bulk density of the soil is generally higher in conservation agriculture than in conventional agriculture, the infiltration capacity of the soil using this technique is often significantly greater (Figure 1.13).

Figure 1.14. *Soybeans grown using direct seeding in grain straw residues on a plot using conservation agriculture in Gers (photo: L. Alletto)*

On the other hand, no-till techniques lead to a very different distribution of organic matter between the two systems. As plowing results in the homogenization of the tilled layer's organic carbon content, the soil surface content is generally lower than in a non-plowing situation, where the crop residues are not buried (Figure 1.16). Therefore, the soil surface's organic carbon content is generally greater in conservation agriculture than in conventional agriculture (Figure 1.15), which gives better structural stability to the soil surface and therefore better resistance to the erosive effect of runoff.

Figure 1.15. *Block of soil extracted with a garden spade from a plot using conservation agriculture. The coloring gradient of organic matter and the macroporosity created by biological activity are clearly visible (photo: L. Alletto). For a color version of this figure, see www.iste.co.uk/coquet/agricultural.zip*

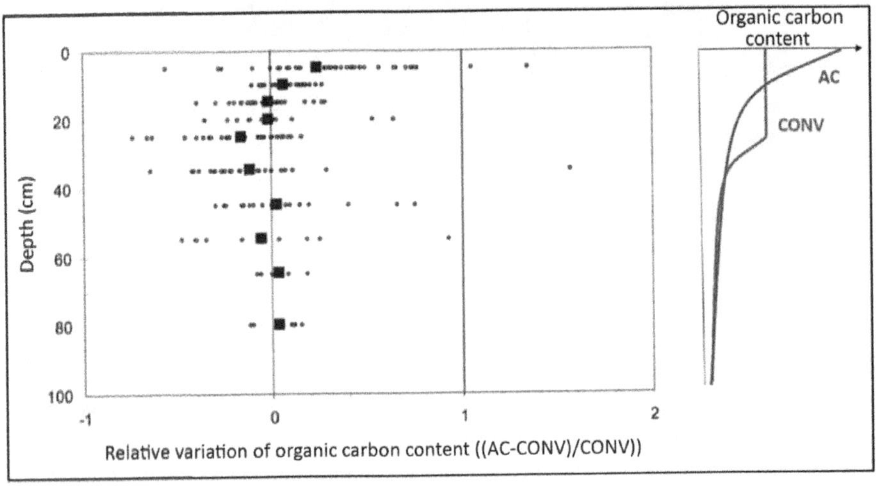

Figure 1.16. *Relative difference in organic C content between no-till (AC) and plowing (CONV) for 237 pairs of plots in 47 sites around the world (according to Angers and Eriksen-Hamel 2008). For a color version of this figure, see www.iste.co.uk/coquet/agricultural.zip*

During the conversion of a cultivation system from conventional agriculture to conservation agriculture, a progressive densification of the soil layer initially tilled can be observed. The term "densification" is used instead of compaction, because this process is observed without any settlement taking place due to agricultural machinery or by the trampling of animals. The agricultural performance of the plots converted to conservation agriculture is often lower and more variable than before conversion during a so-called "transition" period, whose duration varies depending on the type of soil, its initial situation and the agricultural practices implemented. In certain situations, the densification of the soil that occurs in conservation agriculture may require tillage for cracking the soil (using subsoiling, for example, see section 1.2.2.6) in order to recreate porosity, particularly at greater depths.

1.4. Conclusion

Tillage has always been an important component of agricultural activity. Today, given the need to reduce the impact of agriculture on the climate, the evolution of agricultural systems toward techniques with lower energy-consumption appears to be a positive development. From this point of view, the reduction of tillage is notable. By foregoing plowing operations, 15–50 L of diesel per hectare can be saved. But there are also drawbacks to no-till techniques, especially in terms of weed management. In many cases, techniques that do not use plowing give rise to difficulties in controlling weeds, which can no longer be buried. This results in an increased use of herbicides, particularly glyphosate, which has the advantage of being a very effective total herbicide whose agronomic effect is short-lived, and which is used on plots where weed growth is out of hand. Insofar as the European Union wishes to drastically reduce the use of pesticides in agriculture, it is to be expected that this will give rise to a clash that will have to be resolved in the future by agricultural research, which is more grounded in biological regulations and a better understanding of the relationships between soils and plants.

1.5. References

Alletto, L., Coquet, Y., Benoit, P., Heddadj, D., Barriuso, E. (2010). Tillage management effects on pesticide fate in soils. A review. *Agronomy for Sustainable Development*, 30, 367–400.

Alletto, L., Cueff, S., Bréchemier, J., Lachaussée, M., Derrouch, D., Page, A., Gleizes, B., Perrin, P., Bustillo, V. (2022). Physical properties of soils under conservation agriculture: A multi-site experiment on five soil types in south-western France. *Geoderma*, 428, 116228. doi: 10.1016/j.geoderma.2022.116228.

Angers, D.A. and Eriksen-Hamel, N.S. (2008). Full-inversion tillage and organic carbon distribution in soil profiles: A meta-analysis. *Soil Science Society of America Journal*, 72, 1370–1374.

Baize, D. and Jabiol, B. (2011). *Guide pour la description des sols*. Editions Quæ, Versailles.

Bertolino, A.V.F.A., Fernandes, N.F., Miranda, J.P.L., Souza, A.P., Lopes, M.R.S., Palmieri, F. (2010). Effects of plough pan development on surface hydrology and on soil physical properties in Southeastern Brazilian plateau. *Journal of Hydrology*, 393, 94–104.

Boizard, H., Peigné, J., Sasal, M.C., de Fátima Guimarães, M., Piron, D., Tomis, V., Vian, J.F., Cadoux, S., Ralisch, R., Tavares Filho, J. et al. (2017). Developments in the "profil cultural" method for an improved assessment of soil structure under no-till. *Soil and Tillage Research*, 173, 92–103.

Bresson, L.M. and Moran, C.J. (2004). Micromorphological study of slumping in a hardsetting seedbed under various wetting conditions. *Geoderma*, 118, 277–288.

Coquet, Y. (1996). Étude in situ des phénomènes de retrait-gonflement des sols. Application à deux sols tropicaux peu gonflants. PhD Thesis. Université d'Orléans, Orléans.

Desbourdes-Coutadeur, C. (2002). Etude du transport de l'eau dans un sol labouré. Modélisation 2-D de l'infiltration et de la redistribution dans un sol à structure hétérogène. PhD Thesis, Institut National Agronomique Paris-Grignon, Paris.

Gupta, S.C., Sharma, P.P., DeFranchi, S.A. (1989). Compaction effects on soil structure. *Advances in Agronomy*, 42, 311–338.

Jeřábek, J., Zumr, D., Dostál, T. (2017). Identifying the plough pan position on cultivated soils by measurements of electrical resistivity and penetration resistance. *Soil and Tillage Research*, 174, 231–240.

Keller, T., Sandin, M., Colombi, T., Horn, R., Or, D. (2019). Historical increase in agricultural machinery weights enhanced soil stress levels and adversely affected soil functioning. *Soil and Tillage Research*, 194, 104293.

Kuipers, H. (1963). The objectives of soil tillage. *Netherlands Journal of Agricultural Science*, 11(2), 91–96.

Kuznetsova, I.V. and Danilova, V.I. (1988). Loosening of soils by swelling and shrinkage. *Soviet Soil Science*, 20(6), 108–120.

Lamotte, M., Bruand, A., Pédro, G. (1997). Tendance à la lapidification de sols sableux (*hardé* du Nord Cameroun). Une évolution naturelle sous climat semi-aride à fort pouvoir évaporant. *Comptes-Rendus de l'Académie des Sciences de Paris, Sciences de la Terre et des Planètes*, 325, 577–584.

Manichon, H. (1982). Influence des systèmes de culture sur le profil cultural : élaboration d'une méthode de diagnostic basée sur l'observation morphologique. Thesis, INAPG, Paris.

Manichon, H. (1987). Observation morphologique de l'état structural et mise en évidence d'effets de compactage des horizons travaillés. In *Soil Compaction and Regeneration*, Monnier, G. and Goss, M.J. (eds). A.A. Balkema, Rotterdam.

Or, D., Keller, T., Schlesinger, W.H. (2021). Natural and managed soil structure: On the fragile scaffolding for soil functioning. *Soil and Tillage Research*, 208, 104912.

O'Sullivan, M.F. and Ball, B.C. (1993). The shape of the water release characteristic as affected by tillage, compaction and soil type. *Soil and Tillage Research*, 25, 339–349.

Palm, C., Blanco-Canqui, H., DeClerck, F., Gatere, L., Grace, P. (2014). Conservation agriculture and ecosystem services: An overview. *Agriculture Ecosystems & Environment*, 187, 87–105.

Roger-Estrade, J., Richard, G., Manichon, H. (2000). A compartmental model to simulate temporal changes in soil structure under two cropping systems with annual mouldboard ploughing in a silt loam. *Soil and Tillage Research*, 54, 41–53.

Sauzet, O., Cammas, C., Barbillon, P., Etienne, M.P., Montagne, D. (2016). Illuviation intensity and land use change: Quantification via micromorphological analysis. *Geoderma*, 266, 46–57.

Scopel, E., Triomphe, B., Affholder, F., Da Silva, F.A.M., Corbeels, M., Xavier, J.H.V., Lahmar, R., Recous, S., Bernoux, M., Blanchart, E. et al. (2013). Conservation agriculture cropping systems in temperate and tropical conditions, performances and impacts. A review. *Agronomy for Sustainable Development*, 33, 113–130.

Sewell, M.C. (1919). Tillage: A review of the literature. *Journal of the American Society of Agronomy*, 11(7), 269–290.

Sigault, F. (1977). Quelques notions de base en matière de travail du sol dans les anciennes agricultures européennes. *Journal d'agriculture traditionnelle et de botanique appliquée*, 24(2–3), 139–169.

Strudley, M.W., Green, T.R., Ascough, J.C. (2008). Tillage effects on soil hydraulic properties in space and time: State of the science. *Soil and Tillage Research*, 99, 4–48.

Tomis, V., Duparque, A., Boizard, H. (2019). Development of the "mini 3D soil profile" – A visual method derived from the "profil cultural". *Soil and Tillage Research*, 194, 104285.

Unger, P.W. and Kaspar, T.C. (1994). Soil compaction and root growth: A review. *Agronomy Journal*, 86, 759–766.

2

The Biodiversity of Agricultural Soils

Laure Vieublé GONOD[1], Sophie JOIMEL[1] and Audrey NIBOYET[2,3]

[1] *UMR ECOSYS, AgroParisTech, INRAE, Université Paris Saclay, Palaiseau, France*

[2] *UMR iEES Paris, Sorbonne Université, Université Paris Cité, UPEC, CNRS, INRAE, IRD, Paris, France*

[3] *AgroParisTech, Palaiseau, France*

2.1. Soil organisms

The organisms found in soils are characterized by a wide variety of sizes, trophic types, and lifestyles, ranging from strictly aquatic to fully terrestrial. If we consider the size criteria that vary between a few micrometers and several millimeters or centimeters, or even up to several meters, soil organisms can be classified into five main groups: (i) microorganisms composed mainly of bacteria, archaea and fungi; (ii) microfauna (<100 µm) including in particular nematodes, protozoa and tardigrades; (iii) mesofauna (100 µm–2 mm) of which the most representative organisms are mites, Collembola and enchytraeids; (iv) macrofauna (2 mm–8 cm) including in particular earthworms; and (v) megafauna (>8 cm) (Figure 2.1).

The values set out in this chapter for the size limits between the different classes of soil organisms are the conventionally accepted values. Nevertheless, some disparities can be found among different authors. Soil biodiversity includes all of these groups, while soil fauna, also known as pedofauna, focuses on micro-, meso- and macrofauna.

Agricultural Soil Science,
coordinated by Yves COQUET and Joël MICHELIN. © ISTE Ltd 2025.

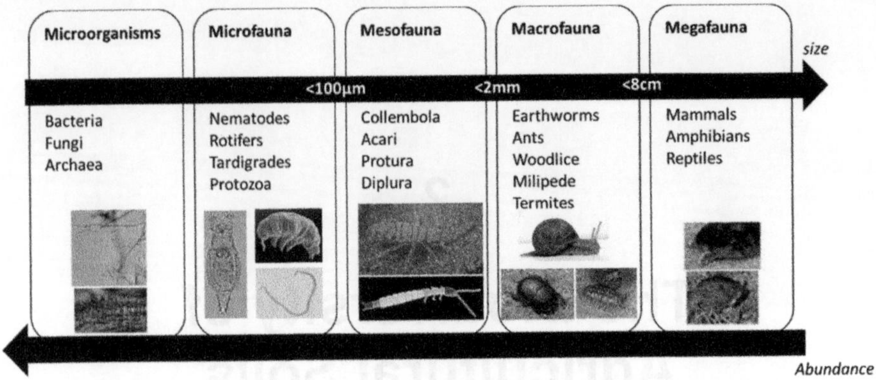

Figure 2.1. *Classification of soil organisms by size (adapted from Tibbett et al. 2020; photos from Jeffery et al. 2013). For a color version of this figure, see www.iste.co.uk/coquet/agricultural.zip*

These organisms can be very numerous in soils with abundances tending to decrease as the size of the organisms increases (Figure 2.1, Table 2.1) and represent several tons per hectare (Table 2.1). The soils are also a reservoir of biodiversity, since they shelter about a quarter of the overall biodiversity on Earth (Decaëns et al. 2006). Several thousands of species of earthworms and Collembola, and tens of thousands of species of bacteria, fungi, protozoa, nematodes and mites have been identified to date (Table 2.1). This diversity, which has long been underestimated (Torsvik et al. 1996), remains yet mostly unknown (Decaëns et al. 2006).

Soil organisms, interacting with each other and with their environment, participate in many different functions, such as the mineralization of organic matter (OM), the recycling of nutrients, the stability of the soil structure, the degradation of pollutants and the regulation of populations. Soil organisms thus contribute directly (or indirectly) to numerous ecosystem services[1], such as soil fertility, crop protection, the regulation of hydrological processes, the fight against erosion, soil detoxification and the protection of water quality, air quality, or climate regulation (carbon (C) sequestration) (see Chapter 7). In some cases, soil organisms can also provide ecosystem disservices, such as being responsible for diseases in plants or via the production of toxic molecules or greenhouse gas emissions (CO_2, CH_4, N_2O).

1 Goods and services provided to humans by ecosystems to ensure their well-being. Developed at the end of the 20th century, this anthropocentric concept was popularized by the Millennium Ecosystem Assessment (2005).

	Group of organisms	Density	Biomass (kg ha^{-1})	Number of species	Total number of known species	% of described species
Microorganisms	Bacteria	10^6–10^9 ind.g^{-1}	1,500	>10,000 g^{-1}	15,000	<1.5
	Fungi	–	3,500	>1,000 g^{-1}	97,000	<6.5
Microfauna	Nematodes	10^6–3×10^7 ind.m^{-2}	60–300	50–100 m^{-2}	25,000	<2.5
	Protozoa	10^5–10^{11} ind.m^{-2}	250	50–100 m^{-2}	21,000	<0.3
Mesofauna	Collembola	2×10^4–4×10^5 ind.m^{-2}	2–40	50–100 m^{-2}	8,500	17
	Acari	2×10^4–4×10^5 ind.m^{-2}	2–40	~150 m^{-2}	40,000	55
Macrofauna	Earthworms	50–400 ind.m^{-2}	200–4,000	~10 m^{-2}	7,000	23

Table 2.1. *Average biomass and species richness (known or estimated) of soil organisms in temperate regions (adapted from Schaefer and Schauermann 1990; Gobat et al. 2010; Orgiazzi et al. 2016)*

There are also other more functional classifications of soil organisms, which are based on eco-morphological criteria such as size, color or life forms (Bouché 1993), such as the eco-morphological classification of earthworms into three functional groups: epigeic (which are located in the litter on the surface of the soil), anecic (which move vertically in the soil, come to feed on the surface and bury the OM in the soil) and endogeic earthworms (which live deeper in the soil). In order to take into account the variability of species within different groups, new approaches based on functional traits[2] have been developed in the last few decades (Pey et al. 2014).

Due to different sizes and habitats, all of these organisms coexist in the soil and interact at different and complementary scales of space and time. Fresh OM, such as crop residues, is first broken down by different organisms in the mesofauna and especially in the macrofauna categories into smaller OMs. These smaller OMs can then be assimilated by soil microorganisms, which can then further break them down and turn them into minerals. Additionally, through bioturbation or after

2 Morphological, physiological, phenological or behavioral characteristics measurable from the cell to the whole organism, without reference to its environment or to any level of organization. Databases on different traits, such as BETSI (https://portail.betsi.cnrs.fr/), can be accessed online.

ingesting fragments of OM and excreting fecal pellets rich in OM, soil fauna participates in the redistribution of OM in the soil profile and promotes the accessibility of OM to microorganisms. Microfauna and mesofauna also play a more indirect role in the degradation of OM by regulating the populations of bacteria and fungi.

Soil biodiversity, which can be described according to the criteria of (i) abundance, density or biomass, (ii) diversity (such as via taxonomic richness, Shannon[3] and evenness[4] indices), and (iii) functionality, is therefore now no longer considered as merely an intrinsic heritage value of soils but as being at the origin of many ecosystem services (Millennium Ecosystem Assessment 2005). This diversity also enables soils to be resistant[5] and resilient[6] when faced with various types of disturbances and stresses (Brussaard et al. 2007). The objective of this chapter is to take stock of knowledge on soil biodiversity in agricultural soils, and the effects of agricultural practices on soil biodiversity.

2.2. Biodiversity of agricultural soils compared with other land uses

The studies that compared the microbial biomass for different land uses showed that forest soils sheltered a greater quantity of microorganisms than vineyard soils, orchard soils and finally, agricultural soils. The microbial biomass in soils is mainly controlled by the physicochemical characteristics of the soil (Dequiedt et al. 2011) and particularly by the clay content, organic carbon content and pH (Horrigue et al. 2016). On the other hand, the soils richest in microorganisms are not necessarily those that show the greatest diversity. In fact, the physicochemical factors that influence microbial diversity, and particularly bacterial diversity, are different from those that control microbial biomass (Constancias et al. 2015; Terrat et al. 2018). Dequiedt et al. (2020) have shown that alkaline soils with a coarser texture and with easily degradable OM are generally home to a greater bacterial diversity. This means that agricultural soils can present higher levels of diversity than forest and grassland soils. However, this level of richness does not provide information on the quality of microbial diversity, and more specifically on the presence of species with

---

3 The Shannon index takes into account the number of species present in a sample as well as the abundance of each species and the total size of the population (Shannon and Weaver 1963).

4 Distribution of the number of individuals by species: ratio between the Shannon H' index and the maximum value that this index would reach if all the species found in the sample were represented equally (Pielou 1975).

5 Resistance refers to the ability to maintain its initial state when facing a disturbance.

6 Resilience is the ability to return to its initial state after experiencing a disturbance.

effects considered beneficial or harmful (pathogenic). Several studies have also observed an increase in bacterial opportunistic and pathogens with the intensity of the agricultural practices (Lienhard et al. 2014).

Regarding the soil fauna, several studies have demonstrated that agricultural soils host a lower abundance and species richness of earthworms and Collembola compared to forest and even urban soils (Scimia et al. 2014; Joimel et al. 2017). However, as for microorganisms, this level of species richness does not provide information on the functionality of these species (Joimel et al. 2021).

Recent research has focused on the networks of biotic interactions between soil microorganisms, with the underlying hypothesis, resulting from work on macroorganisms (Schleuning et al. 2015), that the complexity and structure of interaction networks are essential for the proper functioning of an ecosystem. Morriën et al. (2017) and later Karimi et al. (2019) have shown that (i) bacterial networks are more complex (in terms of the number of links between bacterial genera) in forest soils, followed by grasslands and field crops, and lastly in soils in vineyards and orchards; and (ii) in vineyards soils, certain groups of bacterial taxa were isolated from the rest of the community.

2.3. Effects of agricultural practices on soil biodiversity

Agricultural soils tend to contain fewer organisms, but with potentially larger proportions of opportunists and pathogens, as well as less complex networks. Nevertheless, the results vary widely from one agricultural soil to another, which is explained at least in part by the diversity of agricultural practices and cropping systems. The practices developed, particularly by conventional agriculture, such as tillage, fertilization, the use of phytosanitary products and monoculture are clearly identified as having effects on the physical, chemical and biological components of soils with repercussions on biological activities (Millennium Ecosystem Assessment 2005). Indeed, these practices can impact soil organisms directly through physical damage, causing injury or death, exposing them to predation, stimulating, or inhibiting them (Roger-Estrade et al. 2010) or indirectly via effects on habitats, the location of nutrient resources and the physicochemical characteristics of the soil (temperature, oxygenation and humidity) (Vargas Gil et al. 2011).

The effects of practices on the levels of abundance, biomass, densities, diversities and activities of soil organisms can vary, ranging from negative to positive impacts and depends on the taxa considered. In this chapter, we will focus on the effects of (i) mineral fertilizers, (ii) the addition of exogenous OM, (iii) pesticides, (iv) tillage and (v) cultivated biodiversity.

2.3.1. *Effects of mineral fertilizers*

The effects of mineral fertilizers on soil organisms can be direct or indirect (Clermont-Dauphin et al. 2014), and either beneficial or harmful.

Various studies have highlighted the harmful effects of mineral fertilizers, particularly on earthworm (Hansen and Engelstad 1999) and microbial populations (Treseder 2008; Fierer et al. 2012; Zhang et al. 2015). Repeated high-dose applications of fertilizers (particularly nitrogen and phosphates) very often lead to a reduction in the abundance and diversity of mycorrhizal fungi and N_2-fixing microorganisms (Clermont-Dauphin et al. 2014; de Graaff et al. 2019), probably due to their impacts on soil characteristics (decrease in pH) and a lower dependence of plants under conditions of high nitrogen (N) and phosphorus (P) levels (Treseder 2004; Qin et al. 2015). Similarly, Wei et al. (2012) observed negative effects on the diversity of soil fauna, though for low doses and in the short term (de Graaff et al. 2019). However, the interactions between organisms and mineral fertilizers can be much more complex. For example, Simonin et al. (2015) studied the impact of nitrogenous mineral fertilization on the various microbial groups involved in nitrification, and demonstrated (i) an increase in the abundance of ammonium oxidizing bacteria (AOB) but no significant effect on that of ammonium oxidizing archaea (AOA); and (ii) an increase in the abundance of the *Nitrobacter*-like bacteria, and conversely, a decrease in the *Nitrospira*-like bacteria, which are involved in the oxidation of nitrites into nitrates. These differing responses of the microbial groups involved in nitrification therefore lead to changes in the structure of the nitrifying microbial communities following mineral nitrogen fertilization, with possible functional consequences on the nitrogen cycle. Contrasting responses were also observed for the different nematode trophic groups. Fungivorous nematodes decrease linearly in response to nitrogen inputs (Liang et al. 2009), while bacterivorous nematodes are stimulated for low doses (Wei et al. 2012). Other soil fauna taxa, such as Collembola, seem to be not particularly sensitive to nitrogen fertilization (Coulibaly et al. 2017).

The use of mineral fertilizers may also cause long-term beneficial effects, particularly on soil microorganisms and their activities, resulting in an increase in plant production and thus an increase in OM returning to the soil (e.g. in the form of root exudates and root biomass, via the restitution of crop residues) (Clermont-Dauphin et al. 2014). This OM can be trophic resources that are available directly to soil organisms, and also impacts the physicochemical properties of soils (e.g. structure, gas and water transfer, water retention, cation exchange capacity) and local conditions (water content, temperature, etc.).

Wang et al. (2016) observed that the increase in plant-based carbon following fertilization gave rise to an increase in the diversity of microbial communities and fauna. In some cases, the negative responses to mineral fertilizers could therefore be counterbalanced by positive responses due to the accumulation of OM.

The variability in the results observed with harmful or beneficial effects on soil biodiversity can be explained by differences in the amounts of fertilizers added, the nature of the fertilizers, the duration of the fertilization and the times at which the effects are observed and studied (short, medium and long term) (de Graaff et al. 2019). The effects of fertilizers on the amounts of OM in the soils are only observed after several years of fertilization and numerous inputs. The changes induced by fertilization on the soil's OM content, which can be particularly important in mediating the response of soil organisms, can only be observed over the long term.

2.3.2. *Effects of the addition of exogenous organic matter*

The effects of OM on soil biodiversity (abundance, biomass, diversity and activity) can be direct as trophic resources for heterotrophic organisms or as micro-habitats, as well as indirect through the role of OM on the physical and chemical properties of soils and on the biological component of soils. Indeed, OM plays a major role in the physical structure of the soil and the circulation of water, air and nutrients, in the retention of nutrients, water, pollutants and toxins.

Numerous studies have highlighted the positive effects of different organic amendments on soil microorganisms and their activities (Houot et al. 2014). The increase in microbial biomass is mainly due to the stimulation of indigenous microbial communities following the addition of carbon (Pérez-Piqueres et al. 2006), but may also be due to an intake of microorganisms via organic waste products[7] (OWPs) (Ros et al. 2006). Indeed, the microbial biomass of composts can be up to 1,000 times higher than that of soil (Annabi et al. 2007). Applications of OWPs also stimulate activities such as soil respiration, which reflects the overall activity of the soil microbial communities. The effects on more specific activities, such as enzyme activities involved in the carbon, nitrogen, or phosphorus cycles, are more variable and can be affected by many different factors (e.g. nature of the OM and dose applied).

---

7 The recycling of organic waste products (OWPs) in agriculture, which has seen extensive development in France, enables the recovery of waste (under the status of waste or products after standardization or homologation) used for their fertilizing and/or amending properties with moderate implementation costs.

The durability and intensity of the effects can also vary depending on the type of OWPs provided. For instance, Vieublé Gonod et al. (2009) have shown that the stimulation of microbial biomass, fungal biomass and mineralization activity was dependent on the characteristics of the composts and over time. Thus, a municipal solid waste compost (MSW) stimulated the total microbial and fungal biomass in a very intense but temporary way, due in particular to the contribution of a high fraction of labile C that can be directly assimilated by the microorganisms. Conversely, a co-compost of green waste and sewage sludge (GWS), with a more stable and less rapidly available supply of carbon to microorganisms, exerted more moderate but much more persistent effects (up to six months after the application). Compost application also stimulated carbon mineralization to a greater extent with the GWS compost than with the MSW compost, four to eight months after the addition.

The characteristics of the effects also depend on the number of OWP applications. In her PhD thesis, Obriot (2016) examined the impact of single or repeated short-, medium- and long-term OWP inputs on soil microbial communities and their degradation activities related to carbon and nitrogen cycles, at plot scale. The results showed that a single application of OWP (e.g. biowaste compost, green waste and sludge compost or manure) had moderate and transient effects on soil microbial communities (total microbial biomass, fungi, bacteria, nitrifying and denitrifying populations). These OWP applications also stimulated carbon and nitrogen mineralization and emissions of N_2O. The N_2O emissions after the application of OWP were generally higher than those obtained with mineral fertilization, as the carbon and nitrogen added with OWP stimulate the first stages of denitrification. However, the N_2O emissions also seem to vary according to the characteristics of the OWPs. Repeated applications have more marked effects. At the QualiAgro site[8] (Ile de France), the seven additions of OWPs carried out between 1998 and 2011 increased the total microbial biomass, regardless of the organic treatments (factor 1.4) (Figure 2.2), due to an increase in OM contents in the soils.

Vieublé Gonod et al. (2009) also observed that these stimulating effects of OWPs on soil microorganisms and their activities were spatialized in conventional farmed soils and mainly located in the interfurrows where plant residues and composts are concentrated after plowing, and which are therefore richer in carbon and nitrogen (Roger-Estrade et al. 2000; see section 1.2.3).

Nevertheless, some studies have also highlighted the lack of effects or negative effects, following the application of OWPs, and that excessive fertility through the

8 See: https://www6.inrae.fr/qualiagro.

OM applications could have a negative impact on key functional groups such as mycorrhizal fungi (Orgiazzi et al. 2016).

These contradictory results could be due to the factors previously mentioned (e.g. diversity of OM, doses applied and pedoclimatic context), but also to the measurement methods used to analyze the effects on microbial communities (e.g. fumigation–extraction and molecular biology methods) and/or to the presence of pollutants (e.g. high doses of sewage sludge rich in trace metal elements (TMEs) before the implementation of specific regulations on sludge spreading) (Houot et al. 2014).

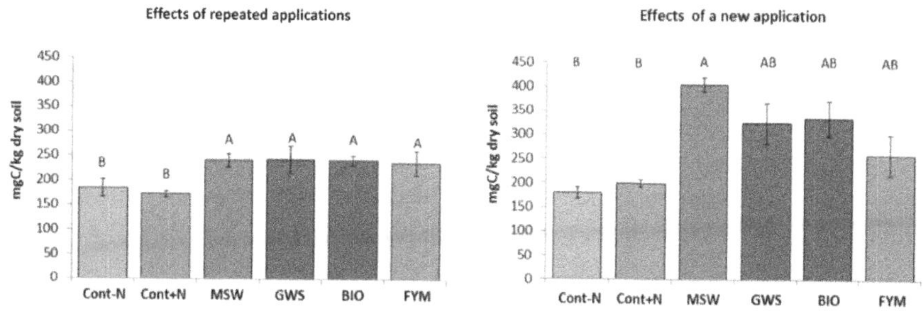

Figure 2.2. *Impact of repeated applications of different types of OWPs in the long term (left) and short term (right) on the total microbial biomass in QualiAgro soils. The control plots either receive (Cont + N) or do not receive (Cont – N) mineral nitrogen fertilization, and the amended plots receive a municipal solid waste compost (MSW), a co-compost of green waste and sewage sludge (GWS), a biowaste compost (BIO), or a farmyard manure (FYM) (Obriot 2016). For a color version of this figure, see www.iste.co.uk/coquet/agricultural.zip*

Far fewer studies have looked at the impacts of OWPs on the abundance, density and diversity of soil fauna and have focused primarily on earthworms, mites, Collembola and nematodes (Houot et al. 2014; Cortet and Hedde 2020). In general, the application of OMs stimulates earthworms (Houot et al. 2014). Pérès et al. (2011) observed that the number of earthworms per m^2 was higher in plots that had been supplied with OWPs for more than 10 years. The authors showed that the effects also depended on the characteristics of the added OM, with the effects in this study being more marked after the addition of manure than after the addition of MSW, co-compost of GWS or biowaste compost. These OM additions also impact the composition of the earthworm population by increasing the proportion of anecic earthworms that dig vertical galleries, feed at the soil surface and bury OM in the soil profile. Thus, the additions of OM to the soil have stimulated the bioturbation activity of the earthworms which feed on it, with consequences on soil structure, and consequently on water and air transfers in particular.

Likewise, OM additions stimulate Collembola and mites, and particularly Oribates, which are the most numerous in soils, but the intensity of the effects varies according to OWP characteristics (Houot et al. 2014; Cortet and Hedde 2020). The addition of OM also tend to stimulate nematodes, and in particular bacterivorous and fungivorous nematodes. The effects may be indirect via the increase in bacterial and fungal populations, as well as direct via an improvement in, for example, water retention.

The overall biodiversity of soil fauna is rarely significantly altered by the addition of OM, although significant changes in the structure of the food web can be observed. For example, there is a relative increase in the fungivorous functional groups (such as Collembola and mites) and predators (such as mites) following the addition of stabilized OM in the form of green waste compost. Conversely, additions of more labile OM favor bacterivores and opportunists. Oribate mites are very sensitive to the chemical quality of organic amendments, and their abundance and diversity can be negatively affected by a decrease in the abundance of fungi and by additions of saline or polluted organic amendments.

For both microorganisms and soil fauna, the OMs applied to soils generally have positive effects on the abundance and activity of soil organisms, but these effects depend on many factors. These factors include the organisms considered, the nature of the OM and in particular its C/N ratio, in relation to the original characteristics of the OM and the treatments applied before spreading (composting, methanization), the dose applied; the number and frequency of additions; and the characteristics of the soil (OM content) or even the climatic conditions (Houot et al. 2014).

More global effects of repeated OWP additions were assessed as part of a multi-criteria evaluation, that considered the various services and disservices, based on soil and crop quality indices, and particularly soil biodiversity and biological activity indices (Obriot et al. 2016). These indices were calculated on the QualiAgro site and used to compare the effects of repeated additions of OWPs on these different indices compared to mineral fertilization. The results highlighted the beneficial effects of the OWPs on biodiversity and biological activities (Figure 2.3). The most biodegradable OWPs, such as MSW and farmyard manure, are the most effective in promoting the biodiversity and biological activity of soils.

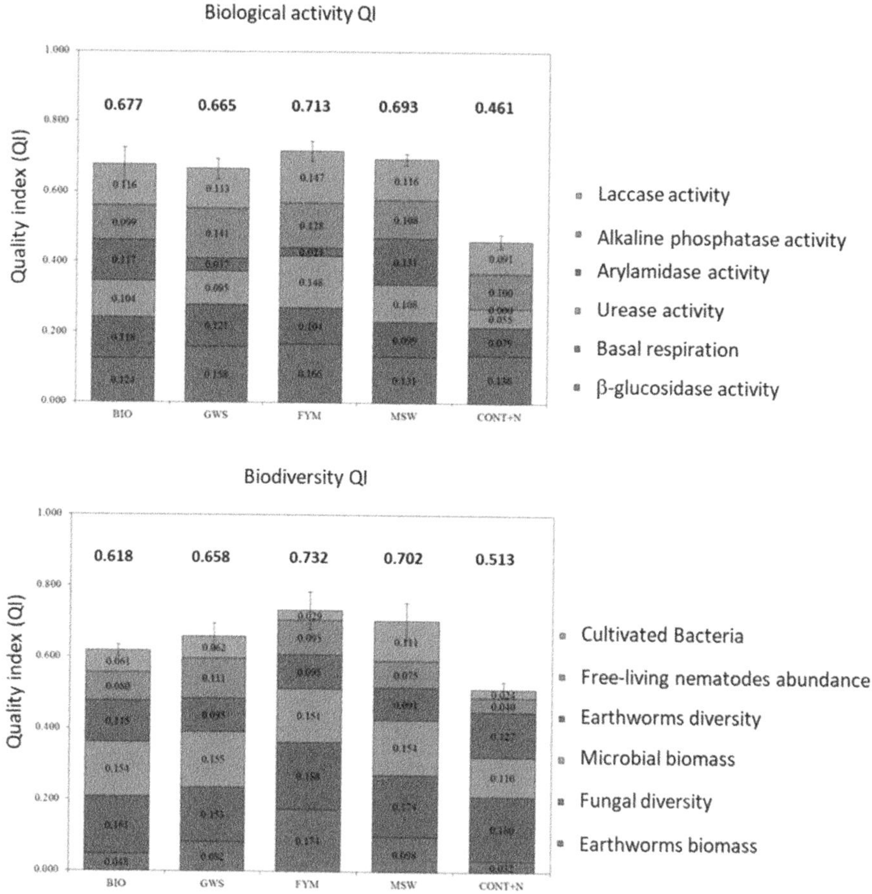

Figure 2.3. *Contributions of different indicators to biological activity and biodiversity indices for the control plots receiving mineral fertilization (CONT + N) and the plots amended with biowaste compost (BIO), green waste and sewage sludge compost (GWS), municipal solid waste compost (MSW) and farmyard manure (FYM) (Obriot et al. 2016). For a color version of this figure, see www.iste.co.uk/coquet/agricultural.zip*

2.3.3. *Effects of pesticides*

While the widespread use of pesticides has contributed to increasing agricultural productivity by reducing the incidence of diseases, pests, parasites and pathogenic organisms, it has also generated several forms of pollution. When pesticides are applied, only a few percent reach their intended targets, and a high proportion may remain in the soil or be transferred to the air or water. Despite efforts to reduce their use, pesticides are still omnipresent in the environment (found in the air, soil and water) and it is possible to find both molecules banned from use for many years (i.e. atrazine), as well as molecules currently used, either in intact form or as metabolites.

Pesticides or their metabolites can have toxic effects on soil organisms and impact their abundance, diversity and/or their activities. In this context, tests to receive approval before being allowed on the market are carried out, despite their possible limitations (Pelosi et al. 2013; Joimel et al. 2022). Nevertheless, the impact of pesticides or their metabolites on soil communities can vary depending on the type of pesticide, the time span of the application and the amount applied (Edwards and Bohlen 1996). The effects generally increase with the concentrations and number of applications (Cycoń and Piotrowska-Seget 2009). In addition, effects can be limited to one class or affect a wide spectrum of organisms, directly or indirectly. The effects may be particularly strong on organisms in the same taxonomic group as the target organisms (fungi for fungicides, insects for insecticides, nematodes for nematicides), though this is not always the case.

Harmful effects of fungicides have been highlighted on soil fauna and microorganisms. Fungicides significantly reduce the specific richness, and in particular the abundance of nematodes (Puissant et al. 2021), and are harmful to arbuscular mycorrhizal (AM) fungi (Plenchette et al. 2005). However, these effects vary depending on the active substance and the application dose. It has also been shown that certain fungicides (e.g. fosetyl-Al, metalaxyl and promamocarb) do not have negative effects, and sometimes even have positive effects on AM fungi. Fungicides applied as a seed coating would likely inhibit the development of AM mycorrhizae more than fungicides applied when the plants are already mycorrhized (Plenchette and Perrin 1992).

Broad-spectrum nematicides do not seem to target specific nematode taxa. Puissant et al. (2021) have shown that the use of such products affects all nematodes by decreasing the total abundance of nematodes (–32.1%), plant parasitic nematodes (–70.2%), taxonomic richness (–16.6%) and by affecting the structure of the food web.

Within soil fauna, herbicides can impact communities of Collembola and detritivorous mites (Cortet and Hedde 2020) or nematodes (Zhang et al. 2010), with effects that can be direct or indirect, by modifying the availability of resources (i.e. a reduction of OM inputs) in the soil. The effects of herbicides can manifest themselves at different scales, from the cellular to the community scale, and also impact functions, as has been demonstrated for earthworms (Pelosi et al. 2014). Intra-group variability has also been reported: for example, Zhao et al. (2013) showed that fungivorous nematodes, which are more present at the surface, could be more exposed to herbicides and more impacted. However, the effects on the groups vary according to the studies. While Zhao et al. (2013) observed an increase in parasitic nematode populations following an herbicide treatment, Puissant et al. (2021) observed no effects on this same group.

The impacts of herbicides on microbial communities are also variable and may be only transient. Vieublé Gonod et al. (2006), for example, demonstrated that the structure of soil bacterial communities was impacted by a soil treatment with 2,4-D, but only for a very short time. These effects in the first days of incubation can be explained both by the growth delays of degrading microorganisms, as has also been previously observed for 2,4-D (Vieublé Gonod et al. 2006) or for other herbicides such as isoproturon (IPU) (Sorensen et al. 2001) and by toxic effects. After seven days of incubation, the significant differences between the control soil and the soil treated with 2,4-D disappeared, and these results can partly be explained by the fact that certain microorganisms are capable of metabolizing herbicides (Vieublé Gonod et al. 2006; Kepler et al. 2020).

Indeed, recurrent treatments of soils with pesticides can lead to the adaptation of microorganisms that acquire capacities for degrading pesticides. Nevertheless, there are very little data in the literature on the temporal evolution of the microbial degradation capacities of pesticides when the selection pressure no longer exists. This is the case in plots converted to organic farming or to zero pesticide use (Arbeli and Fuentes 2007). Allain (2017) compared the biodegradation potential of pesticides in plots cultivated under conventional agriculture (with repeated applications) and in plots converted for a longer or shorter time to organic farming (without pesticides and chemical fertilizers) and zero pesticides (chemical fertilizers allowed). The main hypothesis tested was that microbial biodegradation was greater in conventional farming than in organic/zero pesticide farming, due to the microbial adaptation induced by the repeated application of pesticides. The results showed that, while the biodegradation of certain molecules was indeed greater in conventional systems than in organic/zero pesticide farming (2,4-D, IPU), this was not confirmed for all molecules (glyphosate, epoxyconazole and pendimethalin). These results can be explained by strong adsorption of these molecules or degradation mainly by co-metabolism (Figure 2.4). Moreover, the conversion to organic farming or to zero pesticides, even after almost 20 years, did not necessarily lead to a loss of pesticide mineralization capacities.

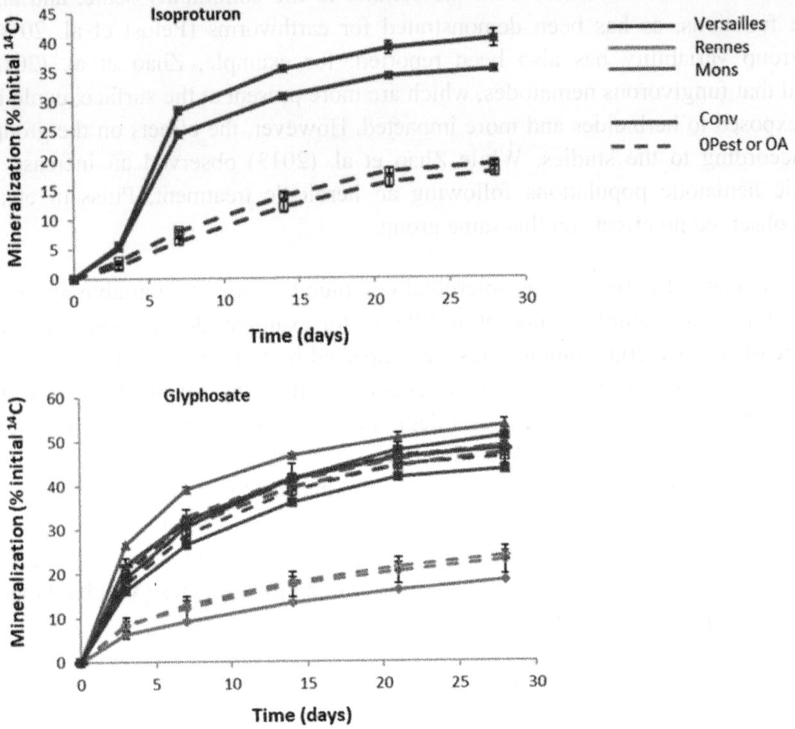

Figure 2.4. *Mineralization of isoproturon and glyphosate in soils cultivated under conventional agriculture (Conv), zero pesticides (0Pest) or organic agriculture (OA) at the Versailles, Rennes and Mons sites (Allain 2017). For a color version of this figure, see www.iste.co.uk/coquet/agricultural.zip*

Despite the abundant scientific literature on the effects of pesticides and their metabolites on soil organisms, on the one hand, and on the biodegradation and fate of pesticides in soils, on the other hand, there is still much to be done due in particular to (i) the broad diversity of molecules which makes case-by-case studies impossible; some studies have proposed classifying organic contaminants according to their intrinsic properties (molecular descriptors) and behavioral properties in the environment (retention, biotic, and abiotic degradation, transfer to air and water) and their toxicological and ecotoxicological effects (TyPol database) (Mamy et al. 2015), which should make it possible to select model molecules and facilitate the generalization of the results; (ii) the effects of the formulations, which remain very poorly known – most of the studies focusing on the active substances, yet the adjuvants present in commercial formulas intended to improve the solubility of the pesticide, the compatibility of the active ingredients, the penetration and translocation of the active ingredients in the target or even its resistance to rain could impact the fate and the effects of the active molecules (Krogh et al. 2003); (iii) the existence of cocktail effects from the application of different types of molecules – a common practice in agriculture – and/or the presence of other organic and inorganic contaminants in the soil; and (iv) the effects of repeated applications of pesticides.

In addition, in a context of agricultural transition and major recent or forthcoming regulations (e.g. ban of certain molecules such as glyphosate), the fate and effects of pesticides in alternative systems is important to consider and necessary for a better understanding of soil and water contaminations.

2.3.4. *Effects of tillage*

The effects of tillage on the soil organisms and their activities have been studied extensively. Tillage (see Chapter 1), by turning over or compacting the soil through the use of machines, impacts both the structure of the soil and its porosity with the potential to cause habitat destruction, as well as the spatial distribution of resources and therefore their accessibility to soil organisms (Bouthier et al. 2014; Coulibaly et al. 2022). Tillage can also lead to increased mortality of soil organisms through direct (mechanical) effects, surface drying or exposure to predation. Finally, soil compaction can also have an impact on larger burrowing fauna, since digging compacted soils requires a high allocation of energy, to the detriment of other activities such as foraging or mating.

More specifically, microbial biomass is significantly impacted by tillage, with effects that increase with the intensity of tillage on a gradient scale: no tillage, limited tillage (simplified cultivation techniques and surface tillage) and deep tillage (decompaction and plowing). Numerous studies have shown that burying crop residues in conventional tillage promoted the development of bacteria, while no-till systems, leading to an accumulation of crop residues and a concentration of OM on the surface, tended to be dominated by fungi (Clermont Dauphin et al. 2014). Thus, the fungi/bacteria ratios are higher in no-till systems, as compared to a system using conventional tillage (Brady and Weil 2002). With conventional tillage, it is common to find food webs based on bacteria, while when no-till occurs, they are instead based on fungi, with effects on nutrient availability, soil aggregate stability and nitrate leaching (Hendrix et al. 1986).

Tillage, in comparison with reduced tillage or no tillage, can stimulate microbial activities in the short term (mineralization of OM, enzyme activities) but generally affects them negatively in the longer term (Clermont Dauphin et al. 2014).

Therefore, the effects are different over the short versus the longer term (van Capelle et al. 2012; Dequiedt et al. 2020).

The results observed may also depend on the soil depth under consideration. van Capelle et al. (2012) showed that the total amount of microorganisms present in the soil was not significantly different in soils with or without tillage. On the other hand, the vertical distribution of microbial biomass in the soil shows significant differences between the different tillage systems, with a homogeneous distribution over the 0–30 cm range in the conventional system and a decrease in microbial biomass with soil depth in the absence of tillage, linked to OM gradients (Figure 1.16).

The impacts of tillage on microbial diversity are also contradictory. For example, Dequiedt et al. (2020) showed that taxonomic richness follows an inverse gradient to microbial biomass, since it tends to increase with tillage. The underlying hypothesis is that tillage may, on the one hand, promote the development of new ecological niches through the destruction of soil aggregates, and, on the other hand, reduce competition between taxa. This hypothesis is based on the theory of intermediate disturbances. Nevertheless, this positive effect of tillage on microbial diversity should be put into perspective, as some of the taxa favored by tillage are microorganisms that have high growth rates and are opportunistic (r strategy) (Dequiedt et al. 2020).

However, de Graaf et al. (2019) showed in their meta-analysis that tillage had negative effects on bacterial biodiversity but no impact on fungal diversity, and that the results were not affected by the duration of the experiment. This lack of fungal response is unexpected, since tillage is often described as being able to physically affect fungal hyphae. In this way, mechanical disturbances could have greater consequences on fungal biomass than on diversity (de Graaf et al. 2019). Nevertheless, the lack of effects in this case could also be explained by the high variability of the responses measured in the individual studies.

Research conducted on soil fauna also shows contrasting effects, which seem to be highly dependent on the size of the organisms considered. While Zhang et al. (2019) observed that tillage had a slightly negative impact on nematodes, Zhang et al. (2012) showed that four years of direct seeding had no significant effect on the abundance and biomass of total nematodes. Similarly, tillage seems to have little effect on the abundance of Collembola, instead having more effect on their richness, which is often already very low in field crops (Cortet and Hedde 2020), even if some studies have also shown the opposite effect (Orgiazzi et al. 2016). Tillage seems to affect the three eco-morphological categories of Collembola.

Euedaphic species are the most affected due to their limited burrowing activity, in particular in fine-textured soils (van Capelle et al. 2012). Mites show wide ranges of responses (Orgiazzi et al. 2016). van Capelle et al. (2012) showed that it was mainly the vertical distribution of mites that varied with tillage intensity, as well as the composition of the mite communities.

In contrast, there appears to be more consensus on the effect of tillage on the soil macrofauna. Earthworms are generally affected over the long term by conventional tillage, in both temperate and tropical zones (Clermont Dauphin et al. 2014), with effects on abundance, biomass and diversity (van Capelle et al. 2012). Here too, the eco-morphological groups are differently affected, with the anecic species, which move the soil vertically in the soil profile, and the epigeic species, which inhabit the litter layer on the soil surface, being generally the most affected. Endogeic species that derive more of their food from the soil rather than from surface litter are less affected by tillage and in some cases may even have easier access to the resource, and thus develop and reproduce faster (Clermont Dauphin et al. 2014; Orgiazzi et al. 2016).

In any case, the effect of tillage is also modulated by the type of soil (van Capelle et al. 2012), the tools used, the date of tillage and therefore the risk of exposure to predation or extreme temperatures, or related practices (Kladivko 2001; Pelosi et al. 2008).

In conclusion, it has been shown that the soil mesofauna and macrofauna, which are larger, reproduce more slowly and have a longer generation time, are more sensitive to the impact of tillage than microfauna and microorganisms (Wardle 1995). The effects of tillage intensity on soil organisms also differ depending on their ability to burrow, their adaptation to certain soil depths, their habitat requirements and their food preferences, with potential consequences on food webs (Briar et al. 2007). Conversely, tillage often leads to a qualitative simplification of the soil food web, with a reduction in the number of trophic groups and interactions.

Tillage, often summarized in three greatly simplified categories (conventional, reduced tillage and no tillage), in fact encompasses a wide range of practices which are diversified by the tools used or the depth of tillage (see Chapter 1), which can also explain the wide variability of results observed.

2.3.5. *Effects of crop diversification*

Crop diversification can be implemented in different ways, via (i) associations of different crops at the same time or (ii) crop rotations and/or the establishment of intermediate crops, where different plant species are cultivated over time. More recently, varietal mixing practices have developed, where in crops such as wheat, different varieties can be combined at the same time (Barot et al. 2017; Borg et al. 2018).

2.3.5.1. *Diversification over time: crop rotations and intermediate crops*

Various studies have highlighted the positive effects of crop rotations (Boswell et al. 1998; Chen et al. 2008; Venter et al. 2016) and cover crops (McDaniel et al. 2014; Kim et al. 2020) on soil microbial abundance, diversity and activity. However, the observed results can vary greatly from one study to another, depending on the pedoclimatic context and the agricultural practices (e.g. tillage) (Kim et al. 2020).

Regarding soil fauna, Puissant et al. (2021) showed that, for nematodes, crop rotation reduced the abundance of phytophagous nematodes (crop pests) when the crop rotation was longer than two years. They also observed that, in particular, cover crops increased the density of total, bacterivorous and phytophagous nematodes. Therefore, the stimulation of the abundance and diversity of organisms in intercropping and rotation systems can play an essential role in the fight against parasites and diseases. Conversely, monoculture tends to decrease the abundance of soil organisms, as has been shown for mites, Collembola and earthworms (Orgiazzi et al. 2016) or the macrofauna of the soil in general (Sileshi et al. 2008).

Similarly, the richness and diversity of soil microorganisms (Chen et al. 2008; Orgiazzi et al. 2016; Venter et al. 2016) and soil fauna, in particular, nematodes (Puissant et al. 2021), Collembola (Orgiazzi et al. 2016) and macrofauna (Sileshi et al. 2008), is generally lower with monocrops than with crop rotations due in particular to the close relationships between above- and below-ground communities.

These results can be explained by greater inputs of OM, and particularly carbon, into the soils, especially when the crop residues are not exported (McDaniel et al. 2014). In addition, intercropping reduces nitrogen leaching and increases nitrogen availability in the soil. Finally, crop rotations lead to a greater abundance and diversity of plant litter, with potential impacts on soil structure, water retention and cation exchange capacity, as well as on local temperature and moisture conditions.

Venter et al. (2016) also suggest that the increase in soil biodiversity with crop diversification could also, at least in part, be due to the differential action of the roots, as the exploration of the soil by successive crops could allow the proliferation of microorganisms in larger soil volumes. The physical changes due to the roots could create a favorable microclimate for the development of soil microorganisms.

However, once again, the results found in the literature linking above- and below-ground biodiversity may be contradictory (Yin et al. 2010; Navarro-Noya et al. 2013). Thus, to date, no specific combinations of crops or varieties have been identified as being more beneficial to soil biodiversity than others. In addition, few studies have measured the impact of crop rotations and intercropping on the activities of organisms and soil functioning.

2.3.5.2. *Spatial diversification: species associations and mixtures of varieties*

In addition to species associations, one way to increase intra-field plant diversity is through the design of varietal mixtures, combining different varieties of the same cultivated species. Increasing intra-specific diversity within fields is likely to have positive effects on the biodiversity of agricultural soils. This is due to selection effects and complementary effects between these varieties, as has been observed at the inter-specific scale (Barot et al. 2017). An experiment conducted on winter wheat (*Triticum aestivum*) allowed for a comparison between the soil biodiversity in plots comprising one, two, four, or eight wheat varieties from a pool of 16 varieties, and one, two, three, or four functional groups defined according to the traits of the wheat varieties (Dubs et al. 2018, 2023). The effects observed on soil biodiversity vary and are dependent on the groups considered. In particular, Salmon et al. (2021) did not observe a significant effect of the number of wheat varieties and of the number of wheat functional groups on Collembola, while Chassé et al. (2019) observed variable effects in earthworms, with a positive effect of diversity on juvenile growth and survival, but a negative effect on adult growth.

2.4. Comparison of cropping systems

Due to the environmental and health impacts of conventional agriculture, new farming systems have emerged in recent decades to reconcile socioeconomic and environmental concerns. These systems are characterized, for example, by a reduction or even elimination of tillage, the ban of phytosanitary products and chemical inputs (e.g. conservation agriculture[9] and organic farming[10]). The stability and sustainability of these alternative cropping systems are based in particular on their ability to promote soil biodiversity and optimize the functioning of the biological component of the soil (Wardle et al. 2004; Brussaard et al. 2007).

---

9 Conservation agriculture is labeled (CA) and is defined through its three pillars defined by the FAO: (i) reduction/elimination of tillage; (ii) permanent soil cover; and (iii) diversification of rotations. See section 1.3.

10 Organic agriculture (OA) is regulated and excludes the use of synthetic chemicals and GMOs, and limits the use of additives.

However, the relationships between farming practices and biodiversity are not simple, as we have previously seen. Some of these alternative systems have revealed dilemmas when applying these practices in real agricultural contexts. For example, in conservation agriculture, the benefits of no-tillage and plant cover, which allow the maintenance of the integrity of soil habitats and a good level of trophic resources for soil organisms, may be counterbalanced by an increase in the doses of herbicides used for weed management (Colbach and Vacher 2014). Similarly, in organic farming, the use of mechanical weeding to compensate for the absence of chemical weeding, and therefore the use of repeated and even superficial tillage, could alter the habitats of soil organisms and affect their abundance, diversity and/or activity. Conversely, the positive effects of the OWP could counteract the negative effects of tillage on biodiversity, as has been shown on the soil physical properties (Treonis et al. 2010). This is why the effects of these factors, often studied independently of each other, must be considered holistically at the scale of cropping systems to better understand their effects on soil biodiversity. Likewise, the dynamics at play within systems in transition are poorly documented (Pelosi et al. 2016).

To study these effects, two investigation methods were implemented. The first of these proposed to prioritize the impact of practices on soil organisms (Nivelle et al. 2016). Puissant et al. (2021), for example, have shown in a meta-analysis that organic/mineral fertilization, cover crops and the use of nematicides were the practices with the strongest effects on nematodes, while the use of herbicides, crop rotations, mineral fertilization and tillage had weaker effects. The second method considered cropping systems as a whole, with in situ studies and some directly at farmers' sites. The objective of these studies was to compare different cropping systems, mainly organic farming versus conventional agriculture, conservation agriculture versus conventional agriculture, and organic farming versus conservation agriculture. A recent study by Christel et al. (2022) showed that the total number of publications concerning the effect of cropping systems on the ecological quality of soils has been increasing sharply since the early 2000s. Nevertheless, research on the effects of conservation agriculture remains small in number compared to that done on the effects of organic farming.

2.4.1. *Organic farming versus conventional agriculture*

Research comparing organic and conventional agriculture have mainly targeted microorganisms, and to a lesser extent soil fauna (in particular, micro-arthropods) (Christel et al. 2022). Various studies, carried out in various pedoclimatic contexts, have shown that the biomass, abundance and density of microorganisms (bacteria and fungi) and microfauna (nematodes), mesofauna (Collembola and mites) and macrofauna (earthworms, ground beetles, spiders, etc.) as well as the biological activity of the soil, were generally higher in organic farming compared to conventional systems (e.g. Bengtsson et al. 2005; Gomiero et al. 2011; Henneron et al. 2015; Lori et al. 2017).

These effects on soil biodiversity can be explained by crop rotations, the use of legumes and the addition of OM in organic farming. However, the effects vary depending on land use (agriculture, orchards and grasslands) and climate (Lori et al. 2017). According to Hartmann et al. (2015), the quantities and the quality of the OMs used in organic farming are the major determinants of microbial biodiversity. As a result, the stimulation of soil microbial biomass is higher in organic farming systems that receive high amounts of organic inputs (Gunapala and Scow 1998; Lundquist et al. 1999), highlighting the fact that one practice (tillage) is compensated for by another (the addition of OM). Puissant et al. (2021) showed that conventional agriculture reduced the composition, taxonomic richness and size of nematode populations, likely due to the absence of crop rotations (monocultures), tillage and fertilization.

Even if many results point to greater soil biodiversity in organic farming than in conventional agriculture, some more nuanced results exist in the literature. In their meta-analysis, de Graaf et al. (2019) observe no significant difference between organic farming and conventional agriculture. These results may be explained by the fact that the practices associated with control (conventional agriculture) and treatment (organic farming) differed greatly from one experiment to the other. In general, conventional and organic farming systems differed in more than one of the following factors: fertilization practices, intensity of tillage, use of herbicides and species or cultivars. In addition, in many conventional systems, farmers can include practices such as reduced tillage, the incorporation of cover crops and/or fertilization with a combination of inorganic and organic amendments.

Finally, some studies have also shown that the biological quality of soils from organic farming plots was not always better, and may even be worse than that of conventional plots, for different soil organisms (Maron and Ranjard 2019; Christel et al. 2022). These results can be explained by the fact that organic farming systems are associated with frequent tillage, and even if they use no synthetic pesticides, they may use substances such as copper that accumulate in the soil (Joimel et al. 2016, 2020) and that can be harmful to soil organisms (Cortet and Hedde 2020).

The effect of organic farming on soil biodiversity varies depending on the pedoclimatic context, as well as on the age of the system's implementation. For example, Puissant et al. (2021) observed that the nematode enrichment index increased with the time since the agricultural system was in place (conventional system/organic farming system), which suggests that soil functions and fertility changes continue over time after the conversion of the system.

These different results thus make it clear that additional studies must be carried out in order to evaluate the mechanisms by which organic farming influences the soil properties and soil biodiversity. If the classification of agricultural systems into conventional or organic constitutes a simple categorization, it is probably not precise enough to evaluate the impact of these practices on soil biodiversity, and the literature on this subject actually covers a variety of different practices.

2.4.2. *Conservation agriculture versus conventional agriculture*

The practices of reducing or even eliminating tillage, permanent soil cover and diversified crop rotations in conservation agriculture, also have numerous effects on the abundance and diversity of microbial communities and soil fauna, and concern both beneficial organisms and pathogens. These practices help the integrity of the soil habitats to be maintained and a good level of trophic resources. However, this agriculture is still highly dependent on the use of synthetic pesticides such as herbicides, which limits its virtuous aspects.

Various studies have demonstrated an increase in biomass, diversity and microbial activities in conservation agriculture, due to a particular microclimate and a greater availability of carbon resulting from less soil disturbance and the retention of crop residues on the soil surface (Liu et al. 2014; Palm et al. 2014; Singh et al. 2018).

Similarly, several studies have noted the positive effects of conservation agriculture on soil fauna. The structure indices of macrofauna communities, such as density, are often higher in conservation agriculture than in conventional agriculture, and more so than for microfauna or microorganisms (Henneron et al. 2015; Mafongoya et al. 2016). On the basis of a meta-analysis, Briones and Schmidt (2017) concluded that conservation agriculture significantly increased the abundance of earthworms (+127%) and their biomass (+101%) compared to a conventional system.

Verhulst et al. (2010) showed that the conservation farming practices could also affect pathogens. No-till practices make it possible to maintain the stability of communities and to conserve species located higher up in the food webs, such as predators and parasites of pathogenic organisms. However, the effects vary depending on the pathogenic organisms (Palm et al. 2014). Conversely, the presence of residues, which give rise to a colder and wetter climate at the soil surface, could promote an increase in diseases. Crop rotations could change the composition of soil communities and reduce the size of pathogenic populations.

As in the case of transitions from conventional agriculture to organic farming, Puissant et al. (2021) observed an increase in the nematode enrichment index during transitions from conventional to conservation agriculture. Various studies have shown that earthworm populations and microbial biomass change rapidly when a farmer moves from a conventional system to another system that is considered to be more sustainable, such as conservation agriculture and organic farming (Ruiz-Camacho et al. 2009; Pérès et al. 2011).

2.4.3. *Organic farming versus conservation agriculture*

Few studies have focused on the comparison of these two types of agriculture. Henneron et al. (2015) demonstrated that after 14 years, conservation agriculture was overall more effective than organic farming in promoting soil biodiversity. Conservation agriculture increased the abundance of bacteria, fungi and all groups of nematodes, anecic and epigeic earthworms and most macroinvertebrates. Organic farming mainly stimulated the abundance of bacteria, bacterivorous nematodes and epigeic, endogeic and anecic earthworms. These results can be explained, at least in part, by the fact that the organic farming system used in this trial received very few or no exogenous organic inputs.

2.5. Soil biodiversity management – keys for promoting action

2.5.1. *Bioindicators*

To clarify the impacts of agricultural practices and cropping systems on biodiversity and soil functioning over the short and long term, appropriate indicators need to be developed. Bioindicators, as compared to chemical or physical indicators, have the advantage of responding quickly to environmental disturbances. For example, Vian and Peigné (2012) showed that different types of tillage in organic farming, after five years, had impacted microbial abundances, as well as the structure of bacterial communities and enzyme activities, while the organic C contents on the first 15 cm were not significantly different in the different treatments. Other studies have shown the sensitivity of soil organisms to agricultural practices and changes in land use (Le Roux 2012; Constancias et al. 2015; Joimel et al. 2017, 2021). Bioindicators thus appear to be relevant tools for assessing the effects of agricultural practices on soil biodiversity. However, the routine use of bioindicators is still limited due to the lack of standards covering different pedoclimatic contexts (Pulleman et al. 2012). These guidelines are under development, but still remain partial at this time.

2.5.2. *Link between biodiversity and soil functioning*

To be able to develop decision-making tools and manage soil biodiversity, knowledge about the links between indicators, functions and services delivered is also essential (Gardi et al. 2009). This knowledge should make it possible to answer questions such as: what is the minimum biomass required to ensure a given function? Is there a causal relationship between biological diversity and soil functioning? How many species are needed to ensure each function and ecosystem service? The answers to these questions are complex, given that a large number of interacting processes and organisms are involved at different spatial and temporal scales. In addition, different species are capable of performing a same function (functional redundancy), some species provide different functions (multifunctionality), and finally, other species contribute comparatively little (or at least their contribution is unknown) to soil functioning. Due to functional redundancy, the loss of taxonomic diversity does not necessarily lead to the loss of soil functions. Nevertheless, even in the case of functional redundancies, organisms capable of performing the same functions are not necessarily the same, depending on the habitat, the time and the process or the pressures considered (Isbell et al. 2011). Thus, even if some species may appear redundant at a given time or place, these species could be necessary to maintain soil functioning and the ecosystem services in environments subjected to disturbances (hypothesis of biodiversity insurance). This can be particularly important in a context of climate change and an increase in the frequency and severity of extreme climate events, such as episodes of drought or intense rainfall (IPCC 2021). In this way, several studies have shown that the resistance of soil microbial communities to an increase in atmospheric CO_2 concentration and an increase in mineral nitrogen inputs depended on the structure of soil microbial communities, with the resistance of microbial communities increasing with the fungi/bacteria ratio and decreasing with the Gram+ bacteria/Gram– bacteria ratio (Simonin et al. 2017).

Many studies have shown that the performance, resistance and resilience of an ecosystem are directly linked to the biodiversity of the communities it supports (Garbeva et al. 2004; Brussaard et al. 2007; Tardy et al. 2014; Oliver et al. 2015). An erosion of diversity can lead to functional losses related to denitrification (Philippot et al. 2013), decomposition of OM (Maron et al. 2018), structural stability or an increase in the survival time of pathogenic microorganisms in the soil (Ranjard et al., personal communication). Nielsen et al. (2010) focused on the relationship between biodiversity and the carbon cycle and showed a positive relationship, especially in low biodiversity systems. The loss of species with specialized functions in the soil may have a greater effect on carbon dynamics. They also showed that when two organisms do not occupy the same ecological niche, the loss of one of the two species can lead to the loss of the function. Nevertheless, this study has some limitations, since the results were obtained from experiments manipulating biodiversity, leading to a very reduced biodiversity compared to what can be observed in the field. Even if biodiversity is often associated with a better use of ecosystem resources (complementarity effects and selection effects), some authors have found contrary relationships (negative selection effects) (Balvanera et al. 2014; Harrison et al. 2014; Lemanceau et al. 2015). The literature on diversity–function relationships raises the question of determining at what level of resolution biodiversity can be considered important for regulating ecological processes in soils (Beare et al. 1997). Another difficulty in establishing relationships between indicators, functions and services arises from the fact that a service usually involves several functions and processes, which in turn can contribute to several services.

2.5.3. *Ecological engineering*

2.5.3.1. *Direct interventions on soil biodiversity*

Direct interventions are based on the introduction of organisms into the soil, with objectives such as (i) stimulating plant growth (i.e. rhizobacteria to promote nitrogen nutrition in plants, mycorrhizal fungi to improve the mineral and water nutrition, or earthworms to improve soil fertility); (ii) controlling pests or diseases (e.g. biological control agents); and (iii) improving the physical properties of soils (e.g. anecic earthworms that dig vertical galleries). Aguilar-Paredes et al. (2020), for example, showed through a literature review that the inoculation of PGPR (plant growth-promoting rhizobacteria) such as *Bacillus*, *Pseudomonas* or *Acetobacter* had demonstrable effects both on the growth and yields of cereal crops such as wheat or rice, and in agroforestry. The marketing of biofertilizers is subject to European regulations that are currently being developed.

2.5.3.2. *Indirect interventions in soil biodiversity*

Indirect interventions consist of manipulating abiotic factors that control soil biological activity (habitat structures, microclimates, nutrients and energy resources). This is the case for most agricultural practices (e.g. addition of OM, tillage, fertilization and diversification of cultivated biodiversity).

2.6. Conclusion

Soil biodiversity is crucial to the functioning of agricultural soils. It is therefore necessary to understand the effects of agricultural practices and cropping systems on the composition and abundance of soil microbial communities and soil fauna communities; this is in order to be able to limit those that degrade the biodiversity of agricultural soils and, conversely, promote those that preserve or even restore the biodiversity of agricultural soils. Agroecology must therefore lead to the development of practices that promote proper soil biological activity and maximize the ecosystem services provided by agricultural soils (i.e. fertility, depollution, erosion reduction and carbon storage), while reducing the ecosystem dis-services provided by agricultural soils (N losses and greenhouse gas emissions). Ultimately, the control of the biological functions of the soil should make it possible to significantly reduce the use of chemical inputs (e.g. fertilizers and pesticides) and lead to the modulation of certain agricultural practices known to be deleterious to soil biodiversity (i.e. intensive tillage and lack of plant cover).

2.7. References

Aguilar-Paredes, A., Valdés, G., Nuti, M. (2020). Ecosystem functions of microbial consortia in sustainable agriculture. *Agronomy*, 10, 1902. doi: 10.3390/agronomy10121902.

Allain, A. (2017). Potentiel de biodégradation des pesticides en agriculture conventionnelle vs agriculture biologique et 0 pesticide. Internship Report, M2 Mention agrosciences, environnement, territoires, paysage, forê (parcours gestion des sols et des services ecosystéiques).

Annabi, M., Houot, S., Francou, C., Poitrenaud, M., Bissonnais, Y.L. (2007). Soil aggregate stability improvement with urban composts of different maturities. *Soil Science Society of America Journal*, 71, 413–423. doi: 10.2136/sssaj2006.0161.

Arbeli, Z. and Fuentes, C.L. (2007). Accelerated biodegradation of pesticides: An overview of the phenomenon, its basis and possible solutions; and a discussion on the tropical dimension. *Crop Protection*, 26, 1733–1746. doi: 10.1016/j.cropro.2007.03.009.

Balvanera, P., Siddique, I., Dee, L., Paquette, A., Isbell, F., Gonzalez, A., Byrnes, J., O'Connor, M.I., Hungate, B.A., Griffin, J.N. (2014). Linking biodiversity and ecosystem services: Current uncertainties and the necessary next steps. *BioScience*, 64, 49–57. doi: 10.1093/biosci/bit003.

Barot, S., Allard, V., Cantarel, A., Enjalbert, J., Gauffreteau, A., Goldringer, I., Lata, J.-C., Le Roux, X., Niboyet, A., Porcher, E. (2017). Designing mixtures of varieties for multifunctional agriculture with the help of ecology. A review. *Agron. Sustain. Dev.*, 37, 13. doi: 10.1007/s13593-017-0418-x.

Beare, M.H., Reddy, M.V., Tian, G., Srivastava, S.C. (1997). Agricultural intensification, soil biodiversity and agroecosystem function in the tropics: The role of decomposer biota. *Applied Soil Ecology, Soil Biodiversity, Agricultural Intensification and Agroecosystem Function*, 6, 87–108. doi: 10.1016/S0929-1393(96)00150-3.

Bengtsson, J., Ahnström, J., Weibull, A.-C. (2005). The effects of organic agriculture on biodiversity and abundance: A meta-analysis. *Journal of Applied Ecology*, 42, 261–269. doi: 10.1111/j.1365-2664.2005.01005.x.

Borg, J., Kiær, L.P., Lecarpentier, C., Goldringer, I., Gauffreteau, A., Saint-Jean, S., Barot, S., Enjalbert, J. (2018). Unfolding the potential of wheat cultivar mixtures: A meta-analysis perspective and identification of knowledge gaps. *Field Crops Research*, 221, 298–313. doi: 10.1016/j.fcr.2017.09.006.

Boswell, E.P., Koide, R.T., Shumway, D.L., Addy, H.D. (1998). Winter wheat cover cropping, VA mycorrhizal fungi and maize growth and yield. *Agriculture, Ecosystems & Environment*, 67, 55–65. doi: 10.1016/S0167-8809(97)00094-7.

Bouché, M.B. (1993). *Lombriciens de France. Ecologie et systématique*. Inra Editions, Paris.

Bouthier, A., Pelosi, C., Villenave, C., Pérès, G., Hedde, M., Ranjard, L., Vian, J.F., Peigné, J., Cortet, J., Piron, D. (2014). Impact du travail du sol sur son fonctionnement biologique. In *Faut-il travailler le sol?* Labreuche, J., Laurent, F., Roger-Estrade, J. (eds). Éditions Quæ, Versailles.

Brady, N.C. and Weil, R.R. (2002). *The Nature and Properties of Soils*, 13th edition. Prentice Hall, Upper Saddle River, NJ.

Briar, S.S., Grewal, P.S., Somasekhar, N., Stinner, D., Miller, S.A. (2007). Soil nematode community, organic matter, microbial biomass and nitrogen dynamics in field plots transitioning from conventional to organic management. *Applied Soil Ecology*, 37, 256–266. doi: 10.1016/j.apsoil.2007.08.004.

Briones, M.J.I. and Schmidt, O. (2017). Conventional tillage decreases the abundance and biomass of earthworms and alters their community structure in a global meta-analysis. *Global Change Biology*, 23, 4396–4419. doi: 10.1111/gcb.13744.

Brussaard, L., de Ruiter, P.C., Brown, G.G. (2007). Soil biodiversity for agricultural sustainability. *Agriculture, Ecosystems & Environment*, 121, 233–244. doi: 10.1016/j.agee.2006.12.013.

van Capelle, C., Schrader, S., Brunotte, J. (2012). Tillage-induced changes in the functional diversity of soil biota – A review with a focus on German data. *European Journal of Soil Biology*, 50, 165–181. doi: 10.1016/j.ejsobi.2012.02.005.

Castellanos-Navarrete, A., Verhulst, N., Govaerts, B., Verachtert, E., Mezzalama, M., Wall Patrick, C., Deckers, J., Sayre, K.D. (2010). Conservation agriculture, improving soil quality for sustainable cropping systems? In *Advances in Soil Science: Food Security and Soil Quality*, Lal, R. and Stewart, B.A. (eds). CRC Press, Boca Raton.

Chassé, P., Pelosi, C., Lata, J.-C., Barot, S. (2019). Impact of crop genetic diversity on a litter consumer. *Basic and Applied Ecology*, 36, 1–11. doi: 10.1016/j.baae.2019.02.002.

Chen, M., Chen, B., Marschner, P. (2008). Plant growth and soil microbial community structure of legumes and grasses grown in monoculture or mixture. *Journal of Environmental Sciences*, 20, 1231–1237. doi: 10.1016/S1001-0742(08)62214-7.

Christel, A., Maron, P.-A., Ranjard, L. (2022). Méta-analyse sur l'impact des modes de production agricole sur la qualité écologique du sol. *Étude et Gestion des Sols*, 29, 117–144.

Clermont-Dauphin, C., Blanchart, E., Loranger-Merciris, G., Meynard, J.-M. (2014). Cropping systems to improve soil biodiversity and ecosystem services: The outlook and lines of research. In *Sustainable Agriculture Reviews 14: Agroecology and Global Change, Sustainable Agriculture Reviews*, Ozier-Lafontaine, H. and Lesueur-Jannoyer, M. (eds). Springer, Cham. doi: 10.1007/978-3-319-06016-3_5.

Colbach, N. and Vacher, C. (2014). Travail du sol et gestion de la flore adventice. In *Faut-il travailler le sol ?* Labreuche, J., Laurent, F., Roger-Estrade, J. (eds). Éditions Quæ, Versailles.

Constancias, F., Saby, N.P.A., Terrat, S., Dequiedt, S., Horrigue, W., Nowak, V., Guillemin, J.-P., Biju-Duval, L., Prévost-Bouré, N.C., Ranjard, L. (2015). Contrasting spatial patterns and ecological attributes of soil bacterial and archaeal taxa across a landscape. *Microbiology Open*, 4, 518–531. doi: 10.1002/mbo3.256.

Cortet, J. and Hedde, M. (2020). La faune du sol pour évaluer l'impact des pratiques agricoles et la santé des sols. *Génie écologique*. doi: 10.51257/a-v1-ge1058.

Coulibaly, S.F.M., Coudrain, V., Hedde, M., Brunet, N., Mary, B., Recous, S., Chauvat, M. (2017). Effect of different crop management practices on soil Collembola assemblages: A 4-year follow-up. *Applied Soil Ecology*, 119, 354–366. doi: 10.1016/j.apsoil.2017.06.013.

Coulibaly, S.F.M., Aubert, M., Brunet, N., Bureau, F., Legras, M., Chauvat, M. (2022). Short-term dynamic responses of soil properties and soil fauna under contrasting tillage systems. *Soil and Tillage Research*, 215, 105191. doi: 10.1016/j.still.2021.105191.

Cycoń, M. and Piotrowska-Seget, Z. (2009). Changes in bacterial diversity and community structure following pesticides addition to soil estimated by cultivation technique. *Ecotoxicology*, 18, 632–642. doi: 10.1007/s10646-009-0321-6.

Decaëns, T., Jiménez, J.J., Gioia, C., Measey, G.J., Lavelle, P. (2006). The values of soil animals for conservation biology. *European Journal of Soil Biology*, 42(Supplement 1), S23–S38. doi: 10.1016/j.ejsobi.2006.07.001.

Dequiedt, S.S., Saby, N.P.A., Lelievre, M., Jolivet, C., Thioulouse, J., Toutain, B., Arrouays, D., Bispo, A., Lemanceau, P., Ranjard, L. (2011). Biogeographical patterns of soil molecular microbial biomass as influenced by soil characteristics and management. *Global Ecology and Biogeography*, 20, 641–652. doi: 10.1111/j.1466-8238.2010.00628.x.

Dequiedt, S.S., Karimi, B., Chemidlin Prévost-Bouré, N., Terrat, S., Horrigue, W., Djemiel, C., Lelievre, M.M., Nowak, V., Wincker, P., Jolivet, C. et al. (2020). Le RMQS au service de l'écologie microbienne des sols français. *Etude et Gestion des Sols*, 27, 51–71.

Dubs, F., Le Roux, X., Allard, V., Andrieu, B., Barot, S., Cantarel, A., Vallavielle-Pope, C., de Gauffreteau, A., Goldringer, I., Montagnier, C. et al. (2018). An experimental design to test the effect of wheat variety mixtures on biodiversity and ecosystem services. Document, HAL [Online]. Available at: https://hal.archives-ouvertes.fr/hal-01843564.

Dubs, F., Enjalbert, J., Barot, S., Porcher, E., Allard, V., de Vallavielle-Pope, C., Gauffreteau, A., Niboyet, A., Pommier, T., Saint-Jean, S. et al. (2023). Unfolding the link between multiple ecosystem services and bundles of functional traits to design multifunctional crop variety mixtures. *Agronomy for Sustainable Development*, 43, 71. doi: 10.1007/s13593-023-00924-6.

Edwards, C.A. and Bohlen, P.J. (1996). *Biology and Ecology of Earthworms*, 3rd edition. Chapman & Hall, London.

Fierer, N., Lauber, C.L., Ramirez, K.S., Zaneveld, J., Bradford, M.A., Knight, R. (2012). Comparative metagenomic, phylogenetic and physiological analyses of soil microbial communities across nitrogen gradients. *ISME J*, 6, 1007–1017. doi: 10.1038/ismej.2011.159.

Garbeva, P., van Veen, J.A., van Elsas, J.D. (2004). Microbial diversity in soil: Selection microbial populations by plant and soil type and implications for disease suppressiveness. *Annu. Rev. Phytopathol.*, 42, 243–270. doi: 10.1146/annurev.phyto.42.012604.135455.

Gardi, C., Montanarella, L., Arrouays, D., Bispo, A., Lemanceau, P., Jolivet, C., Mulder, C., Ranjard, L., Römbke, J., Rutgers, M. et al. (2009). Soil biodiversity monitoring in Europe: Ongoing activities and challenges. *European Journal of Soil Science*, 60, 807–819. doi: 10.1111/j.1365-2389.2009.01177.x.

Gobat, J.-M., Aragno, M., Mathey, W. (2010). *Le sol vivant. Bases de pédologie, biologie des sols*, 3rd edition. PPUR, Lausanne.

Gomiero, T., Pimentel, D., Paoletti, M.G. (2011). Is there a need for a more sustainable agriculture? *Critical Reviews in Plant Sciences*, 30, 6–23. doi: 10.1080/07352689.2011.553515.

de Graaff, M.-A., Hornslein, N., Throop, H.L., Kardol, P., van Diepen, L.T.A. (2019). Effects of agricultural intensification on soil biodiversity and implications for ecosystem functioning: A meta-analysis. In *Advances in Agronomy*, Sparks, D.L. (ed.). Academic Press, Cambridge, MA. doi: 10.1016/bs.agron.2019.01.001.

Gunapala, N. and Scow, K.M. (1998). Dynamics of soil microbial biomass and activity in conventional and organic farming systems. *Soil Biology and Biochemistry*, 30, 805–816. doi: 10.1016/S0038-0717(97)00162-4.

Hansen, S. and Engelstad, F. (1999). Earthworm populations in a cool and wet district as affected by tractor traffic and fertilisation. *Applied Soil Ecology*, 13, 237–250. doi: 10.1016/S0929-1393(99)00037-2.

Harrison, P.A., Berry, P.M., Simpson, G., Haslett, J.R., Blicharska, M., Bucur, M., Dunford, R., Egoh, B., Garcia-Llorente, M., Geamănă, N. et al. (2014). Linkages between biodiversity attributes and ecosystem services: A systematic review. *Ecosystem Services*, 9, 191–203. doi: 10.1016/j.ecoser.2014.05.006.

Hartmann, M., Frey, B., Mayer, J., Mäder, P., Widmer, F. (2015). Distinct soil microbial diversity under long-term organic and conventional farming. *ISME J*, 9, 1177–1194. doi: 10.1038/ismej.2014.210.

Hendrix, P.F., Parmelee, R.W., Crossley, D.A., Coleman, D.C., Odum, E.P., Groffman, P.M. (1986). Detritus food webs in conventional and no-tillage agroecosystems. *BioScience*, 36, 374–380. doi: 10.2307/1310259.

Henneron, L., Bernard, L., Hedde, M., Pelosi, C., Villenave, C., Chenu, C., Bertrand, M., Girardin, C., Blanchart, E. (2015). Fourteen years of evidence for positive effects of conservation agriculture and organic farming on soil life. *Agron. Sustain. Dev.*, 35, 169–181. doi: 10.1007/s13593-014-0215-8.

Horrigue, W., Dequiedt, S., Chemidlin Prévost-Bouré, N., Jolivet, C., Saby, N.P.A., Arrouays, D., Bispo, A., Maron, P.-A., Ranjard, L. (2016). Predictive model of soil molecular microbial biomass. *Ecological Indicators*, 64, 203–211. doi: 10.1016/j.ecolind.2015.12.004.

Houot, S., Pons, M.-N., Pradel, M., Savini, I., Tibi, A. (2014). Valorisation des matières fertilisantes d'origine résiduaire sur les sols à usage agricole ou forestier. Report, Ministère de l'Agriculture.

IPCC (2021). Climate Change 2021: The Physical Science Basis. Contribution of Working Group I to the Sixth Assessment Report of the Intergovernmental Panel on Climate Change. Report, Cambridge University Press, Cambridge.

Isbell, F., Calcagno, V., Hector, A., Connolly, J., Harpole, W.S., Reich, P.B., Scherer-Lorenzen, M., Schmid, B., Tilman, D., van Ruijven, J. et al. (2011). High plant diversity is needed to maintain ecosystem services. *Nature*, 477, 199–202. doi: 10.1038/nature10282.

Jeffery, S., Gardi, C., Jones, A., European Commission, Joint Research Centre, Institute for Environment and Sustainability. (2013). *European Atlas of Soil Biodiversity*. Publications Office, Luxembourg.

Joimel, S., Cortet, J., Jolivet, C.C., Saby, N.P.A., Chenot, E.D., Branchu, P., Consalès, J.N., Lefort, C., Morel, J.L., Schwartz, C. (2016). Physico-chemical characteristics of topsoil for contrasted forest, agricultural, urban and industrial land uses in France. *Science of The Total Environment*, 545–546, 40–47. doi: 10.1016/j.scitotenv.2015.12.035.

Joimel, S., Schwartz, C., Hedde, M., Kiyota, S., Krogh, P.H., Nahmani, J., Pérès, G., Vergnes, A., Cortet, J. (2017). Urban and industrial land uses have a higher soil biological quality than expected from physicochemical quality. *Science of The Total Environment*, 584–585, 614–621. doi: 10.1016/j.scitotenv.2017.01.086.

Joimel, S., Cortet, J., Consalès, J.N., Branchu, P., Haudin, C.-S., Morel, J.L., Schwartz, C. (2020). Contribution of chemical inputs on the trace elements concentrations of surface soils in urban allotment gardens. *J. Soils Sediments*, 21, 328–337. doi: 10.1007/s11368-020-02784-z.

Joimel, S., Schwartz, C., Bonfanti, J., Hedde, M., Krogh, P.H., Pérès, G., Pernin, C., Rakoto, A., Salmon, S., Santorufo, L. et al. (2021). Functional and taxonomic diversity of collembola as complementary tools to assess land use effects on soils biodiversity. *Front. Ecol. Evol.*, 9, 630919. doi: 10.3389/fevo.2021.630919.

Joimel, S., Chassain, J., Artru, M., Faburé, J. (2022). Collembola are among the most pesticide sensitive soil fauna groups: A meta-analysis. *Environ. Toxicol. Chem.*, 1–9. doi: 10.1002/etc.5428.

Karimi, B., Dequiedt, S., Terrat, S., Jolivet, C., Arrouays, D., Wincker, P., Cruaud, C., Bispo, A., Chemidlin Prévost-Bouré, N., Ranjard, L. (2019). Biogeography of soil bacterial networks along a gradient of cropping intensity. *Scientific Reports*, 9, 3812. doi: 10.1038/s41598-019-40422-y.

Kepler, R.M., Epp Schmidt, D.J., Yarwood, S.A., Cavigelli, M.A., Reddy, K.N., Duke, S.O., Bradley, C.A., Williams, M.M., Buyer, J.S., Maul, J.E. (2020). Soil microbial communities in diverse agroecosystems exposed to the herbicide glyphosate. *Applied and Environmental Microbiology*, 86, e01744-19. doi: 10.1128/AEM.01744-19.

Kim, N., Zabaloy, M.C., Guan, K., Villamil, M.B. (2020). Do cover crops benefit soil microbiome? A meta-analysis of current research. *Soil Biology and Biochemistry*, 142, 107701. doi: 10.1016/j.soilbio.2019.107701

Kladivko, E.J. (2001). Tillage systems and soil ecology. *Soil and Tillage Research*, 61, 61–76. doi: 10.1016/S0167-1987(01)00179-9.

Krogh, K.A., Halling-Sørensen, B., Mogensen, B.B., Vejrup, K.V. (2003). Environmental properties and effects of nonionic surfactant adjuvants in pesticides: A review. *Chemosphere*, 50, 871–901. doi: 10.1016/S0045-6535(02)00648-3.

Le Roux, X. (2012). *Agriculture et biodiversité valoriser les synergies : expertise scientifique collective Inra Juillet 2008*. Éditions Quæ, Versailles.

Lemanceau, P., Maron, P.-A., Mazurier, S., Mougel, C., Pivato, B., Plassart, P., Ranjard, L., Revellin, C., Tardy, V., Wipf, D. (2015). Understanding and managing soil biodiversity: A major challenge in agroecology. *Agron. Sustain. Dev.*, 35, 67–81. doi: 10.1007/s13593-014-0247-0.

Liang, W., Lou, Y., Li, Q., Zhong, S., Zhang, X., Wang, J. (2009). Nematode faunal response to long-term application of nitrogen fertilizer and organic manure in Northeast China. *Soil Biology and Biochemistry, Science Goes Underground in China*, 41, 883–890. doi: 10.1016/j.soilbio.2008.06.018.

Lienhard, P., Terrat, S., Prévost-Bouré, N.C., Nowak, V., Régnier, T., Sayphoummie, S., Panyasiri, K., Tivet, F., Mathieu, O., Levêque, J. et al. (2014). Pyrosequencing evidences the impact of cropping on soil bacterial and fungal diversity in Laos tropical grassland. *Agron. Sustain. Dev.*, 34, 525–533. doi: 10.1007/s13593-013-0162-9.

Liu, C., Lu, M., Cui, J., Li, B., Fang, C. (2014). Effects of straw carbon input on carbon dynamics in agricultural soils: A meta-analysis. *Global Change Biology*, 20, 1366–1381. doi: 10.1111/gcb.12517.

Lori, M., Symnaczik, S., Mäder, P., Deyn, G.D., Gattinger, A. (2017). Organic farming enhances soil microbial abundance and activity – A meta-analysis and meta-regression. *PLOS ONE*, 12, e0180442. doi: 10.1371/journal.pone.0180442.

Lundquist, E.J., Scow, K.M., Jackson, L.E., Uesugi, S.L., Johnson, C.R. (1999). Rapid response of soil microbial communities from conventional, low input, and organic farming systems to a wet/dry cycle. *Soil Biology and Biochemistry*, 31, 1661–1675. doi: 10.1016/S0038-0717(99)00080-2.

Mafongoya, P., Rusinamhodzi, L., Siziba, S., Thierfelder, C., Mvumi, B.M., Nhau, B., Hove, L., Chivenge, P. (2016). Maize productivity and profitability in conservation agriculture systems across agro-ecological regions in Zimbabwe: A review of knowledge and practice. *Agriculture, Ecosystems & Environment*, 220, 211–225. doi: 10.1016/j.agee.2016.01.017.

Mamy, L., Patureau, D., Barriuso, E., Bedos, C., Bessac, F., Louchart, X., Martin-laurent, F., Miege, C., Benoit, P. (2015). Prediction of the fate of organic compounds in the environment from their molecular properties: A review. *Critical Reviews in Environmental Science and Technology*, 45, 1277–1377. doi: 10.1080/10643389.2014.955627.

Maron, P.-A. and Ranjard, L. (2019). Qualité écologique des sols. Reference Article, *Techniques de l'ingénieur*. doi: 10.51257/a-v1-ge1051.

Maron, P.-A., Sarr, A., Kaisermann, A., Lévêque, J., Mathieu, O., Guigue, J., Karimi, B., Bernard, L., Dequiedt, S., Terrat, S. et al. (2018). High microbial diversity promotes soil ecosystem functioning. *Applied and Environmental Microbiology*, 84, e02738-17. doi: 10.1128/AEM.02738-17.

McDaniel, M.D., Tiemann, L.K., Grandy, A.S. (2014). Does agricultural crop diversity enhance soil microbial biomass and organic matter dynamics? A meta-analysis. *Ecological Applications*, 24, 560–570. doi: 10.1890/13-0616.1.

Millennium Ecosystem Assessment. (2005). *Ecosystems and Human Well-Being: Synthesis*. Island Press, Washington, D.C.

Morriën, E., Hannula, S.E., Snoek, L.B., Helmsing, N.R., Zweers, H., de Hollander, M., Soto, R.L., Bouffaud, M.-L., Buée, M., Dimmers, W. et al. (2017). Soil networks become more connected and take up more carbon as nature restoration progresses. *Nat. Commun.*, 8, 14349. doi: 10.1038/ncomms14349.

Navarro-Noya, Y.E., Gómez-Acata, S., Montoya-Ciriaco, N., Rojas-Valdez, A., Suárez-Arriaga, M.C., Valenzuela-Encinas, C., Jiménez-Bueno, N., Verhulst, N., Govaerts, B., Dendooven, L. (2013). Relative impacts of tillage, residue management and crop-rotation on soil bacterial communities in a semi-arid agroecosystem. *Soil Biology and Biochemistry*, 65, 86–95. doi: 10.1016/j.soilbio.2013.05.009.

Nielsen, U.N., Osler, G.H.R., Campbell, C.D., Burslem, D.F.R.P., van der Wal, R. (2010). The influence of vegetation type, soil properties and precipitation on the composition of soil mite and microbial communities at the landscape scale. *Journal of Biogeography*, 37, 1317–1328. doi: 10.1111/j.1365-2699.2010.02281.x.

Nivelle, E., Verzeaux, J., Habbib, H., Kuzyakov, Y., Decocq, G., Roger, D., Lacoux, J., Duclercq, J., Spicher, F., Nava-Saucedo, J.-E. et al. (2016). Functional response of soil microbial communities to tillage, cover crops and nitrogen fertilization. *Applied Soil Ecology*, 108, 147–155. doi: 10.1016/j.apsoil.2016.08.004.

Obriot, F. (2016). Epandage de produits résiduaires organiques et fonctionnement biologique des sols : de la quantification des impacts sur les cycles carbone et azote à l'évaluation multicritère de la pratique à l'echelle de la parcelle. Thesis, AgroParisTech, Paris.

Obriot, F., Stauffer, M., Goubard, Y., Cheviron, N., Pérès, G., Eden, M., Revallier, A., Vieublé-Gonod, L., Houot, S. (2016). Multi-criteria indices to evaluate the effects of repeated organic amendment applications on soil and crop quality. *Agriculture, Ecosystems & Environment*, 232, 165–178. doi: 10.1016/j.agee.2016.08.004.

Oliver, T.H., Heard, M.S., Isaac, N.J.B., Roy, D.B., Procter, D., Eigenbrod, F., Freckleton, R., Hector, A., Orme, C.D.L., Petchey, O.L. et al. (2015). Biodiversity and resilience of ecosystem functions. *Trends in Ecology & Evolution*, 30, 673–684. doi: 10.1016/j.tree. 2015.08.009.

Orgiazzi, A., Bardgett, R.D., Barrios, E., Behan-Pelletier, V., Briones, M.J.I., Chotte, J.L., Deyn, G.B., de Eggleton, P., Fierer, N., Fraser, T. et al. (2016). Global soil biodiversity atlas. *European Union*, doi: 10.2788/2613.

Palm, C., Blanco-Canqui, H., DeClerck, F., Gatere, L., Grace, P. (2014). Conservation agriculture and ecosystem services: An overview. *Agriculture, Ecosystems & Environment*, 187, 87–105. doi: 10.1016/j.agee.2013.10.010.

Pelosi, C., Bertrand, M., Makowski, D., Roger-Estrade, J. (2008). WORMDYN: A model of Lumbricus terrestris population dynamics in agricultural fields. *Ecological Modelling*, 218, 219–234. doi: 10.1016/j.ecolmodel.2008.07.002.

Pelosi, C., Joimel, S., Makowski, D. (2013). Searching for a more sensitive earthworm species to be used in pesticide homologation tests – A meta-analysis. *Chemosphere*, 90, 895–900. doi: 10.1016/j.chemosphere.2012.09.034.

Pelosi, C., Barot, S., Capowiez, Y., Hedde, M., Vandenbulcke, F. (2014). Pesticides and earthworms. A review. *Agron. Sustain. Dev.*, 34, 199–228. doi: 10.1007/s13593-013-0151-z.

Pelosi, C., Pey, B., Caro, G., Cluzeau, D., Peigné, J., Bertrand, M., Hedde, M. (2016). Dynamics of earthworm taxonomic and functional diversity in ploughed and no-tilled cropping systems. *Soil and Tillage Research*, 156, 25–32. doi: 10.1016/j.still.2015.07.016.

Pérès, G., Vandenbulcke, F., Guernion, M., Hedde, M., Beguiristain, T., Douay, F., Houot, S., Piron, D., Richard, A., Bispo, A. et al. (2011). Earthworm indicators as tools for soil monitoring, characterization and risk assessment. An example from the national Bioindicator programme (France). *Pedobiologia*, 54(Supplement), S77–S87. doi: 10.1016/j.pedobi. 2011.09.015.

Pérez-Piqueres, A., Edel-Hermann, V., Alabouvette, C., Steinberg, C. (2006). Response of soil microbial communities to compost amendments. *Soil Biology and Biochemistry*, 38, 460–470. doi: 10.1016/j.soilbio.2005.05.025.

Pey, B., Nahmani, J., Auclerc, A., Capowiez, Y., Cluzeau, D., Cortet, J., Decaëns, T., Deharveng, L., Dubs, F., Joimel, S. et al. (2014). Current use of and future needs for soil invertebrate functional traits in community ecology. *Basic and Applied Ecology*, 15, 194–206. doi: 10.1016/j.baae.2014.03.007.

Philippot, L., Spor, A., Hénault, C., Bru, D., Bizouard, F., Jones, C.M., Sarr, A., Maron, P.-A. (2013). Loss in microbial diversity affects nitrogen cycling in soil. *ISME J.*, 7, 1609–1619. doi: 10.1038/ismej.2013.34.

Pielou, E.C. (1975). *Ecological Diversity*. John Wiley & Sons, New York.

Plenchette, C. and Perrin, R. (1992). Evaluation in the greenhouse of the effects of fungicides on the development of mycorrhiza on leek and wheat. *Mycorrhiza*, 1, 59–62. doi: 10.1007/BF00206137.

Plenchette, C., Clermont-Dauphin, C., Meynard, J.M., Fortin, J.A. (2005). Managing arbuscular mycorrhizal fungi in cropping systems. *Can. J. Plant Sci.*, 85, 31–40. doi: 10.4141/P03-159.

Puissant, J., Villenave, C., Chauvin, C., Plassard, C., Blanchart, E., Trap, J. (2021). Quantification of the global impact of agricultural practices on soil nematodes: A meta-analysis. *Soil Biology and Biochemistry*, 161, 108383. doi: 10.1016/j.soilbio.2021.108383.

Pulleman, M., Creamer, R., Hamer, U., Helder, J., Pelosi, C., Pérès, G., Rutgers, M. (2012). Soil biodiversity, biological indicators and soil ecosystem services – An overview of European approaches. *Current Opinion in Environmental Sustainability*, 4, 529–538. doi: 10.1016/j.cosust.2012.10.009.

Qin, H., Lu, K., Strong, P.J., Xu, Q., Wu, Q., Xu, Z., Xu, J., Wang, H. (2015). Long-term fertilizer application effects on the soil, root arbuscular mycorrhizal fungi and community composition in rotation agriculture. *Applied Soil Ecology*, 89, 35–43. doi: 10.1016/j.apsoil.2015.01.008.

Roger-Estrade, J., Richard, G., Boizard, H., Boiffin, J., Caneill, J., Manichon, H. (2000). Modelling structural changes in tilled topsoil over time as a function of cropping systems. *European Journal of Soil Science*, 51, 455–474. doi: 10.1046/j.1365-2389.2000.00323.x.

Roger-Estrade, J., Anger, C., Bertrand, M., Richard, G. (2010). Tillage and soil ecology: Partners for sustainable agriculture. *Soil and Tillage Research*, 111, 33–40. doi: 10.1016/j.still.2010.08.010.

Ros, M., Klammer, S., Knapp, B., Aichberger, K., Insam, H. (2006). Long-term effects of compost amendment of soil on functional and structural diversity and microbial activity. *Soil Use and Management*, 22, 209–218. doi: 10.1111/j.1475-2743.2006.00027.x.

Ruiz-Camacho, N., Velasquez, E., Pando, A., Decaëns, T., Dubs, F., Lavelle, P. (2009). Indicateurs synthétiques de la qualité du sol. *Etude et Gestion des Sols*, 16, 323–338.

Salmon, S., Vittier, T., Barot, S., Ponge, J.-F., Ben Assoula, F., Lusley, P. (2021). Responses of Collembola communities to mixtures of wheat varieties: A trait-based approach. *Pedobiologia*, 87–88, 150755. doi: 10.1016/j.pedobi.2021.150755.

Schaefer, M. and Schauermann, J. (1990). The soil fauna of beech forests: Comparison between a mull and a moder soil. *Pedobiologia*, 34, 299–314.

Schleuning, M., Fründ, J., García, D. (2015). Predicting ecosystem functions from biodiversity and mutualistic networks: An extension of trait-based concepts to plant–animal interactions. *Ecography*, 38, 380–392. doi: 10.1111/ecog.00983.

Scimia, J., Hoeffner, K., Branchu, P., Joimel, S., Schwartz, C., Bechet, B., Dumat, C., Guernion, M., Poupelin, M., Cluzeau, D. (2014). Surprisingly evaluation of earthworm diversity in associative gardens of metropolitan areas. In *The First Global Soil Biodiversity Conference*, Dijon.

Shannon, C.E. and Weaver, W. (1963). *The Mathematical Theory of Communication*. University of Illinois Press, Champaign, IL.

Sileshi, G., Mafongoya, P.L., Chintu, R., Akinnifesi, F.K. (2008). Mixed-species legume fallows affect faunal abundance and richness and N cycling compared to single species in maize–fallow rotations. *Soil Biology and Biochemistry*, 40, 3065–3075. doi: 10.1016/j.soilbio.2008.09.007.

Simonin, M., Le Roux, X., Poly, F., Lerondelle, C., Hungate, B.A., Nunan, N., Niboyet, A. (2015). Coupling between and among ammonia oxidizers and nitrite oxidizers in grassland mesocosms submitted to elevated CO_2 and nitrogen supply. *Microb. Ecol.*, 70, 809–818. doi: 10.1007/s00248-015-0604-9.

Simonin, M., Nunan, N., Bloor, J.M.G., Pouteau, V., Niboyet, A. (2017). Short-term responses and resistance of soil microbial community structure to elevated CO_2 and N addition in grassland mesocosms. *FEMS Microbiol. Lett.*, 364. doi: 10.1093/femsle/fnx077.

Singh, G., Bhattacharyya, R., Das, T.K., Sharma, A.R., Ghosh, A., Das, S., Jha, P. (2018). Crop rotation and residue management effects on soil enzyme activities, glomalin and aggregate stability under zero tillage in the Indo-Gangetic Plains. *Soil and Tillage Research*, 184, 291–300. doi: 10.1016/j.still.2018.08.006.

Sorensen, S.R., Ronen, Z., Aamand, J. (2001). Isolation from agricultural soil and characterization of a *Sphingomonas* sp. able to mineralize the phenylurea herbicide isoproturon. *Applied Environmental Microbiology*, 67, 5403–5409. doi: 10.1128/AEM.67.12.5403-5409.2001.

Tardy, V., Mathieu, O., Lévêque, J., Terrat, S., Chabbi, A., Lemanceau, P., Ranjard, L., Maron, P.-A. (2014). Stability of soil microbial structure and activity depends on microbial diversity. *Environ. Microbiol. Rep.*, 6, 173–183. doi: 10.1111/1758-2229.12126.

Terrat, S., Karimi, B., Chemidlin, N., Horrigue, W., Djemiel, C., Ranjard, L. (2018). La caractérisation des communautés microbiennes du sol à l'échelle de la France pour évaluer l'effet de l'usage des sols. *Innovations Agronomiques*, 69, 27–37.

Tibbett, M., Fraser, T.D., Duddigan, S. (2020). Identifying potential threats to soil biodiversity. *PeerJ*, 8, e9271. doi: 10.7717/peerj.9271.

Torsvik, V., Sørheim, R., Goksøyr, J. (1996). Total bacterial diversity in soil and sediment communities – A review. *Journal of Industrial Microbiology and Biotechnology*, 17, 170–178. doi: 10.1007/BF01574690.

Treonis, A.M., Austin, E.E., Buyer, J.S., Maul, J.E., Spicer, L., Zasada, I.A. (2010). Effects of organic amendment and tillage on soil microorganisms and microfauna. *Applied Soil Ecology*, 46, 103–110. doi: 10.1016/j.apsoil.2010.06.017.

Treseder, K.K. (2004). A meta-analysis of mycorrhizal responses to nitrogen, phosphorus, and atmospheric CO_2 in field studies. *New Phytologist*, 164, 347–355. doi: 10.1111/j.1469-8137.2004.01159.x.

Treseder, K.K. (2008). Nitrogen additions and microbial biomass: A meta-analysis of ecosystem studies. *Ecology Letters*, 11, 1111–1120. doi: 10.1111/j.1461-0248.2008.01230.x.

Vargas Gil, S., Meriles, J., Conforto, C., Basanta, M., Radl, V., Hagn, A., Schloter, M., March, G.J. (2011). Response of soil microbial communities to different management practices in surface soils of a soybean agroecosystem in Argentina. *European Journal of Soil Biology*, 47, 55–60. doi: 10.1016/j.ejsobi.2010.11.006.

Venter, Z.S., Jacobs, K., Hawkinsc. H.J. (2016). The impact of crop rotation on soil microbial diversity: A meta-analysis. *Pedobiologia*, 59, 215–223. doi: 10.1016/j.pedobi.2016.04.001.

Verhulst, N., Govaerts, B., Verachtert, E., Castellanos-Navarette, A., Mezzalama, M., Wall, P.C., Chocobar, A., Deckers, J., Sayre, K.D. (2010). Conservation agriculture, improving soil quality for sustainable production systems? In *Food Security and Soil Quality*, Lal, R. and Stewart, B.A. (eds). CRC Press, Boca Raton, FL.

Vian, J.F. and Peigné, J. (2012). Influence du travail du sol sur différents bioindicateurs du sol. Résultats des mesures des bioindicateurs sur le site de Thil. Program, Bioindicateurs de la qualité des sols, phase II. ADEME.

Vieublé Gonod, L., Martin-Laurent, F., Chenu, C. (2006). 2,4-D impact on bacterial communities, and the activity and genetic potential of 2,4-D degrading communities in soil. *FEMS Microbiol. Ecol.*, 58, 529–537. doi: 10.1111/j.1574-6941.2006.00159.x.

Vieublé Gonod, L., Benoit, P., Cohen, N., Houot, S. (2009). Spatial and temporal heterogeneity of soil microorganisms and isoproturon degrading activity in a tilled soil amended with urban waste composts. *Soil Biology and Biochemistry*, 41, 2558–2567. doi: 10.1016/j.soilbio.2009.09.017.

Wang, Z., Chen, Q., Liu, L., Wen, X., Liao, Y. (2016). Responses of soil fungi to 5-year conservation tillage treatments in the drylands of northern China. *Applied Soil Ecology*, 101, 132–140. doi: 10.1016/j.apsoil.2016.02.002.

Wardle, D.A. (1995). Impacts of disturbance on detritus food webs in agro-ecosystems of contrasting tillage and weed management practices. In *Advances in Ecological Research*, Begon, M. and Fitter, A.H. (eds). Academic Press, Cambridge. doi: 10.1016/S0065-2504(08)60065-3.

Wardle, D.A., Bardgett, R.D., Klironomos, J.N., Setälä, H., van der Putten, W.H. , Wall, D.H. (2004). Ecological linkages between aboveground and belowground biota. *Science*, 304, 1629–1633. doi: 10.1126/science.1094875.

Wei, C., Zheng, H., Li, Q., Lü, X., Yu, Q., Zhang, H., Chen, Q., He, N., Kardol, P., Liang, W. et al. (2012). Nitrogen addition regulates soil nematode community composition through ammonium suppression. *PLoS ONE*, 7, e43384. doi: 10.1371/journal.pone.0043384.

Yin, C., Jones, K.L., Peterson, D.E., Garrett, K.A., Hulbert, S.H., Paulitz, T.C. (2010). Members of soil bacterial communities sensitive to tillage and crop rotation. *Soil Biology and Biochemistry*, 42, 2111–2118. doi: 10.1016/j.soilbio.2010.08.006.

Zhang, J., Li, Q., Liang, W. (2010). Effect of acetochlor and carbofuran on soil nematode communities in a Chinese soybean field. *AJAR*, 5, 2787–2794. doi: 10.5897/AJAR. 9000195.

Zhang, X., Li, Q., Zhu, A., Liang, W., Zhang, J., Steinberger, Y. (2012). Effects of tillage and residue management on soil nematode communities in North China. *Ecological Indicators*, 13, 75–81. doi: 10.1016/j.ecolind.2011.05.009.

Zhang, X., Liu, W., Zhang, G., Jiang, L., Han, X. (2015). Mechanisms of soil acidification reducing bacterial diversity. *Soil Biology and Biochemistry*, 81, 275–281. doi: 10.1016/j.soilbio.2014.11.004.

Zhang, S., Cui, S., McLaughlin, N.B., Liu, P., Hu, N., Liang, W., Wu, D., Liang, A. (2019). Tillage effects outweigh seasonal effects on soil nematode community structure. *Soil and Tillage Research*, 192, 233–239. doi: 10.1016/j.still.2019.05.017.

Zhao, J., Neher, D.A., Fu, S., Li, Z., Wang, K. (2013). Non-target effects of herbicides on soil nematode assemblages. *Pest Management Science*, 69, 679–684. doi: 10.1002/ps.3505.

3

Spatial Variability and Mapping of Agricultural Soils

Jean-Marc GILLIOT
UMR ECOSYS, AgroParisTech, INRAE, Université Paris Saclay, Palaiseau, France

3.1. Introduction

In the development of sustainable agricultural practices, the consideration of the spatial variability of soils is an essential factor. From the intra-plot scale (in the case of precision agriculture) to the global scale, knowledge of the spatial variability of soils is necessary to address the challenges of climate change, such as carbon storage or the sustainable management of water resources. Classic soil mapping based on information from auger surveys, pits and laboratory analyses has gradually begun to incorporate some of the techniques developed in the digital revolution (Figure 3.1). All this information is now stored and analyzed spatially in a database using a geographic information system (GIS). The proliferation of GNSS (Global Navigation Satellite Systems) positioning systems facilitates the integration of field data in the form of maps, with increasing ease of use (via smartphones) and accuracy thanks to differential models (DGNSS). More and more digital sensors provide auxiliary data, which are combined with the "classic" data in an approach known as "digital soil mapping" (DSM) (McBratney et al. 2003). Modeling techniques using advanced processing methods, such as "machine learning", are becoming increasingly common to spatially predict soil properties, densify them or generalize them geographically.

Agricultural Soil Science,
coordinated by Yves COQUET and Joël MICHELIN. © ISTE Ltd 2025.

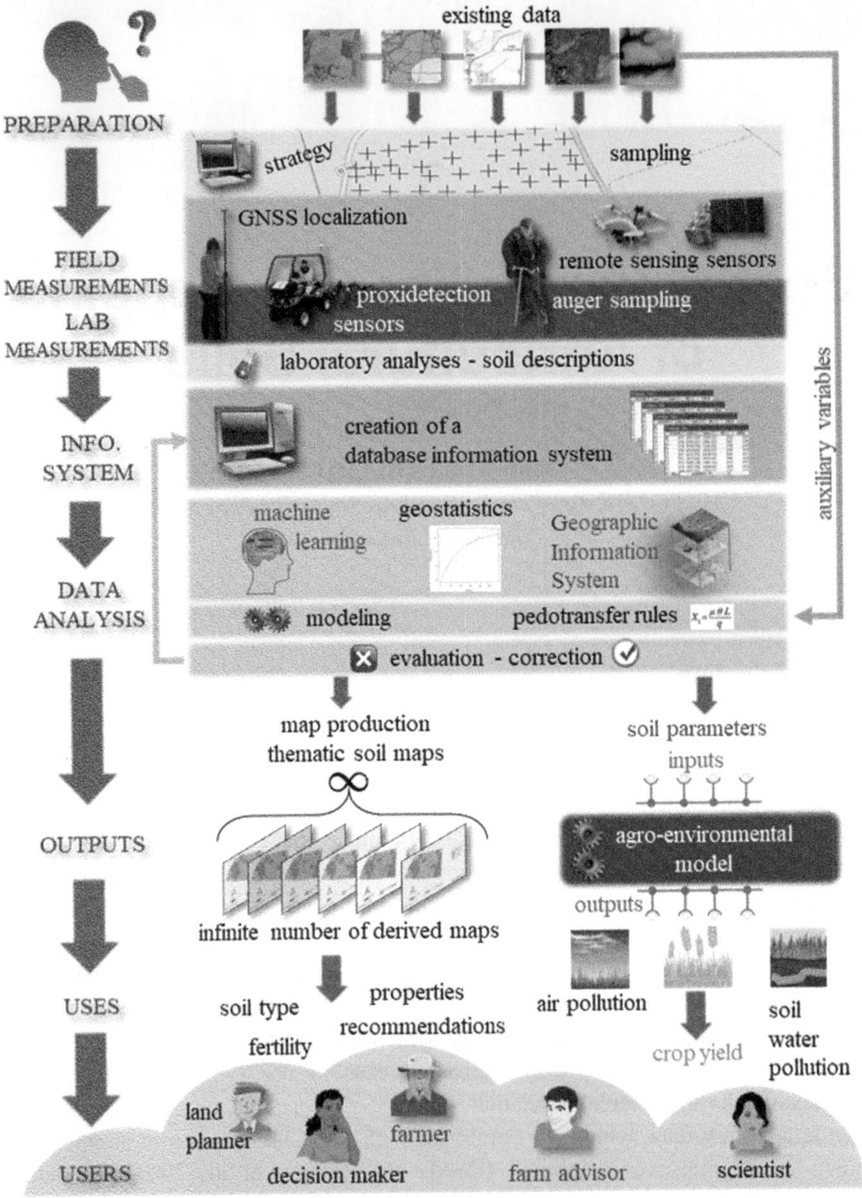

Figure 3.1. *Approach and use of soil mapping. For a color version of this figure, see www.iste.co.uk/coquet/agricultural.zip*

Instead of using a single fixed paper map, the creation of a database containing all of the above information makes it possible to produce a very large number of derived maps, or thematic maps, responding to different uses, for both researchers and farmers. These soil maps are also often used as one of the inputs for agro-environmental models, which make it possible to monitor and predict the functioning of agricultural systems. These models have proven to be very useful planning tools for testing the adaptations of cropping systems to climate change.

3.2. Background: the origins of soil mapping in France

Drawing maps has been a very important area of activity since the beginning of human history. Examples have been found of certain prehistoric petroglyphs whose geometric figures are sometimes interpreted as representations of "agrarian land registers". The most famous of these stone engravings is the "map" of Bedolina in Italy (Brocard 2005), which dates back to the Bronze Age. During Antiquity, the foundations of scientific cartography were defined by the Greeks, particularly Hipparchus, who introduced the concept of cartographic projection, and Ptolemy with his treatise on geography in the year 150 (Guerrier 2016). In France, in 1748, Louis XV assigned the task of triangulating his battlefields in Flanders to Cassini, who made maps so precise that the king later said to him: "I want the map of my kingdom to be prepared in the same way, and I am entrusting that task to you" (Gallois 1909). Cassini's map then laid the groundwork for the general and systematic cartography of the national territory of France (Bonnerot et al. 2021), using the method of geodetic triangulation. Later, by order of Napoleon I, Cassini's map would then be modernized between 1820 and 1866, giving rise to the ordinance survey map, at a scale of 1:80,000. In 1940, the departments that produced the ordinance survey map were replaced by the *Institut géographique national*, or IGN (French National Geographic Institute), which would then develop topographic maps at different scales, as we know them today. With regard to the mapping of the subsoil, the work of the French committee for the history of geology (Savaton 2003) shows that, although Jean-Etienne Guettard created a map and a mineralogical atlas of France in the middle of the 18th century, France did not yet have a geological map of the entire territory at the beginning of the 19th century. The geognostic map of the surroundings of Paris (Cuvier and Brongniart 1811) is considered the first French geological map. Departmental maps were then published between 1825 and 1875, and then replaced beginning in 1868 by the detailed geological coverage at 1/80,000th scale, carried out by the new geological mapping service of France created through an initiative of Napoleon III (BRGM 2021). In 1967, the BRGM and the geological mapping service of France merged, which led to the creation of the modern version of the geological map at 1/50,000th scale. The first traces in

France of the use of maps for agricultural uses, which could not yet quite qualify as soil maps, date back to the creation of departmental agronomic maps; the first of these was created by the Marquis Arcisse de Caumont in 1843 for the department of Calvados (Hitier 1894). However, in 1883, a new arrival from Russia would set off a revolution. With the publication of his works on the soils of the vast plains of Russia in his book "The Russian Chernozem", Vasily Vasilyevich Dokuchayev put forward his global, temporal and geographical approach to soil, thus founding the field of pedology (Boulaine 1983). Another Russian, V. Agafonoff, became known as the creator of the first schematic pedological map of France, at a scale of 1:2,500,000 (Geze 1965). He presented this map in his publication entitled "The Soil Zones of France (including a map)" (Agafonoff 1927). In 1935, the French Association for the Study of Soils (*Association Française pour l'Etude des Sols*, AFES) was founded, which on the basis of the works of Agafonoff sought to promote the creation of a soil map of France at 1/1,000,000th scale, though without immediate success. The efforts in the field of pedology at that time seemed more focused to the overseas territories than on mainland France. Still, a few soil maps were produced in the years that followed. Among these are the map at 1/1,000,000th scale of northwest France, published in 1952 by A. Oudin, G. Aubert, and P. Duchaufour, as well as the map of the Aisne soils at 1/25,000th scale done in 1963 by J. Hebert and M. Jamagne (Geze 1965). The first map of the soils of the territory of France at one-millionth scale was finally produced and published in 1966, under the direction of Dupuis. At the request of Jean Boulaine, and under the impetus of the Inspector General of the INRA, G. Drouineau, the French Pedology and Soil Mapping Commission was formed. Between 1964 and 1967, the commission brought together specialists in agricultural research, who convened at the National Superior School of Agronomy in Grignon. As a result of these works, a classification of soils was created, which was used for more than 20 years, before being improved via the Pedological Reference System by Baize and Girard (AFES, 1992). Finally, in 1968, Drouineau oversaw the creation of the Soil Study Service and the Pedological Map of France (*Service d'Etude des Sols et de la Carte Pédologique de France* (SESCPF)), whose direction was then entrusted to Marcel Jamagne at the INRA center in Versailles (Boulaine 2012). In 1982, the SESCPF moved to the premises of INRA Orléans and would become a driving force in French soil mapping, with Marcel Jamagne often being considered one of the fathers of soil mapping in France. The initial mission of the SESCPF was to coordinate the surveys and the publication of the soil maps of France at 1/100,000th scale (Arrouays et al. 2004). The INRAE Info&Sols unit in Orléans has today succeeded the SESCPF and the Group of Scientific Interest on Soils (*Groupement d'intérêt Scientifique sur les Sols* (GIS SOL)) has been coordinating the acquisition and dissemination of information on the soils of France since 2001, including cartography (Gis Sol 2021a).

3.3. The digital age, from paper maps to spatialized databases: the available information for soil mapping

3.3.1. *Augerholes, pits and soil analyses*

One of the most frequent tasks in soil mapping is to seek to spatialize the information obtained during a "classic" soil prospecting mission using descriptions of horizons recorded from auger surveys or soil pits, and laboratory analyses carried out on soil samples from these horizons. To do this, the position of the augerholes and pits is determined by measuring their geographical coordinates using a GNSS-type receiver; in this method, it is important to note the geographical reference system used to set the receiver, most often "WGS84" for the international latitude/longitude system, or Lambert 93 for the official French system. Prior to going to the field, for the variables used to describe the samples, it is recommended to define a data dictionary; in other words, to codify all the permitted values for these variables. When filling in the description sheets of the samples in the field, this dictionary will be referenced to, so that the codification of the variables is completely standardized. This standardization is very important, so that the information can be entered into a database: the same value must not be coded differently between the records. In the same way, leaving unfilled boxes in the form should be avoided, as there could be ambiguous interpretations of these at the time during which they are input into a computerized system. Everything must be made explicit.

3.3.2. *The accessible French databases*

3.3.2.1. *Répertoire national des études cartographiques de sol, Refersols (The French National Directory of Soil Cartographic Studies)*

Refersols (2021) is an online accessible database, which contains an inventory of the metadata of all the studies referenced in the national directory of cartographic studies by Gis Sol. In 2021, 6,239 studies were referenced in this database, of which 2,100 were scanned and 228 were vectorized.

3.3.2.2. *Base de Données Géographiques des Sols de France, BDGSF (The Geographical Database of the Soils of France)*

The BDGSF has been created through the digitization of the soil map of France at 1/1,000,000th scale. To this day, it is still the only soil database to be completely exhaustive in its coverage of the national territory of France. In addition, it is also used by the European soil database for its coverage of France (see section 3.3.3.1). It is widely used for applications on global scales, but its low geographical and semantic resolution means that it is less advantageous for use on a larger scale (Voltz et al. 2018). The database can be downloaded from the INRAE "DATA" portal (INRAE 2018)

3.3.2.3. *Inventaire, Gestion et Conservation des Sols, IGCS (Soil Inventory, Management, and Conservation)*

As was noted in the section on the history of soil mapping (section 3.2), at its creation, the SESCPF's mission was to publish soil maps at 1/100,000th scale, but at the end of the 1980s, it became clear that this result was not achievable in a reasonable time frame. Thus, the French Ministry of Agriculture and INRA decided to launch the IGCS program, with a target scale of 1/250,000 (Arrouays et al. 2004). IGCS comes in three main spatial scales (Gis Sol 2021b): (i) the "reference sectors" (SR) with scales less than 1/50,000th; (ii) the "Pedological Knowledge of France" (CPF) inventories with scales between 1/50,000th and 1/100,000th; and (iii) the "Regional Pedological Reference System" (RRP) whose main cartographic representation at 1/250,000th will eventually cover the entire French territory. The map of the dominant soils of the RRP can be viewed on the IGN Geoportal. We can thus estimate the progress made: we can note an absence of data for 10 departments. The Gis Sol plans to subsequently disseminate further information on soils on the Geoportal (Messant et al. 2021)

3.3.2.4. *Base de Données d'Analyses des Terres, BDAT (Soil Analysis Database)*

For almost 25 years, the BDAT program has maintained a database containing the results of soil analyses requested by farmers across the national territory of France (Saby et al. 2014). This mainly concerns analyses on the surface horizons of cultivated soils, carried out by laboratories approved by the French Ministry of Agriculture. The first synthesis of soil analysis data can be traced back to 1988 in Brittany; later, Christian Walter with AFES broadened its scope to all of France (Walter et al. 1997). Since 2001, the InfoSol unit of the INRAE of Orléans has been responsible for managing these analyses within the framework of the "BDAT" program. The location of the data is done "at the municipality level". The analyses are grouped by five-year periods: 1991–1994, 1995–1999, 2000–2004, 2005–2009, 2010–2014. Individual analyses cannot be made public in order to protect the confidentiality of the sources of the information. The data are therefore made known through a statistical aggregation at four possible geographical levels: administrative department, small agricultural region (SAR), department and region. The GEOSOL cartographic tool within the Gis Sol allows us to create our own online maps, based on the data from the BDAT (Figure 3.2). The data tables of the BDAT can be downloaded from INRAE's data sharing portal, "DATA" (INRAE 2018) (subject to authorization). The database contains 31.8 million unit analyses from 2.6 million samples collected between 1990 and 2014.

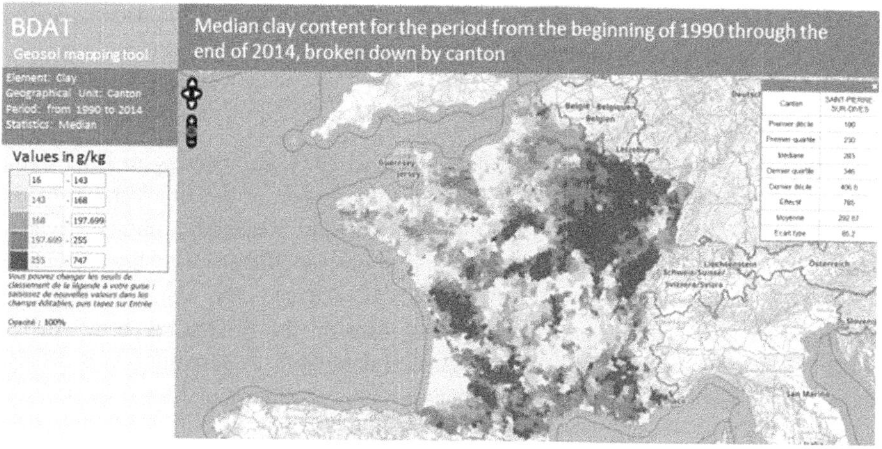

Figure 3.2. *The "GEOSOL" online mapping tool from the BDAT on the Gis Sol website: an example of mapping the median value by administrative department of the clay content of the soils, between 1990 and 2014. For a color version of this figure, see www.iste.co.uk/coquet/agricultural.zip*

3.3.2.5. *Réseau de Mesures de la Qualité des Sols, RMQS (Soil Quality Measurement Network)*

The RMQS is a network for monitoring soil quality, established in the year 2000. It consists of 2,240 measurement sites in France, spread over a regular grid where each edge measures 16 km (Gis Sol 2021c). Soil samples are taken every 15 years. The first RMQS 2000–2009 campaign, which focused on soil contamination, made it possible to map the contents of nine trace metal elements, as well as to reassess the carbon and the microbial biomass amounts in the soils (Gis Sol 2021c).

3.3.2.6. *Base de données sur les sites et sols pollués, BASOL (database on polluted sites and soils)*

An inventory of sites with polluted or potentially polluted soils has been carried out by the Ministry of the Environment since 1994, using the BASOL database (Géorisques 2021). Two other databases also contain information on potentially polluted soils: the database of inventories on former industrial sites and service activities (*base de données d'inventaire des anciens sites industriels et activités de services* (BASIAS)) and the soil information sectors (*secteurs d'informations sur les sols* (SIS)). It may be useful to consult these databases before, for example, establishing an urban agriculture site, in order to verify that there is no history of pollution on the site.

3.3.3. *The available international databases*

3.3.3.1. *The European Soil Database (ESDB)*

The work to unify European soil data began in 1974, culminating in 1985 in the digitization of paper maps and in 1986 with the creation of the first version of the ESDB, with a scale of 1:1,000,000. The JRC (Joint Research Centre), in Ispra (Italy), has been in charge of maintaining it, more specifically the "European Soil Data Center" (Panagos et al. 2012; ESDAC 2021). The database may be downloaded after filling out an application form on the ESDAC website (ESDB 2021) in vector GIS format with an associated data table. A derivative version also exists, in the form of a series of raster images with a resolution of 1 km.

3.3.3.2. *The FAO World Soil Map: Digital Soil Map of the Word (DSMW)*

The DSMW is the digitized version of the paper-based FAO-UNESCO map of the world's soils, at 1/5,000,000th scale (FAO DSMW 2021). The DSMW has currently been replaced by the HWSD database (see section 3.3.3.3).

3.3.3.3. *The Harmonized World Soil Database (HWSD)*

The HWSD contains the information from the FAO-UNESCO World Soil Map at a 1/5,000,000th scale. This database contains more than 20,000 rows of data described by numerous information columns on the properties of soil (FAO HWSD 2021). The database can be downloaded in Microsoft Access format. It can easily be read using a GIS software such as QGIS or ArcGIS. A raster GIS layer, in ".bil" format, is provided with the database, which provides the delimitation of the cartographic units of soils in order to spatialize the pedological data. It is also possible to interconnect with the cartographic units of the DBGSE as they contain the same polygons. HWSD is the result of a collaboration between the FAO, IIASA, ISRIC-World Soil Information, the Institute of Soil Sciences of the Chinese Academy of Sciences (ISSCAS) and the JRC.

3.3.3.4. *GlobalSoilMap*

The GlobalSoilMap project was launched in 2006 by initiative of the "Digital Soil Mapping" Working Group of the International Union of Soil Science (IUSS) (Hempel et al. 2013). It is a predictive database of certain soil properties in raster format, at a resolution of 100 m covering the entire planet, produced from digital geographical data by DSM approaches. The project no longer seems to be active today, but some of its specifications have been applied in other projects.

3.3.3.5. *World Soil Information Service (WoSIS)*

WoSIS is an SQL database of soil profiles, developed by the ISRIC Foundation (International Soil Reference and Information Centre). The georeferenced soil profiles are standardized to be integrated into the database. The database emphasizes point data with the analytical and physical properties of the soil taken into account in the GlobalSoilMap specifications (Batjes et al. 2017). More than 400,000 profiles have been deposited in the database, of which about 200,000 have been standardized and can therefore be used directly. The data can be downloaded or accessed online via the GIS Internet protocol "WFS".

3.3.3.6. *SoilGrids*

SoilGrids is a global DSM system which uses state-of-the-art machine learning methods (Hengl et al. 2017). The SoilGrids prediction models are adjusted using the 200,000 observations of soil profiles from the WoSIS database and a series of environmental covariates. The SoilGrids outputs are raster maps of soil properties at six depth intervals, at a spatial resolution of 250 m, over the entire world. The covariates come from more than 400 layers of products derived from Earth observation and other environmental information. The latest update dates to May 2020 and the data are available for download on the ISRIC website or accessible via the GIS Internet protocol "WMS".

3.3.4. *Auxiliary cartographic data*

At different stages of the soil mapping process, auxiliary data, other than soil field information, can be used. For example, to help establish soil mapping units from surveys, we can rely on boundaries appearing on an aerial photo or from the contour lines of an IGN topographic map. With a DSM approach, we can combine soil information with digital geographic layers in GIS. The vast majority of these data are now available in digital format and can be used in GIS. These data have become much more accessible, thanks in part to the European directive of March 14, 2007 known as the "Inspire" directive, which applies to digital geographical data held by public authorities. The directive requires these public authorities to make these data publicly available on the Internet. This has led to outcomes such as a large part of the IGN data being made freely available for download on the IGN "Geoservices" website (IGN 2021). At most data providers, you can either download the data in the form of GIS format files, or access a geographical data stream on the Internet following one of the standardized protocols (WFS, WMS and

WMC), which are now supported by GIS software. The latter of these protocols allows us to access a large volume of data online, without having to download it beforehand and store it on the hard drive. This is particularly suitable for the simple visualization of so-called "background" geographical layers, such as IGN orthophotographs. The auxiliary layers can also be produced on request by design offices or laboratories using different sensors and measurement vectors.

3.3.4.1. *Cartographic backgrounds or basemaps*

Some maps are used as geographical "backgrounds", providing markers to position information about the soil; these are called basemaps. The most common are the IGN topographic maps and OpenStreetMap, which is an online, open and global collaborative mapping project, and very widely used in GIS and in applications that use mapping.

3.3.4.2. *Remote sensing data*

Images from remote sensing devices, using drones, aerial devices or satellites are often used. On bare soil, these images make it possible to perceive the spatial variations of certain surface properties of soils. Under vegetation or crops, the variations of the coverings can reflect the response of the plants to certain properties of the soil. Various sensors are used to acquire these images, mainly: (i) "standard" optical sensors such as cameras; (ii) multispectral sensors, which also give information on wavelengths not visible to the human eye, most often within the infrared spectrum; and (iii) radar sensors. The Copernicus program, the environmental monitoring program of the European Union, is based in part on the use of Earth observation satellites, in particular the Sentinel family of satellites. Their data are accessible free of charge and without limitation. This is truly revolutionary for access to satellite imagery data (Copernicus 2021). In the same way, American satellite data from the Landsat family are also freely accessible and can be accessed from sources such as the EarthExplorer website from the USGS (USGS 2021).

3.3.4.3. *Data on the morphology of the terrain*

The basic information for analyzing the morphology of the terrain is altitude data, generally provided in the form of a digital elevation model (DEM) or digital terrain model (DTM). Derived variables such as the slope (intensity and direction) can then be calculated in a GIS from the DEM. The IGN provides DTMs on its Geoservices website, including (IGN 2021): the ALTI database at a resolution of 25 m, for medium-scale analyses, and the ALTI RGE (large-scale reference base) at a resolution of 1 m, for large-scale analyses. The SRTM (Shuttle Radar Topography Mission) is a global DTM carried out by NASA, from the Space Shuttle Endeavour, using radar surveys. It can be downloaded freely. SRTM3 is provided at a resolution of 3 arc s (about 90 m at the equator). A detailed DEM can be carried out on the scale of an agricultural plot from GNSS surveys, LIDAR measurements or "stereo" pairs of aerial images (Gilliot et al. 2020).

3.3.4.4. *Geophysical data: soil resistivity/conductivity*

Many of the sensors that are used to acquire information about soils are optical sensors, and therefore can only provide data on area most at the surface of the soil. By using geophysical methods, it is possible to explore the soil in greater depth. Among these methods, one of the most used in agriculture is to measure the electrical resistance of the soils. A pair of electrodes is used to supply an electric current into the soil and other pairs of electrodes measure the current that has passed through the soil at different depths. From the measurements obtained, the electrical resistance of the soil can be deduced. The resistance is a general indicator of the texture, the thickness and the water content of the soil profiles. The electrodes are "dragged" behind an all-terrain vehicle or a tractor and coupled with a GNSS receiver, which makes it possible to obtain maps of the resistance at different depths of the plot. There are several study offices that can carry out this type of acquisition. One of the best known in France is Geocarta (2021). Some sensors, such as the EM38 sensor from the Canadian Geonics, measure electrical conductivity, which is the inverse of resistivity.

3.3.4.5. *Other auxiliary data for soils*

The Veris Iscan scanner, which is pulled across the ground, brings together two sensors: a near infrared sensor, which is used to estimate the organic matter content of the soil, and an electrical conductivity sensor to simultaneously estimate the texture of the soil between 0 and 60 cm. The information is geolocated with a GNSS receiver to produce maps. In France, the company PRECIFIELD offers services using the Iscan Audit (PRECIFIELD 2021). Ground penetrating radar (GPR), or georadar, is mainly used to detect objects in the subsoil, such as for detecting drains in a drainage network in agricultural contexts. GPR remains underused for analyzing soil properties. Researchers from the Université catholique de Louvain and the Liège Space Center, in their RAPAS project, have placed a GPR on a drone to map soil moisture, in order to control the irrigation of an agricultural plot. SpiRiT Sol is a portable near-infrared spectroscopic sensor developed by Arvalis, which makes it possible to estimate the properties by contact with a soil sample directly in the field. Experiments have been conducted using airborne gamma spectrometry for soil mapping. This technique is based on the detection of gamma photons emitted during radioactive decays of isotopes contained in the soil and the source rock. A big advantage of this technique is that it is usable even when vegetation cover is present.

An airborne study conducted by the BRGM and INRA has shown an interesting potential for soil mapping, even though data analysis can be complex (Laroche et al. 2013). GNSS reflectometry, or GNNS-R, is a method of analyzing the waves emitted by the satellites of satellite positioning systems (GNSS) after reflection by the ground, using an antenna directed toward the ground. On this basis, the researchers from the European MISTRALE project have designed a GNSS-R sensor embedded on a drone to produce high-resolution maps of the humidity of agricultural soils (MISTRALE-UE 2021).

3.4. Some of the general concepts and methods of soil mapping

3.4.1. *The concepts of scale, resolution and sampling*

Scale refers to the ratio of a distance shown on a map (on paper or a screen) to the same distance on the ground, for which distances must be given in the same units. For example, a scale of 1/5,000th would mean that a distance of 1 cm on a map represents a distance of 5,000 cm, or 50 m, on the ground. With the advent of digital data, it is conceivable that the notion of scale no longer makes sense in GIS software. With this format, it is possible to zoom in or out without limits, or to change the display scale or rendering scale (Figure 3.3), unlike a paper map where this scale can no longer be changed after printing. However, the notion of scale is also a factor at other stages of the mapping process, such as during the acquisition of information. In the case of using a digital camera, the acquisition scale will depend on the resolution of the camera (number of pixels) and the altitude of the vector that is carrying the sensor. The ground resolution, or ground sample distance (GSD) is the size in meters of the pixel side of the final image matrix; this GSD determines the acquisition scale. For example, the resolution of a multispectral image from the Sentinel 2 satellite, in orbit at an altitude of 786 km, is 10 m, thus giving an area of 100 m^2. It is therefore the representative surface of each elementary reflectance measurement.

In the case of field point measurements, such as soil auger holes, the measurement sampling is to be defined. A sampling strategy defines the density of the auger holes and the method of locating them in space. The choice of the sampling strategy depends essentially on the prior knowledge of the terrain to be mapped and on the human and financial resources available for the study. Without special knowledge of the terrain, if sufficient means are available, it is possible to carry out a significant number of auger holes distributed over a regular grid (Figure 3.4(a)); this is known as “grid soil sampling”, the systematic approach. If the available means are more limited, fewer auger holes will be made. In such a case, there would no longer be any point in distributing them over a regular grid, as the spacing of the grid would be too large. Then, the position will generally be set randomly (Figures 3.4(b) and (c)). If we have a prior knowledge of a particular direction of a variation in the soil, we can orient the position of the auger holes in this direction so as to capture the maximum variability (Figures 3.4(d) and (e)).

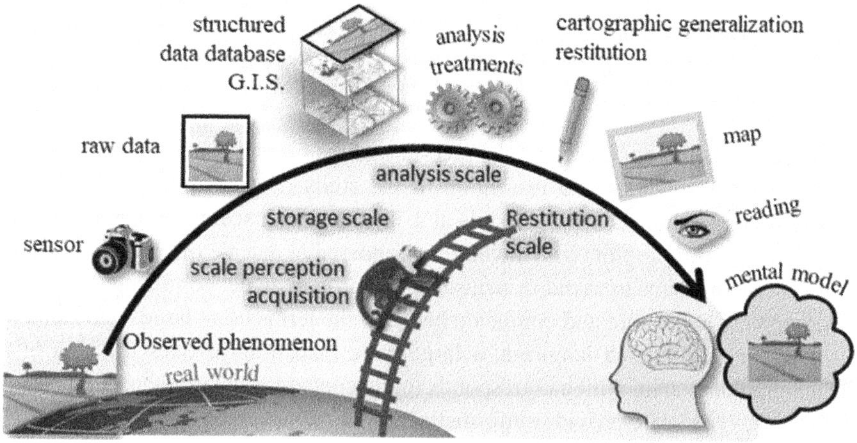

Figure 3.3. *The notion of scale in the different stages of the cartographic process. For a color version of this figure, see www.iste.co.uk/coquet/agricultural.zip*

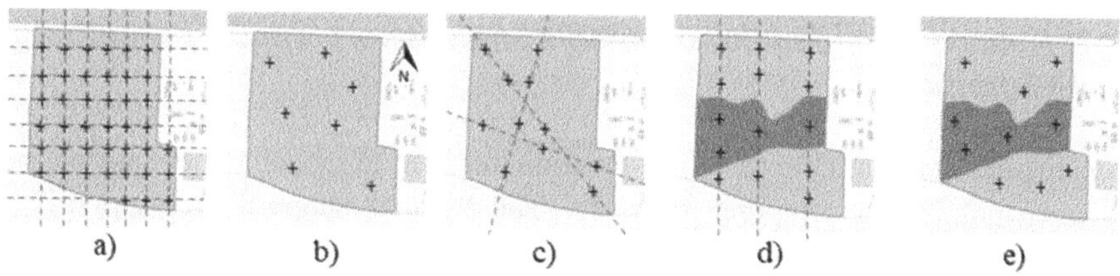

Figure 3.4. *Different sampling strategies for soil surveys at the scale of an agricultural plot: (a) On a regular grid; (b) on random points; (c) on random transects; (d) on selected transects; and (e) on selected positions. For a color version of this figure, see www.iste.co.uk/coquet/agricultural.zip*

Sampling density will depend on the scale of restitution chosen. One possibility is to follow the recommendations of the AFNOR CARTO NF X31560 standard (Table 3.1) and adapt it according to the assumed variability of the soils. At intra-plot scales, in the case of research studies, it is possible to go up to a density of one sampling for every 0.5 ha.

Restitution scale	Density of auger holes	Density of pits
1/250,000th	1 hole per 200–600 ha	1 pit per 2,000–6,000 ha
1/50,000th	1 hole per 10–30 ha	1 pit per 200–300 ha
1/10,000th	1 hole per 2–3 ha	1 pit per 10–50 ha

Table 3.1. *Examples of the recommended densities of soil measurements based on the scale of map rendering, as defined by the AFNOR CARTO NF X31560 standard*

A soil map should not be used if the precision of the information is "incompatible" with the scale of analysis and restitution of the study. For example, it would not make sense to use the BDGSF, which is at a 1/1,000,000th scale, for a study on the scale of an agricultural plot, which would be rendered at a scale of 1/5,000th. There are two classic mishaps to avoid in terms of the notion of scale: confusing the notion of geographical extension, and confusion between the terms of a "large" and "small" scale. In everyday French language, a large-scale map often evokes the idea of a large geographical area, which corresponds to the concept of geographical extension and not scale. The scale provides information about the accuracy of the data and not about the extent of their coverage – these are two different notions. The qualifiers of "large" and "small" scale refer to the results of the fraction and not to the denominator, so a map at a 1/1,000,000th scale is a small-scale map.

3.4.2. *Notions of regionalized variable and support: spatial organization of soil information*

A regionalized variable (RV) is a function intended to model a phenomenon that has a spatial and/or temporal structure. The "spatial support" or "support" is composed of the portions of the geographical space known as "sites", which will carry the RV information. The RV theory was initially developed in Geostatistics for applications in estimating mineral resources, but today, it is more widely used to describe spatially structured data. Once integrated into a GIS, soil data can be presented in different types of support, essentially depending on how it is acquired. There are two main families of geometrical forms the media can take: vectors, which can be points, lines, or polygons, and rasters, which are regular data matrices (grids). Thus, the auger holes and the pits are represented in the form of isolated points (Figure 3.5(a)). The data from the sensors that are pulled through the plot in trailing mode come in the form of a large series of measurement points, which follow the trajectory of the tractor (Figure 3.5(b)); the soil cartographic units of a pedological map are in the form of polygons (Figure 3.5(c)) and the remote sensing imaging data are in the form of raster grids (Figure 3.5(d)).

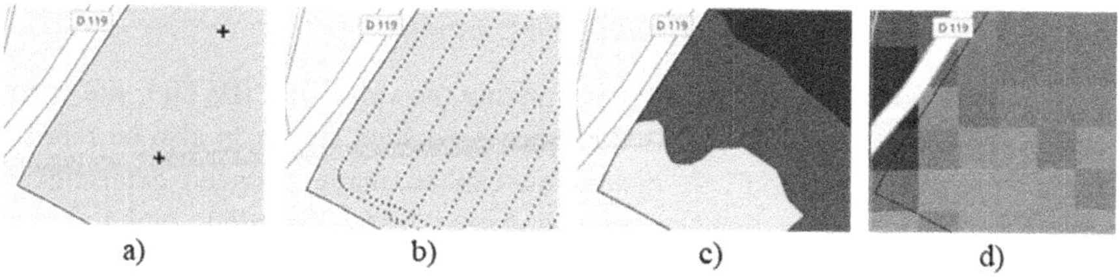

Figure 3.5. *The most common spatial media types (geometries) for soil data in a GIS: (a) points; (b) series of points; (c) polygons; and (d) rasters. For a color version of this figure, see www.iste.co.uk/coquet/agricultural.zip*

In certain cases, three-dimensional geometries can also be found, such as 3D point clouds, which are produced by LIDAR acquisitions or by 3D photogrammetry from a series of "stereo" images (Gilliot et al. 2017).

3.4.3. *Changes to soil information support formats*

It is quite common to want to transfer soil information from its original support to a new destination support in order to increase the spatial density of the information, to inform another data layer, to calculate zonal statistics or to change scale by spatial aggregation or disaggregation. The destination format can have the same geometric nature as the original format (Figure 3.6(b)), or it may be different (Figures 3.6(c) and (d)). The value assigned to a destination site can come from a single origin site, often the closest spatially (this is called the "nearest neighbor" method), or it can be the result of a calculation made from the values of several sites, such as by averaging them.

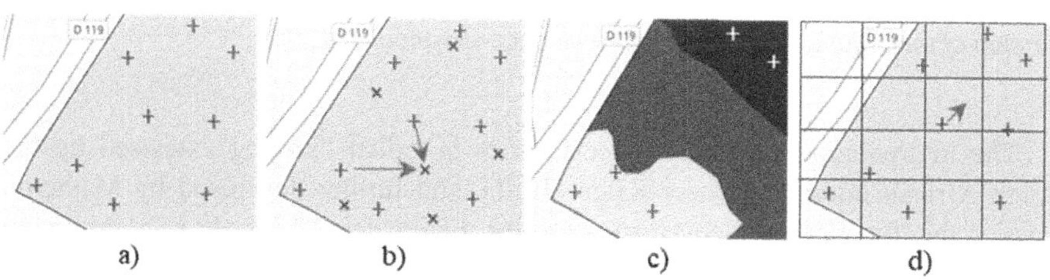

Figure 3.6. *Examples of format change. (a) Original points format. (b) Destination points. (c) Destination polygons. (d) Destination raster grid. For a color version of this figure, see www.iste.co.uk/coquet/agricultural.zip*

3.4.3.1. *Spatial interpolation of quantitative variables*

Spatial interpolation is a method of estimating the value of a RV on a site where no measurement has been taken based on measured sites. This can also be referred to as a "prediction method". A distinction can be made between deterministic methods, which use a predefined equation, and geostatistical methods, which seek to fit a spatial model onto the original sites.

Interpolation using the inverse of distances, or IDW (inverse distance weighting) is a commonly used deterministic method: Z_e, the estimated value, is calculated as the weighted average of the Z_i, the closest measured values (Figure 3.7). The selection of the closest values is made either by using those which are at a distance below a given threshold, or by setting the number of closest values retained. The weighting coefficient is defined as the inverse of the distance d_i between Z_e and Z_i, raised to a power p. The higher p is, the more significant the weight of the closest points is in the calculation of the average. In many software programs, the destination format will be a raster grid, for which the resolution will be defined.

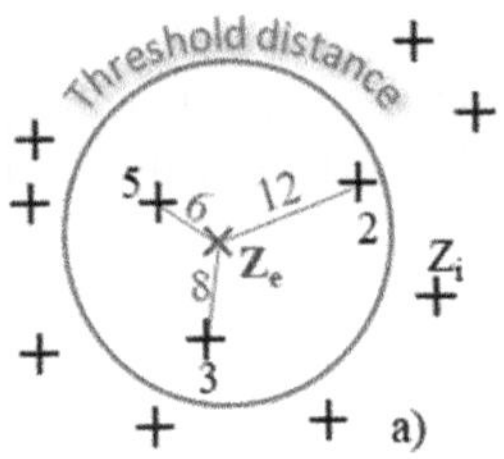

a)

$$Z_e = \frac{\sum_{i=1}^{n} \frac{Z_i}{d_i^p}}{\sum_{i=1}^{n} \frac{1}{d_i^p}}$$

b)

$$= \frac{\frac{5}{6^2} + \frac{2}{12^2} + \frac{3}{8^2}}{\frac{1}{6^2} + \frac{1}{12^2} + \frac{1}{8^2}} = 3{,}96$$

c)

Figure 3.7. *IDW interpolation using the inverse of the distances. (a) Z_e the estimated value from the three measurement points selected. (b) The formula of the IDW calculation. (c) The result of the calculation using the example in (a). For a color version of this figure, see www.iste.co.uk/coquet/agricultural.zip*

The interpolation method in Geostatistics is called "kriging", devised by the South African mining engineer Krige (1951), and further developed by Matheron (1962), the founder of Geostatistics, at the Ecole des Mines. It is a minimum variance estimation method that takes into account the spatial structure of the data. Kriging is based on calculating a semi-variogram (Figure 3.8). If we consider all the measurement points taken in pairs, such that if $Z_i(x)$ is the value measured at a geographical site x, and Zi(x+h) is the value at a site x+h at a distance h from the first site, then the experimental semi-variogram is given by the half-variance of the difference between $Z_i(x+h)$ and $Z_i(h)$, calculated for increasing distances of h:

$$\gamma(h) = \frac{1}{2}Var(Zi(x+h) - Zi(x)) \quad [3.1]$$

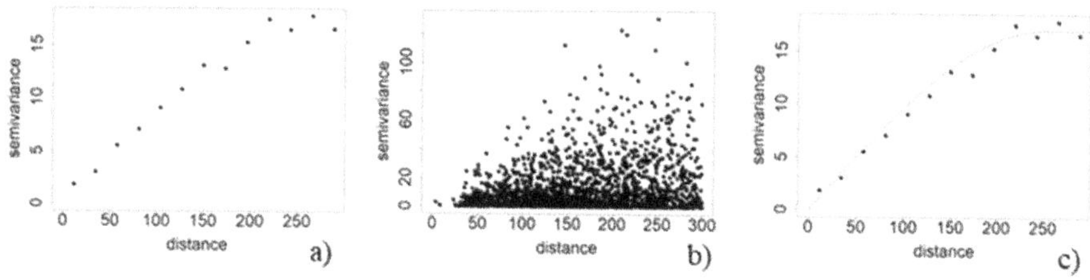

Figure 3.8. *Example of the adjustment of a variographic model: the clay content of the surface samples in an AgroParisTech plot in Grignon (Yvelines, France). (a) Experimental variogram. (b) Variographic cloud. (c) Fitted spherical model*

The experimental variogram is then modeled by adjusting a function (Figure 3.8(c)), called the "theoretical variogram", which will then be input into the kriging process to carry out the interpolation. The big advantage of kriging over the IDW method is that it takes into account the spatial structure of the data. Another advantage is that it produces an error map, which gives the quality of the prediction over the entire plot. Co-kriging is an extension of kriging, where a RV is interpolated on the basis of another RV. This is a fairly direct way of "injecting" an auxiliary variable when we know that it is correlated to the property of the soil that we are trying to interpolate.

3.4.3.2. *Spatialization of the qualitative variables to the nearest neighbor by partitioning the space into tessellations*

The interpolation methods seen in the previous point cannot be applied to qualitative variables, as the notion of a mean average cannot be applied to such variables. In this case, we can use a tessellation, which is a division of space into geometric primitives, which are either regular or irregular, and without spaces or overlap between them, called "tesserae". The algorithm for calculating the tesserae may (or may not be) based on a set of pre-existing points, such as auger holes. One of the most commonly used tessellations is to break down points into Thiessen polygons, also known as a Voronoi diagram (Figure 3.9). From a set of points, each polygon created represents the "zone of influence" of a point with respect to its neighbors.

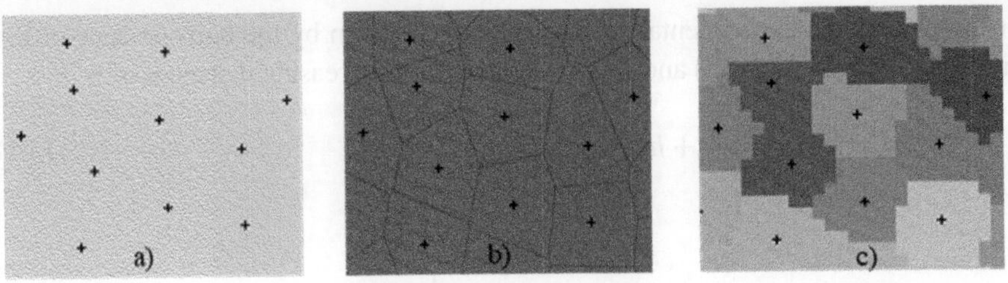

Figure 3.9. *Tessellation by Thiessen polygons. (a) Probe points used to start with. (b) Thiessen polygons. (c) Euclidean allocation in raster mode. For a color version of this figure, see www.iste.co.uk/coquet/agricultural.zip*

Euclidean allocation, or "interpolation" to the nearest neighbor, is an analogous algorithm in raster mode (Figure 3.9(c)). In these nearest neighbor approaches, each polygon or pixel is assigned the value of the geographically closest sample. These are the approaches most often used to spatialize qualitative variables, such as soil type.

3.4.3.3. *Layers overlay*

It is quite common to produce thematic maps to combine several layers of information together. These are known as "layers overlay" in GIS terminology. For instance, to produce a map of the potential erosion of agricultural soils, the texture of the surface horizon and the slope resulting from a DEM can be overlayed. In vector mode, overlay is carried out by a geometric intersection between polygons, whereas in raster mode it is obtained by recombining the values of the pixels between the layers.

3.4.3.4. *Spatial aggregation and disaggregation: changes of scale*

It is quite common to want information to be displayed on a different scale than its original scale – this is known as a "change of scale". Moving to a broader scale is an "aggregation". This conversion is usually not a problem. For example, switching from the number of auger holes per canton to the number of auger holes per department amounts to simple summation (Figure 3.10), and the result is exact. Going in the other direction, a disaggregation is more problematic since we only know the number of auger holes at the departmental level (1,378). An average value can be assigned to each administrative department (53), and it can be seen that the deviation from the real value can be significant in some administrative departments. In doing so, we are assuming homogeneity, which is almost always false. If a second variable is known at the administrative department scale and is correlated to the variable to be disaggregated, it can be used as a distribution "key" (weighting coefficient) in order to improve the estimation.

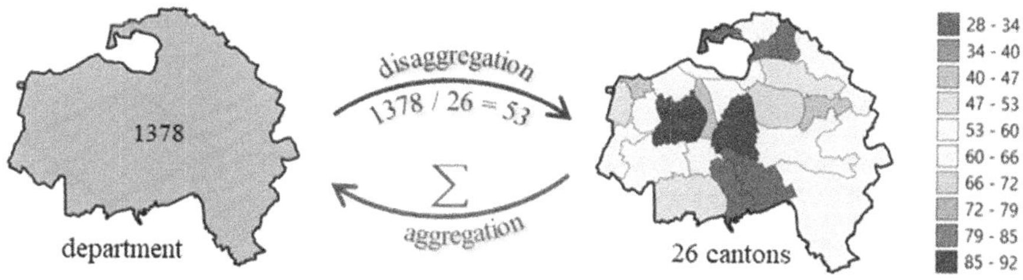

Figure 3.10. *Spatial aggregation and disaggregation of the number of auger holes. Note that 1,378 is the total number of auger holes in the department, 53 is the mean calculated value per canton and the map shows real number of auger holes per canton. For a color version of this figure, see www.iste.co.uk/coquet/agricultural.zip*

3.4.4. *Pedotransfer functions*

Pedotransfer functions (PTFs) are relationships that allow us to predict certain soil properties that are difficult to observe or measure from other properties that are simpler to obtain. They are often used to indirectly map certain soil properties. The relationship can be established either by a statistical regression between properties to be predicted and observed properties, or by expert knowledge. The first PTFs were established for the prediction of soil water properties. For example, the PTF of the hydraulic conductivity at saturation of soil horizons was established by Wösten et al (1999) from the contents of clay, silt and organic matter, as well as bulk density.

3.5. Examples of uses of soil maps

3.5.1. *Precision agriculture using drones*

Unlike traditional agriculture, where agricultural practices are conducted uniformly over the entire extent of a plot, precision agriculture is an approach in which a farmer will vary practices to adapt them to variations within plots. The term "modulation" is used for the spatial variations in an agricultural practice. Though many practices have been tested in precision agriculture, the modulation of fertilizer applications is the technique that is most commonly used. A crucial step in implementing a fertilization modulation is to map different areas in the plot where the doses of fertilizer will be varied: this can be called "zoning" a map. The final map that shows the doses of fertilizer at each point of the plot is called the "recommendation map". One zoning approach maps the development of the crops and estimates their biomass, shortly before the fertilizer is added. This is usually done using a visible near-infrared optical sensor mounted on a tractor (a Yara N-Sensor), a drone or a satellite (FARMSTAR) (Figure 3.11).

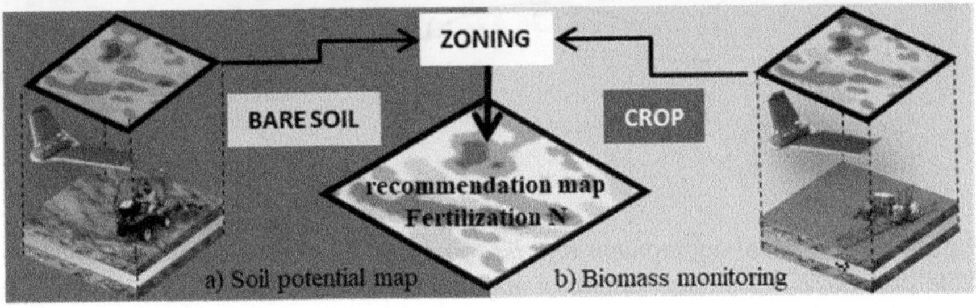

Figure 3.11. *Drone-assisted zoning in precision agriculture based on (a) differences in the soil and (b) the biomass of the crops. For a color version of this figure, see www.iste.co.uk/coquet/agricultural.zip*

Another zoning approach is based on the characterization of the agronomic potential of soils. The principle of controlling agricultural fertilization based on soil zoning is not very new. Alphonse Carnot, the brother of Sadi Carnot (a president of France during its Third Republic) was a Professor of General Chemistry at the Ecole des Mines, of which he was the director in 1901 (Breton 1920). In his report submitted in 1893 to the French National Agricultural Society (Hitier 1894), he proposed a two-step method to help farmers manage the fertilizer applications to a farm plot, based on knowledge of the physical and chemical qualities of its land: "first: make a careful examination of the soil at a certain number of points in order to determine its qualities and defects; second: find a way to generalize these results so as to determine, or at least to sufficiently approximate, the properties of the soil in the sections left between these test points". As we have seen previously, several types of sensors can be used to produce a continuous mapping of the many variations in the soils, which should make it possible to carry out Alphonse Carnot's proposal. In a study conducted by AgroParisTech, on an agricultural plot in Yvelines in 2016 (Gilliot et al. 2017a), an AirInov MultiSPEC 4C camera detecting visible/near infrared multispectral rays was used on board a Sensefly eBee fixed-wing drone (Figure 3.12) to map the surface properties of the soil, and more particularly the organic carbon (OC) and clay content of the soil to be plowed. These maps served as the basis for a precision agriculture zoning.

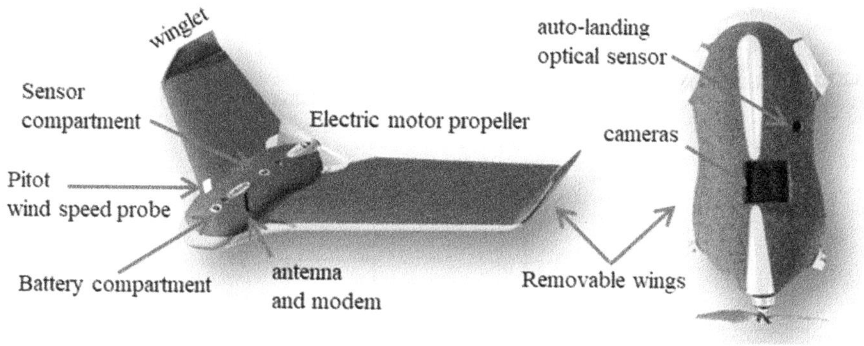

Figure 3.12. *The Sensefly eBee drone, used during the AgroParisTech study*

In the 28-ha plot, a detailed study sub-area of 3.6 ha was defined. Forty-two points distributed regularly in the area were monitored. On each point, an auger hole made it possible to describe the soil, and a sample of the surface horizon was analyzed in the laboratory. The drone flight was carried out on bare soil at an altitude of 60 m, just after soil tillage (Chapter 1), which has the advantage of homogenizing the soil before the data are acquired. Soil reflectance measurements were carried out at the time of the drone flight with a Spectral Evolution SR 3500 field spectroradiometer in order to verify the radiometric calibration of the drone's multispectral camera. The points were located using a Trimble Geo 7x differential GNSS receiver, with decimeter accuracy. Targets placed in the study field, located with the DGNSS receiver, were used for the precise georeferencing of the final drone imaging (Figure 3.13). The images geotagged by the internal GPS of the drone were then processed with the Agisoft Metashape photogrammetry software to produce a 3D model of the plot soils, in the form of a 3D point cloud. A mosaic of reflectance images was exported from this cloud in four wavelengths: green at 550 nm (G), red at 660 nm (R), red edge at 735 nm (RE) and near infrared at 790 nm (NIR). The reflectances of the 42 points were extracted from the image mosaic, then compared with the results of the soil analyses. Partial least square regressions (PLSR) were performed using the software R to establish models for predicting soil properties from reflectances using only 20 points. The remaining points were used to validate the model. The prediction equations obtained were applied in a GIS to the entire image of the plot to produce soil property maps. For each model, the Pearson correlation coefficient (R^2) and the mean squared prediction error (RMSEP) were calculated from the 22 validation points.

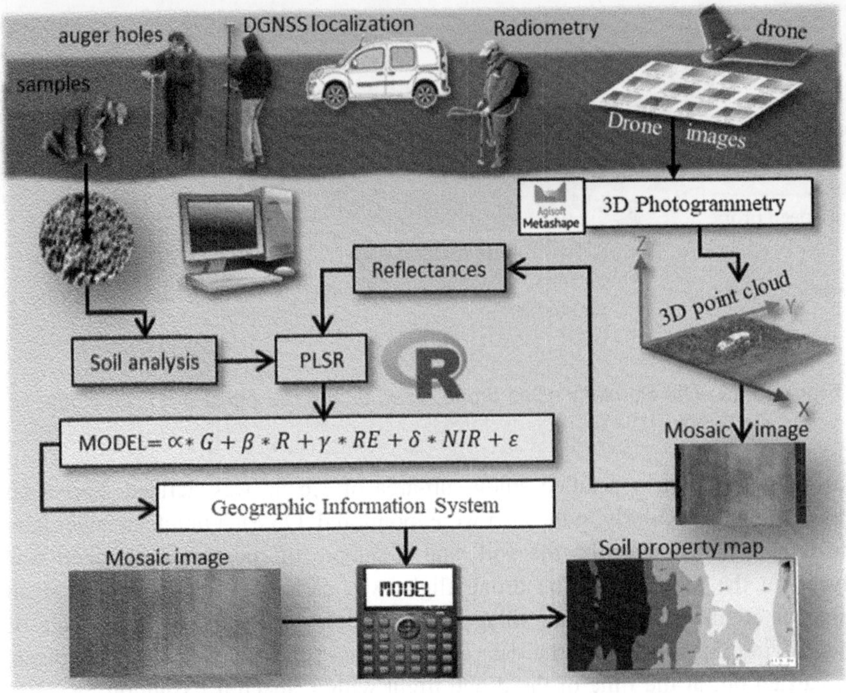

Figure 3.13. *The approach used for calculating maps of soil surface properties using drone multispectral imaging. For a color version of this figure, see www.iste.co.uk/coquet/agricultural.zip*

Other models have also been calculated using fewer points for the calibration of the equation in order to estimate the impact of this number on the quality of the prediction (Table 3.2). For example, for organic carbon (OC), the correlation between the prediction and the validation data is very good, with an R^2 of between 0.69 and 0.81.

With the OC varying from 7 to 21 $g.kg^{-1}$ over the study area, a very satisfactory RMSEP error is obtained, varying from 1.6 to 2 $g.kg^{-1}$, to be compared with the accuracy of the analysis method of the OC in the laboratory by dry combustion (NF ISO 10694), which is generally estimated at 1‰. Thus, the drone imaging approach seems to be quite capable of mapping the different OC levels of the soil surface at the scale of an agricultural plot.

No. of calibration points	R^2 OC	RMSEP OC	R^2 Arg	RMSEP Arg
3	0.69	2	0.84	36
5	0.67	2.1	0.85	34
10	0.75	1.8	0.91	27
20	0.81	1.6	0.91	27

Table 3.2. *Impact of the number of points used to calibrate the prediction relationship on the quality of the prediction (the contents are in g.kg^{-1})*

3.5.2. *Modeling of ammonia emissions arising from agricultural fertilization*

Note that 95% of ammonia (NH_3) emissions into the atmosphere come from agricultural activities, and of these emissions, about 40% are due to applications of nitrogen fertilizers on fields. These emissions are problematic in the following areas: (i) human health, as ammonia is a precursor to the formation of fine particles, resulting in respiratory and cardiovascular risks; (ii) the environment, by contributing to the eutrophication of different environments. The UMR INRAE AgroParisTech ECOSYS has developed the Cadastre_NH_3 software tool (Figure 3.14), which makes it possible to use modeling to map NH_3 emissions across France, also called "emission registers". The model takes into account all the parameters of nitrogen fertilization in the field for a given crop growth period. One of the objectives of this tool is to be able to define regionalized scenarios for adapting fertilization practices, making it possible to reduce NH_3 emissions. Cadastre_NH_3 is a program written in the Python programming language based on the use of the Volt'Air mechanistic model (Le Cadre 2004), which simulates ammonia emissions at the scale of the agricultural plot based on (i) the nature of the nitrogen fertilizer used; (ii) the practices used to apply it; (iii) the weather conditions broken down into hour-long intervals; and (iv) the nature of the soils. Several soil parameters, such as pH levels, influence the intensity of NH_3 emissions. An SQL database is filled in within Cadastre_NH_3 so as to provide all the information necessary to populate the Volt'Air simulations, which must be representative of all fertilization practices throughout the territory of metropolitan France for a given growing period. The sources of information that are used to build the database are (i) the agricultural practices survey (PK) approved by the French Ministry of Agriculture; (ii) the Graphical Parcel Register (RPG) created from farmers' PAC declarations, which provides the areas and locations of several million French agricultural plots; (iii) the weather data in hour-long time intervals from the SAFRAN model of Météo France; (iv) the European soil map (ESDB) at a 1/1,000,000th scale, used for soil spatialization, supplemented by the FAO HWSD database. The variables used to describe the surface composition of the soils are pH, texture (clay, silt and sand), organic carbon content and bulk density. Pedotransfer rules make it possible to estimate the primary soil water properties (hydraulic conductivity at saturation, volumetric water contents, etc.).

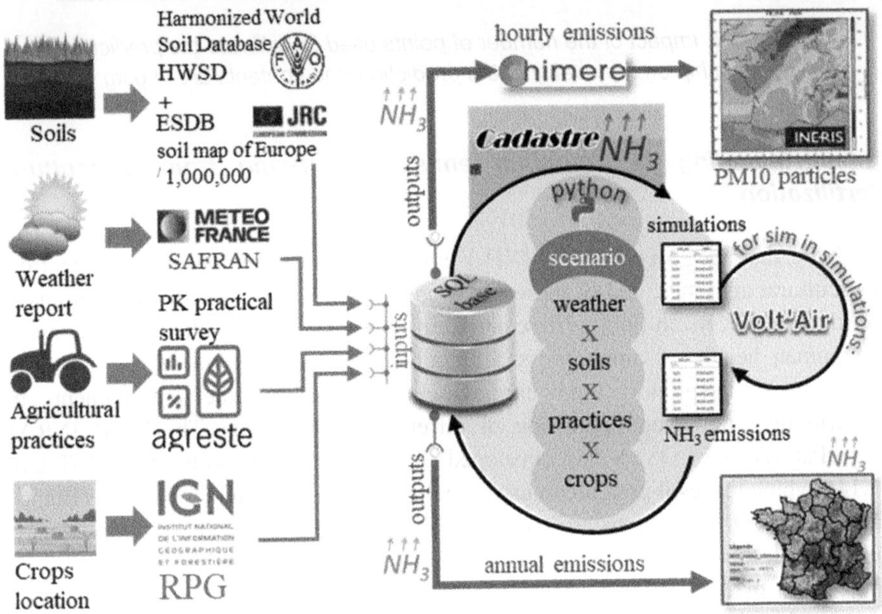

Figure 3.14. *General operating diagram for the INRAE AgroParisTech Cadastre_NH3 tool. For a color version of this figure, see www.iste.co.uk/coquet/agricultural.zip*

All this information is provided primarily at the scale of SARs. Once the database has been filled in, an SQL query by SAR will generate all existing combinations of crop × practice × soil × weather in the "simulations" table. Then, a python loop iterates through this table, calling upon the Volt'Air program to calculate the NH_3 emissions, which will be accumulated in the emissions table. In the latest version of Cadastre_NH_3 (for the 2016–2017 crop period), this represents 191,208 simulations, which are calculated for all of mainland France. Although the algorithm was run in parallel on a workstation with 20 cores and 128 GB of RAM, the calculation time of the simulations was about 36 h. The outputs are mapped and are also used to supply the INERIS CHIMERE atmosphere chemistry model in order to participate in the calculation of the content of fine particles in the atmosphere.

3.6. References

AFES (1992). *Référentiel Pédologique. Principaux sols d'Europe*. INRA Éditions, Paris.

Agafonoff, V. (1927). Les zones des sols de France. *Journal d'agriculture traditionnelle et de botanique appliquée*, 7(72), 513–517.

Arrouays, D., Hardy, R., Schnebelen, N., Bas, C.L., Eimberck, M., Roque, J., Grolleau, E., Pelletier, A., Doux, J., Lehmann, S. et al. (2004). Le programme inventaire gestion et conservation des sols de France. *Étude et Gestion des Sols*, 12.

Batjes, N.H., Ribeiro, E., van Oostrum, A., Leenaars, J., Hengl, T., Mendes de Jesus, J. (2017). WoSIS: Providing standardised soil profile data for the world. *Earth System Science Data*, 9(1), 1–14.

Bonnerot, G., Ducom, E., Joly, F. (2021). Cartographie. *Encyclopedia Universalis* [Online]. Available at: http://www.universalis-edu.com/encyclopedie/cartographie/ [Accessed 29 July 2021].

Boulaine, J. (1983). V. V. Dokouchaev et les débuts de la pédologie. *Revue d'histoire des sciences*, 36(3), 285–306.

Boulaine, J. (2012). Création du service de la carte pédologique de France. *Étude et Gestion des Sols*, 19(1), 29–30.

Breton, J.-L. (1920). Notice sur la vie et les travaux d'Adolphe Carnot. *Académie des sciences* [Online]. Available at: https://www.academie-sciences.fr/pdf/eloges/carnot_vol3273.pdf.

BRGM (2021). Aux origines du BRGM [Online]. Available at: https://histoire.brgm.fr [Accessed 29 July 2021].

Brocard, M. (2005). Les gravures rupestres à parcellaire. *Études rurales*, 2005/3, 175–176, 9–28.

Copernicus (2021). Open access hub [Online]. Available at: https://scihub.copernicus.eu/ [Accessed 24 August 2021].

Cuvier, G. and Brongniart, A. (1811). *Essai sur la géographie minéralogique des environs de Paris, avec une carte géognostique et des coupes de terrain*. Imprimeur de l'Institut Impérial de France, Baudoin [Online]. Available at: https://gallica.bnf.fr/ark:/12148/bpt6k6467343s [Accessed 29 July 2021].

ESDAC (2021). ESDAC – European Commission – Joint Research Centre [Online]. Available at: https://esdac.jrc.ec.europa.eu/ [Accessed 12 August 2021].

ESDB (2021). ESDB – European Commission [Online]. Available at: https://esdac.jrc.ec.europa.eu/content/european-soil-database-v20-vector-and-attribute-data [Accessed 12 August 2021].

FAO DSMW (2021). FAO Digital Soil Map of the World (DSMW) [Online]. Available at: http://www.fao.org/land-water/land/land-governance/land-resources-planning-toolbox/category/details/en/c/1026564/ [Accessed 12 August 2021].

FAO HWSD (2021). Base harmonisée mondiale de données sur les sols [Online]. Available at: http://www.fao.org/soils-portal/soil-survey/cartes-historiques-et-bases-de-donnees-des-sols/base-harmonisee-mondiale-de-donnees-sur-les-sols-version-12/fr/ [Accessed 12 August 2021].

Gallois, L. (1909). L'Académie des Sciences et les origines de la carte de Cassini. *Annales de géographie*, 18(99), 193–204.

Geocarta (2021). Geocarta | Geocarta – cartographie du sol pour l'agriculture, l'aménagement, l'archéologie [Online]. Available at: https://geocarta.net/ [Accessed 19 August 2021].

Géorisques (2021). BASOL : base de données sur les sites et sols pollués (ou potentiellement pollués) [Online]. Available at: https://www.georisques.gouv.fr/articles-risques/basol [Accessed 24 August 2021].

Geze, B. (1965). La cartographie pédologique en France. *Bulletin de l'association française pour l'étude du quaternaire*, 2(1), 1–4.

Gilliot, J.M., Michelin, J., Becu, M., Cissé, M., Hadjar, D., Vaudour, E. (2017a). The potential of UAS imagery for soil mapping at the agricultural plot scale. In *European Geophysical Union General Assembly 2017*, Vienna.

Gilliot, J.M., Vaudour, E., Michelin, J. (2017b). Soil surface roughness measurement: A new fully automatic photogrammetric approach applied to agricultural bare fields. *Computers and Electronics in Agriculture*, 134, 63–78.

Gilliot, J.M., Michelin, J., Hadjard, D., Houot, S. (2020). An accurate method for predicting spatial variability of maize yield from UAV-based plant height estimation: A tool for monitoring agronomic field experiments. *Precision Agriculture*, 22(3), 897–921.

Gis Sol (2021a). Partageons la connaissance des sols [Online]. Available at: https://www.gissol.fr/ [Accessed 30 July 2021].

Gis Sol (2021b). Inventaire, Gestion et Conservation des Sols – IGCS. Program [Online]. Available at: https://www.gissol.fr/le-gis/programmes/inventaire-gestion-et-conservation-des-sols-igcs-67 [Accessed 1 August 2021].

Gis Sol (2021c). Réseau de Mesures de la Qualité des Sols – RMQS. Program [Online]. Available at: https://www.gissol.fr/le-gis/programmes/rmqs-34 [Accessed 2 August 2021].

Guerrier, A. (2016). De l'Antiquité à Google Maps, la cartographie miroir du pouvoir. Article [Online]. Available at: https://www.franceculture.fr/histoire/de-lantiquite-google-maps-la-cartographie-miroir-du-pouvoir [Accessed 29 July 2021].

Hempel, J.W., McBratney, A.B., McKenzie, N.J., Hartemink, A.E., MacMillan, R., Lagacherie, P., Arrouays, D. (2013). Vers une cartographie numérique des propriétés des sols du monde. *Étude et Gestion des Sols*, 20(1), 7–14.

Hengl, T., Mendes de Jesus, J., Heuvelink, G.B.M., Ruiperez Gonzalez, M., Kilibarda, M., Blagotić, A., Shangguan, W., Wright, M.N., Geng, X., Bauer-Marschallinger, B. et al. (2017). SoilGrids250m: Global gridded soil information based on machine learning. *PLoS ONE*, 12(2), e0169748.

Hitier, H. (1894). Les nouvelles cartes agronomiques. *Annales de géographie*, 4(14), 101–106.

IGN (2021). Géoservices [Online]. Available at: https://geoservices.ign.fr/ [Accessed 14 August 2021].

INRAE (2018). Data INRAE [Online]. Available at: https://data.inrae.fr [Accessed 1 August 2021].

Krige, D.G. (1951). A statistical approach to some basic mine valuation problems on the Witwatersrand. *Journal of the Southern African Institute of Mining and Metallurgy*, 52(6), 119–139.

Laroche, B., DeParis, J., Messner, F., Wetterlind, J., Moulin, J., Froger, D. (2013). Potentiel de la spectrométrie gamma aéroportée pour la cartographie des sols et du régolithe : une mini-revue et des premiers exemples en régions Centre et Limousin. *Etude et Gestion des Sols*, 2013, 14.

Le Cadre, E. (2004). Modélisation de la volatilisation d'ammoniac en interaction avec les processus chimiques et biologiques du sol : le modèle VOLT'AIR. PhD Thesis, INA P-G, Paris.

Matheron, G. (1962). *Traité de géostatistique appliquée.* Éditions Technip, Paris.

McBratney, A.B., Mendonca Santos, M.L., Minasny, B. (2003). On digital soil mapping. *Geoderma*, 117(1–2), 3–52.

Messant, A., Lehmann, S., Moulin, J., Lagacherie, P., Jalabert, S., Noraz, A., Lemercier, B., Chafchafi, A. (2021). Diffusion des référentiels régionaux pédologiques sous la forme d'une carte des sols dominants (France métropolitaine Hors – Corse) accessible sur le Géoportail. *Etude et Gestion des Sols*, 28, 57–69.

MISTRALE-UE (2021). Utiliser la réflectométrie GNSS pour cartographier l'humidité des sols. Article, European Commission [Online]. Available at: https://cordis.europa.eu/article/id/202830-using-gnss-reflectometry-to-map-soil-moisture/fr [Accessed 20 August 2021].

Panagos, P., Van Liedekerke, M., Jones, A., Montanarella, L. (2012). European soil data centre: Response to European policy support and public data requirements. *Land Use Policy*, 29(2), 329–338.

PRECIFIELD (2021). Agriculture De Précision [Online]. Available at: https://www.precifield.com [Accessed 19 August 2021].

Refersols (2021). Répertoire national des études cartographiques de sol [Online]. Available at: https://webapps.gissol.fr/georefersols/ [Accessed 2 August 2021].

Saby, N.P.A., Lemercier, B., Arrouays, D., Leménager, S., Louis, B.P., Millet, F. (2014). Le programme Base de Données des Analyses de Terre (BDAT) : bilan de 20 ans de collecte de résultats d'analyses. *Etude et Gestion des Sols*, 10.

Savaton, P. (2003). Les cartes avant les cartes. Les cartes géologiques départementales : la première cartographie détaillée de la France. *Comité français d'Histoire de la Géologie*, 17, 53–74.

USGS (2021). EarthExplorer [Online]. Available at: https://earthexplorer.usgs.gov/ [Accessed 24 August 2021].

Voltz, M., Lemercier, B., Sauter, J., Richer de Forges, A., Lagacherie, P., Arrouays, D., Bispo, A., Slak, M.-F., Laroche, B. (2018). La cartographie des sols en France : état des lieux et perspectives. Report [Online]. Available at: https://hal.inrae.fr/hal-02788992 [Accessed 1 August 2021].

Walter, C., Schvartz, C., Claudot, B., Bouedo, T., Aurousseau, P. (1997). Synthèse nationale des analyses de terre réalisées entre 1990 et 1994 : descriptions statistique et cartographique de la variabilité des horizons de surface des sols cultivés. *Étude et Gestion des Sols*, 4(3), 205–219.

Wösten, J.H.M., Lilly, A., Nemes, A., Le Bas, C. (1999). Development and use of a database of hydraulic properties of European soils. *Geoderma*, 90(3), 169–185.

4

Runoff and Soil Erosion

Philippe MARTIN
UMR SADAPT, AgroParisTech, INRAE, Université Paris Saclay, Palaiseau, France

4.1. Preliminary considerations

Soil erosion is a series of processes by which soil particles are broken off, transported and deposited in a different location. This erosion can result from three primary causes: water, wind and human activity.

In this chapter, we will focus primarily on water erosion in agricultural areas. Before moving onto the central focus of this chapter, we would like to first briefly address the other types of erosion that will not be developed later.

4.1.1. *Forms of erosion*

Soil erosion by water can result from rains and the appearance of surface runoff that, once it gains speed, can cut through soils. Another type of erosion occurs in stream banks at the concave bends, which then deposit on convex sections. In mountainous areas, on soils with a clay or marl substrate, landslides can occur following rains. Solifluction is another type of erosion that occurs on steeply sloping soils. In soils where freezing occurs across the depth of the soil, thawing on the surface leads to downward movements of the upper part of the soil. On the other hand, wind erosion mainly occurs in regions with strong winds and bare soils without vegetation. These conditions are more common in desert areas (whether cold or hot), but they can also be found on agricultural land if the soil is left with no

Agricultural Soil Science,
coordinated by Yves COQUET and Joël MICHELIN. © ISTE Ltd 2025.

vegetation for too long in windy areas. Mechanical erosion occurs due to the descent of soils down a slope due to gravity after being acted on by tillage tools. It should be noted that in some regions where livestock farming is prevalent, mechanical erosion can be generated by the movements of animals instead of tillage tools, which is all the more significant as the animals contribute to the reduction of the soil cover (Boutrais 2000). It should be noted that the different types of erosion can also occur in conjunction with each other, leading to transfers covering long distances. Erosion in mountainous areas can lead to foothill deposits that can then be affected by surface erosion and moved into waterways.

4.1.2. *Connections with human activity*

Erosion processes are often a problem for humans, since top soil removal harms agricultural production due to a reduction in the production potential of the soils. As for soil movements and deposits, they can lead to destructive incidents involving human infrastructure, as was the case in the 1985 Armero disaster in Colombia, which left a death toll of over 20,000 (Voight 1990). However, in some cases, erosion processes are also highly dependent on human activities. This has been observed by archaeologists, who have found a connection between high erosion periods and the advent of ancient civilizations around the Mediterranean (Neboit 1984; Notebaert et al. 2014).

However, erosive processes have also been found to offer great gains in wealth. The best known example can be found in ancient Egypt, the “granary of Rome”, whose production depended heavily on the floods of the Nile and its supply of fertile alluvium, which they diverted to agricultural plots using a network of man-made canals (Burkhalter 1997). The fact that erosion processes do not have uniformly negative consequences is also true within the natural environment. Some shrub formations are only maintained through processes of erosion. The tiger bush, which can be found in Niger, is an interesting example (Seghieri et al. 1994). In this area, runoff is stopped by lines of vegetation located immediately downstream, which thus receives more water than natural precipitation alone would allow, and therefore keeps this particular vegetation alive.

4.1.3. *Regulations put in place*

The negative effects of erosion may be regulated by local agreements, as was the case in the 19th century in the vineyards of Saône-et-Loire, where vineyard tenants who did not remove all the soils that were washed down during a thunderstorm would be required to pay a fine to the owner (Durousset 1994). However, there are also national regulations. For example, France's second empire government enacted legislation to make it mandatory to reforest in the mountains on perimeters deemed to be at risk through the RTM (Restoration of Terrain in the Mountains) (Lilin 1986). But the most famous anti-erosion policy is clearly the policy implemented in the United States in the 1930s, following the occurrence of the dust bowl (Worster 1979; Trimble 2010). This disaster then led to the establishment of a soil conservation service (the Soil Conservation Act, 1935), which provided technical and financial support to farmers who volunteered to change their agricultural practices and landscape planning.

Back in France, the erosive problems did not stop with the implementation of the RTM. Despite certain fears that such an event might occur (Hénin and Gobillot 1950), France has never experienced an erosion crisis comparable to that of the United States. The implementation of agricultural guidance laws and the Common Agricultural Policy (PAC) in the 1960s quickly gave farmers the means to increase agricultural production by compensating for losses in soil fertility, which were more limited than in the United States, through the addition of chemical fertilizers. At the end of the 1970s, the people most harmed by these were not farmers but rather inhabitants downstream of the agricultural areas, who suffered from mud flows that could even prove fatal in some cases. This concern led to the establishment of pilot operations in field cultivation areas affected by the phenomena (Lilin and Paulet 1987) in order to better understand the processes and find technical solutions. These operations ceased at the end of the 1980s, except in the Pays de Caux region, where the operation has continued to this day under the name of AREAS (*Association Régionale d'Etude et d'Amélioration des sols*, the French Regional Association for the Study and Enhancement of Soils). This operation has given rise to a significant amount of research that forms the basis of many of the elements presented in the rest of this chapter.

4.2. The essential processes at work on agricultural land

Erosion by surface water results from the appearance of surface runoff, which can then be loaded with sediments. Runoff and sediment load can result from different processes, and it is important to know how to identify them in order to apply the appropriate solutions.

4.2.1. *Runoff*

We can distinguish two main types of runoff: Hortonian runoff, or runoff by saturation of the profile. Hortonian runoff begins when a soil's surface infiltration capacity is lower than the intensity of rainfall. This is often due to soil surface seals that are only a few millimeters thick[1] (see section 1.2.2.3), though below these seals the soil can have a very good infiltration capacity. Profile saturation runoff, for its part, is when blockages to underground infiltration occur (such as a clay horizon that is not sufficiently deep). The first rains achieve a sufficient infiltration level, but once the profile is saturated, any further precipitation will no longer be able to infiltrate and runoff may occur.

Depending on the region, it may be possible to have primarily Hortonian runoff or runoff through soil profile saturation. However, as Cayot (1993) reminds us, there can also be temporal variations in these mechanisms. As an example of this, in Brittany, thin silts on impermeable substrates (granite) may experience Hortonian runoff in the summer (where soils with little water saturation are exposed to occasional thunderstorms), while saturation runoff will predominate in the winter.

When the runoff is of Hortonian type, the objective will be the fragmentation of the crust (mechanically or biologically) in order to restore the infiltration capacity for the following rains. This can be done using surface tillage (see section 1.2.2.6). However, this surface tillage will have no effect on a plot that produces runoff by profile saturation. On surfaces of this type, the reduction of runoff on agricultural land will instead occur by management types involving underground drainage.

The crusting process can be categorized into two stages. Boiffin (1985) showed that freshly tilled loamy soils had an infiltration capacity of the order of 60 mm/h. To give an idea of what this means in practice, it should be worth noting that a 10-year rain event (one that is likely to occur once every 10 years) is on the order of 45–55 mm in Normandy. Therefore, the infiltration capacity of loamy soils in Normandy is significant, given the local climate context. However, the kinetic energy of the rains will break off soil particles from the surface aggregates; particles that will then splash and gradually close the gaps between the aggregates. This progressive clogging leads to the formation of a so-called "structural" or F1 crust. The infiltration capacity of a structural crust is much lower than a freshly tilled surface, on the order of 2–6 mm/h. As a result, even for ordinary rains, there may be some slight puddling that occurs in the depressions on the surface of the ground.

1 In the following section, we give a description of the ways in which this slaking crust is formed.

Some of the soil particles that continue to be pulled off by the raindrops will then land in these puddles and form sediments of very fine beds with very low porosity. This is the sedimentation that forms slaking crusts, or F2. Beginning in the lowest areas, the slaking crusts will gradually fill them in and expand as the particles torn off from the surface of the soil feed into their expansion. Once the slaking crust is formed, the infiltration capacity of the soil drops to less than 1 mm/h, meaning that even very light rains will generate a puddle on the surface, which can initiate runoff. Depending on the weather conditions and the soil characteristics (such as organic matter content, calcium state, etc.), in loamy soils, 60–120 mm of cumulative precipitation (i.e. on the order of 1–2 months of rainfall in Normandy) must fall in order to result in a well-developed slaking crust of type F2. Of course, these values are given as estimates for average regular rainfall amounts. In the event of a high-intensity thunderstorm, the total rainfall required will be lower.

The appearance of runoff requires connections to form between the puddles created following the formation of the slaking crust. Any obstacle to this connection will mean that runoff may be slowed down. Among other things, these include rough areas on the surface, provided that such areas are not oriented in the direction of the slope. The presence of residues on the ground will also contribute to limiting the interconnection between the puddles and to slowing down the flows, once these connections have been established. Living vegetation will be able to store a few millimeters of water in its aerial sections, and will be able to facilitate water infiltration at the level of contact between the stem and the root.

4.2.2. *Sediment loads*

Immediately after the runoff has appeared, it may be loaded with solid particles through the "splash effect", which tears off particles of soil under the impact of raindrops and projects them into the thin streams of running water. This is then considered to be diffuse erosion, without marked traces on the soil surface. With the help of the slope, if the runoff manages to acquire enough speed, which varies according to the type of materials in place (Figure 4.1), then the soil will erode in different characteristic patterns. On plots cultivated on steep slopes (generally greater than 10%), a large number of channels will form along lines parallel to the main slope, which are usually shallow. If the slope is lower, the flow velocity will not reach a speed that is sufficient to erode the slopes. On the other hand, the runoff may concentrate in the bottoms of valleys, also called "talwegs" (valley bottom paths). The concentration of the runoff that forms in this way will allow the runoff to acquire higher speeds, even though the inclination in the talweg is most often lower than that of the slopes themselves. It is then possible for this runoff to reach a speed such that the talweg erodes. This is then referred to as "concentrated erosion of the valley floor", which gives rise to temporary gullies, which can later be erased by subsequent tillage for crops, or may become permanent if they are too large and no cultivation can be achieved along the route they take.

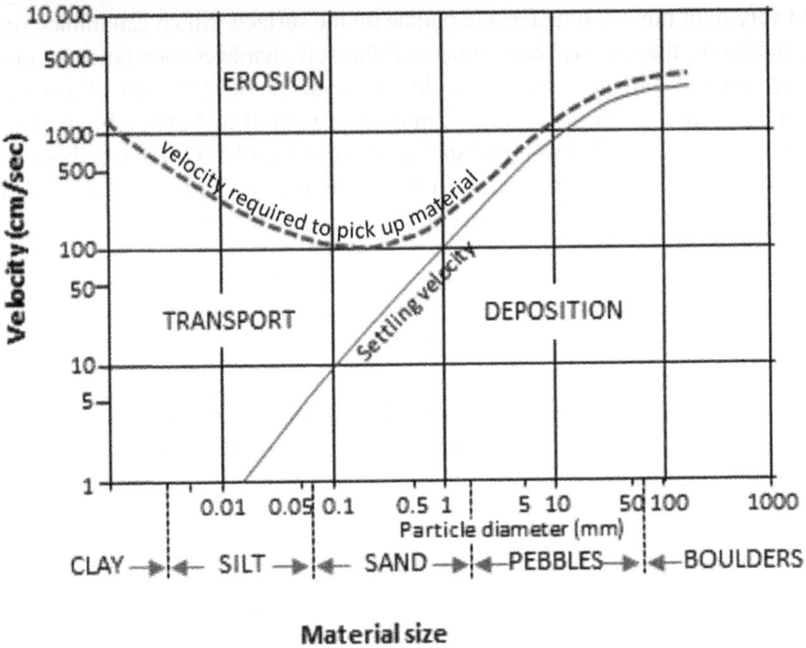

Figure 4.1. *Adaptation of Hjulström's (1935) diagram*

Several important points should be considered regarding this general description we have just provided. The first is that the different types of erosion may combine in different ways across different spaces. For instance, there can be diffuse erosion upstream and slope erosion a little further along, which is then prolonged by concentrated erosion in the talwegs (Figure 4.2). A second important point is that the particle flow rate is not a linear function of the water flow rate (Crawford 1991). In other words, the particle load of a flow of water depends on the flow velocity. As soon as this speed decreases, the transport capacity also decreases, which can result in occasional soil deposits. If the flow regains any further speed, it will in turn regain the corresponding erosive capacities. This last point is important, because it means that we can prevent the soil from leaving at an upstream point of a plot or a watershed with a development that slows down the flows. The loss of speed will trigger a deposit of sediment upstream of the development. On the other hand, if nothing is done to slow down the runoff downstream of this point, the stream of water thus discharged regains all its erosive capacity and may therefore cause more damage downstream than if it had not been discharged upstream of its solid load. In other words, any action to reduce erosive processes upstream must take into account the consequences induced in the downstream section of the system.

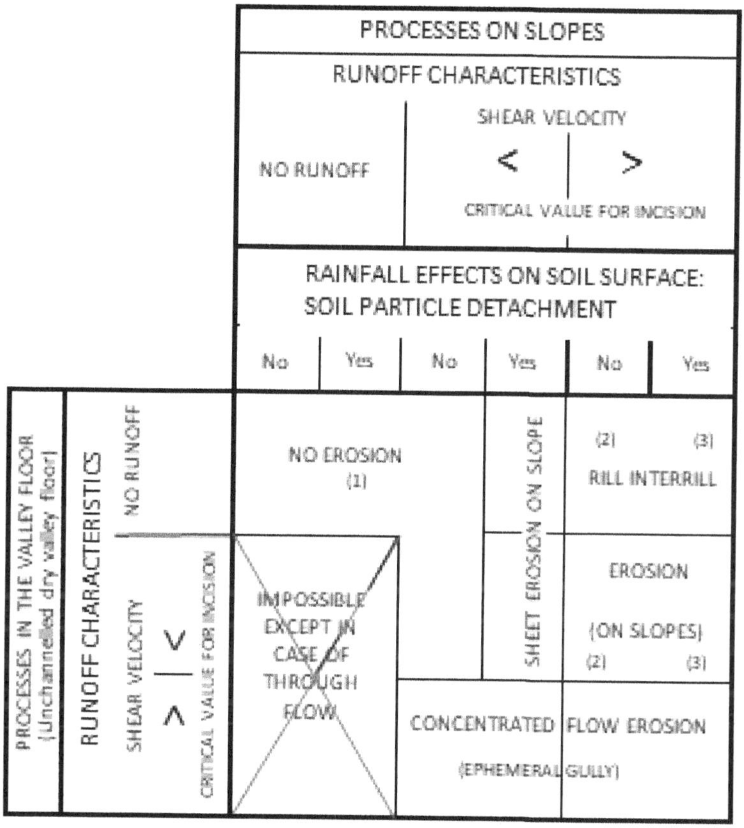

(1) Crusting: soil particle displacement limited to ~10 cm

(2) The interrill area is not a sediment source

(3) The interrill area is a sediment source

Figure 4.2. *Illustration of the different ways in which erosion types may combine within a watershed (according to Auzet et al. 1990)*

4.3. Impact of agricultural activities on processes

In this section, we will first examine how the cropping systems implemented by farmers can guide the processes presented in the previous part. Here, our examination is at the level of cropping systems. We will then look at some long-term effects, that is, long-term developments in agricultural activities.

4.3.1. *Effects of cropping systems*

To illustrate our point, we will look at the case of a cultivated plot located halfway down a gently sloping watershed and crossed by a talweg. This plot is therefore likely to produce diffuse runoff that can concentrate at the level of the talweg, which will erode the talweg in the plots located farther down. This plot can itself be eroded by the talweg which crosses it if concentrated runoff is emitted by the parcels located upstream, or if it itself produces diffuse runoff which concentrates at this point with sufficient speed.

To fully understand the effect of cultivation systems on these processes, we will examine two contrasting situations. The first situation is that of a permanent meadow, that is to say one in which the land cover remains unchanged from one year to the next with the same amount of grass cover. The second situation involves a succession of annual crops, each of which has a phase of sowing, cultivation management and harvesting, and then an inter-crop phase before the sowing of the next crop.

The *permanent meadow* may have been sown a very long time ago. Following the sowing, a slaking crust might have appeared before the grass cover protected the surface. This plot had then been farmed for several months. However, with the development of the covering, the crust of the threshing floor has gradually been fragmented, perforated by the development of the covering and even more so through the development of the biological activity arising due to the appearance of the meadow (earthworms, etc.; see section 2.3, Chapter 2). The meadow thus regains its initial infiltration capacity, which it will then retain due to the protection of the soil surface by the grass covering. If we consider the meadow plot at the time after it regains its infiltration capacity, we can note that given its high infiltration capacity, this surface will produce very little runoff or none at all. It will therefore not pose any threat to the plots downstream where the talweg passes through. Similarly, if a concentrated runoff stream flows into an area upstream of the meadow plot, the talweg will see little or no erosion, because the network of roots that has developed for several years under the meadow will retain the soil effectively. The last aspect is that for the portion of plot located in a sloped orientation, the diffuse runoff that crosses through it to reach the talweg will be slowed down by the vegetation cover, which will lead to the deposition of part of the solid load, or even all of it if the runoff penetrates into the grassland surface. These considerations make it possible to conclude that permanent grasslands play an essential role in reducing erosion problems due to Hortonian runoff.

For *annual crops* on loamy soils prone to crusting, farmers most often resort to plowing (see Chapter 1) and carry out seed sowing very soon after that; in some cases, even on the same day. Under these conditions, just after the sowing occurs, the soil surface is fragmented and has a high infiltration capacity (see section 4.2.1) on the order of 60 mm/h. This means that immediately after sowing, very little runoff is likely to occur in the plot. However, plowing will loosen the soil a great deal, making it very susceptible to erosion concentrated within the tilling depth (25–30 cm). In fact, it is common to see erosion patterns limited to the talweg with a depth equal to that of plowing. After sowing, a "race" occurs between the degradation effect of the rains and the protective action of the developing vegetation cover. In the loamy plains of north-western Europe, characterized by low temperatures and sustained rainfall, it is usually the development of the slaking crust that wins against the development of the plant cover. The plot will therefore see high runoff one or two months after sowing. At the same time, the soil will have become packed up (compacted under the action of its own weight), which will have reduced the risks of erosion in the talweg; risks that will also gradually decrease due to the root development of the crops, which will help to retain the soil.

It is important to note that maintaining a high rate of humus and calcium in the soil will have a strong effect on the structural stability of aggregates, and thus the sensitivity to crusting. High organic matter content levels, as well as good calcium maintenance, make it possible to reduce the sensitivity to slaking. Other, more cyclical factors can also influence the processes. The moister the soil is at the time of sowing, the faster the slaking crust will develop, and the soil will not have time to dry before the first rains. Moreover, a very fine level of seedbed on the surface will create a situation that is more sensitive to slaking than a surface with larger aggregates.

Once the crop has grown, farmers can intervene on the plot to fertilize and protect the crops against weeds, diseases and predators (insects). In conventional agriculture, these various interventions are carried out using spreaders or sprayers that are 24 m or 36 m wide, using the same wheel paths for each intervention. The imprints from these wheels (two 30-cm wheels, 2 m apart) lead to local settlements which reduce porosity and thus water infiltration (section 1.2.2.5), increasing the local runoff, but the overall impact on the plot remains limited (0.6 m/36 = 1.7% of the total surface). On the other hand, these wheel imprints constitute surface agricultural formations that can capture the runoff from the surrounding surfaces and then cause it to accelerate and thus aid in soil incision in local areas.

If a farmer opts for mechanical weed removal, this action will lead to a fragmentation of the crust that gives rise to a loosening of the soil surface along a depth of 2–3 cm. The fragmentation of the slaked crust will restore the infiltration capacity. For its part, the loosening of the profile will increase the risks of erosion in the talweg position if runoff flows in from the upstream plots. However, the risks of erosion are lower compared to the risk levels just after sowing, due to a lower tillage depth (2–3 cm, compared to 25–30 cm).

Harvesting is a stage that often results in significant changes in surface conditions, which vary between different crops. These differences can be illustrated by comparing a harvest of a crop of rapeseeds with that of potatoes. Potatoes are a spring crop, generally grown on mounds and harvested in the autumn. At the time of the harvest, there is almost no vegetation left on the surface. The potato harvester lifts the mounds to collect the potatoes and leaves the fine soil particles on site. With barren soil and numerous wheel prints, the potato plots after harvest often see runoff occur, which encounters no obstacles to concentrating and flowing quickly and often leads to the formation of gullies due to the soil being greatly loosened from the implantation and harvesting of the potatoes. Rapeseed, by contrast, is harvested in the summer. The rapeseed straw is generally crushed, and thus forms a mulch which then sees a rapid development of rapeseed regrowth. This surface sees very little runoff and erosion.

The intercrop period comes between the harvest of one crop and the sowing of the next. Some intercrops are very short (wheat can be sown a few days after the harvest of potatoes or sugar beets). Others can be very long, such as for sowing corn (in May of year n+1 when wheat was harvested in July of year n). To reduce runoff during the intercrop period, farmers can till the soil superficially (stubble cultivation) or sow cover crops. Surface tillage is generally only effective for short periods (two or three months) since roughness ends up being filled by sediment deposits linked to the development of the slaking crust. When possible, the sowing of cover crops after a coarse tillage is an effective solution. Sown at the end of summer (before the fall sowing period), these cover crops develop quickly and thus protect the soil surface before the rains degrade it. The fact that the nitrate directive has made cover crops mandatory in most silty plains (for nitrate trapping purposes) has greatly contributed to reducing the problem of winter agricultural runoff.

4.3.2. *Past and current long-term dynamics*

The evolution of the agricultural context means that cultivation systems evolve over time with possible consequences on the erosion processes. For example, since the end of the 1970s, dairy cattle farming, which has had little support from the PAC, has disappeared in many regions. Also, in those areas where it remains, this is often achieved by intensifying the system with silage corn and soybean meal. This results in a significant reversal of permanent grasslands, especially in loamy plains where annual crops are often more profitable. Given the role played by meadows, these reversals lead to an increase in the risks of runoff and soil loss. These changes take place in two stages. Initially, there is a limited increase in runoff since the high organic matter contents after a permanent meadow make it possible to delay (or even entirely prevent) the formation of a slaking crust. However, over the years, organic matter levels decrease because the contributions specific to crops and organic residues do not allow the losses to be compensated. As a result, the plot becomes increasingly sensitive to slaking, and therefore to runoff (Chenu et al. 2000).

The specialization of certain agricultural regions can also pose problems. Although certain regions of silty plains specialize in winter crops (wheat, rapeseed and barley), most areas will see their highest runoff levels in winter, thus creating a high risk of muddy floods during winter rains (Boardman 1988). The same is true for specializations in spring crops, with regard to the runoff occurring during thunderstorms in the spring (Elyakime and Bruno 2000). The diverse array of crops present in a watershed constitutes an important factor in limiting the risks at outlet points. This means that in the winter, the runoff from wheat or barley plots can be absorbed lower down by plots occupied by intermediate crops, which will see the sowing of spring crops a few months later. However, these possibilities for spatial organization can be undermined if we increase the size of the plots, because the most extreme potential outcome is that we may end up with a single plot with the same crop covering the entirety of a watershed. The increase in the size of plots is also often accompanied by the destruction of bordering elements, such as hedges and embankments, which increases the erosive risks even more.

No-till cultivation techniques, known as soil conservation (see section 1.3.2), were successfully developed in Brazil, beginning in the 1990s, and are now seeing development in France as well, even if this development is more in stony lands with low crop potential than in fertile plains with deep loam. These techniques make it possible to have a soil with a stronger compactness, which makes it less sensitive to surface erosion. On the other hand, reductions in runoff are not always so obvious, and they seem to require successions of cultivation leaving significant residues on the surface. In regions where crops often leave little residue (with predominantly potatoes, flax or beetroot), it is preferable to limit these techniques to steeply sloping areas as well as to plots that are crossed by a talweg. In recent years, other complementary techniques have begun to evolve with the development of direct sowing after cover crop, which is either destroyed and left on the ground in the form of mulch, or kept alive but whose growth is regulated mechanically or chemically. These techniques are often effective, but the regulation of weeds and related cover often requires an increased use of chemical weed-killers, which can pose a problem in terms of water quality.

Without necessarily going as far as no-till, farmers can adopt tillage tools that leave more consideration for the runoff/erosion problem. This is particularly the case for potatoes. As we explained earlier, when this crop is harvested, a very runoff-prone surface is obtained, but this is also the case during the cultivation period due to the fact that potatoes are grown on mounds, and therefore leave spaces between the mounds that have high potential for leading to runoff. An effective technical solution has been found. This solution is to add to the potato planter, devices which scrape the soil at the bottom of the spaces between mounds and which are raised at regular intervals. This thus generates a small pile of soil which very effectively blocks the flow of water in the space between mounds (a "micro-dam"). It should be noted that in a period of climate change, which is likely to lead to a greater frequency of summer droughts, this type of device makes it possible to take better advantage of summer rains to supply the water reserve of the soil, rather than losing this water through runoff (Agassi and Levy 1993).

The general public's demand for organic farming products is leading an increasing number of farmers to convert their farms. Even if the conversions in the very fertile areas of the loess deposits still remain marginal, this translates into overall positive effects for the control of runoff due to a greater use of organic matter (see section 2.4.1) and therefore a significant increase in the organic matter content in the soils (Guyomard et al. 2013). On the other hand, the heavy use of tillage can promote the loss of soil in areas with concentrated runoff and on steep slopes.

4.4. Toward integrated watershed management

Controlling erosion problems is achieved at watershed scale. Each watershed is defined by an outlet, which in our case, is often the place where more or less muddy flows have the greatest impact on human activities. The watershed associated with this outlet represents all the surfaces that converge topographically toward this outlet. Generally, where the outlet is in an inhabited area that we want to protect, it is important to consider three main areas of action.

The first area is the reduction of runoff on agricultural plots. Recall that without runoff, there is no water erosion[2]. This point has been explained at length in the preceding pages. However, it should be remembered that it will be necessary to pay particular attention to tillage, vegetation cover, and also to the management of physical states (settlement), chemical states (calcium content) and biological states (organic matter content and biodiversity) of the soil.

The second area aims to limit the concentration of runoff into the talwegs, the main areas where soil is taken away in these erosion systems. To do so, devices will be set up allowing either the infiltration of the runoff or the regulation of the flows entering into the talwegs, to avoid flows that are too fast or too large, which are highly erosive.

The third area lies in the protection of the talwegs in order to strengthen their consistency and ensure that the soil does not wash away, even in high flows of concentrated runoff.

Below, we will briefly examine these last two areas, which are essential for controlling erosion hazards, and then return to the necessary management actions to address vulnerabilities downstream of the agricultural zones.

4.4.1. *Limiting concentration levels flowing toward the talwegs*

It is desirable to limit the volume and the speed of the runoff as it reaches the talwegs.

If runoff has in fact been produced on plots within the watershed, this runoff can be mitigated by placing highly absorbent surfaces along its path. This can be done by taking special care in the organization of crops within the watershed space (section 4.3.2). In the winter, for example, the runoff from wheat plots can be absorbed by intercrop plots with intermediate crops or by permanent grassland plots. Joannon (2004) has shown that such spatial coordination has plenty of advantages, but it is not easy to put in place in situations where the watershed is drawn on by many different farmers who are not necessarily used to working together.

2 This statement does not apply to the creation of the slaking crust. This is indeed an erosive process (broken off soil, (aerial) transport, and deposition of particles) but which does not mobilize runoff.

The coordination between farmers is complicated by the fact that, over one year-long growing period, the assigning of crops to plots is not decided by farmers using the same time scale (Schaller 2011). In any case, it is essential to have watershed facilitators help support this type of coordination. It may be safer to reduce the amount of runoff with pervious surfaces that remain in place from one year to the next. This would make it possible to propose to farmers that they move crops far upstream toward the middle of sloping areas, so as to be able to absorb runoff from annual crop. This can be done within the same farm or by exchanging plots between two farms. These land exchanges most often require a watershed management structure, exercising a right of land preemption to acquire areas outside the basin that would be exchanged with plots located strategically within the basin. Once these plots have been acquired, the management agency can seek out a farmer who would agree to keep grass on the plot as part of an environmental lease. The framework of "real environmental obligation" can also be used by an owner to maintain grass on a plot via a contract with a community.

To limit the amounts of concentrated water, one option is to put earthen dikes in place upstream of the talwegs, thus providing for water storage. This acts as buffer storage, which is to say that the dikes are equipped with a calibrated runoff flow rate so that the flow rate of the water leaving the structure makes it possible to empty the basin as quickly as possible, without this generating risks of erosion in the talweg.

Prior measures to reduce the amounts of runoff flowing into the talwegs all at once may sometimes prove insufficient. It is then necessary to protect the talwegs.

4.4.2. *Sustainable protection of the talwegs*

The protection of talwegs generally involves sowing grass, which can sometimes be done in conjunction with preliminary earthworks. This ensures a more effective spreading of the water to avoid talweg erosion, or conversely, a very concentrated flow within a grassy ditch.

To limit the flow speed, it is possible to enforce obstacles that slow down the flows. One such type of obstacle are "fascines". These are interlacing branches held in place on the ground by wooden stakes. Placed across the talweg, they make it possible to slow down the flows, and therefore to reduce their erosive effects.

It is important to note that, in the context of concentrated runoff on a low slope, any development along the route of the talwegs can generate changes in the flows over the medium or long term. Since the fascines slow down the flows, they can lead to the depositing of the soil particles contained in the waters. These sediments then accumulate upstream of the fascines to the point of constituting a major obstacle to the flow. This means that it is possible for a new water flow path to be created, as well as erosions in the talweg right next to the fascine that was intended to prevent them. We can see the same type of difficulty with the grass strips that protect the talwegs but which also slow down the flows leading to soil deposits on the surface that cover the grass. The grass often grows back, developing on these sediments and later leading to new deposits. After a few years, grassy strips can thus be found a few centimeters above the initial soil level, thus effectively creating new talwegs, which are sensitive to erosion on either side of the grassy strip. These two illustrations show the benefits of not only reducing and slowing down the flows concentrated in the talwegs, but also of limiting the amount of diffusely eroded soil upstream of the talwegs at the risk that deposits of these sediments do not disturb the protective arrangements of the talwegs too much. This can be done in ways such as by placing fascines upstream of the talwegs to ensure soil particles are deposited before the runoff becomes too concentrated. This depositing will mean that flows that are less heavily loaded with sediments will have a greater erosion capacity, and that it will be especially important to reduce the quantities by infiltration and the flow speed via the creation of hydraulic obstacles.

4.4.3. *Reducing vulnerability in downstream agricultural territories*

The greatest risks occur at the outlet of the watershed, due to greater vulnerability in the infrastructure in these areas (which are generally inhabited). Protecting these areas most often involves infrastructures that make it possible to contain the hazards arising from the agricultural territory. Although much larger in size, these infrastructures are, in principle, very similar to the dikes created to reduce the concentration of runoff in the talweg.

4.4.4. *Population protection measures*

Catastrophic events are still occurring in France at the beginning of the 21st century. Since 1982, the natural disaster law has made it possible to take better care of victims of catastrophic events who are insured. Through the payment of a mandatory surcharge, they pay into a national solidarity fund that is put to use when a “natural disaster” event (particularly floods with high mud concentrations) occurs at a given point within the country’s territory. This system has made it possible to plug the gap in protecting less affluent populations, since land in more “dangerous” areas is often cheaper. The natural disaster law could create an incentive for some communities to rely on this nationally funded scheme and thus not to make sufficient investments in risk prevention at a local level. To avoid this, the law increases the amounts of the deductibles paid for reimbursements made by insurance companies when the frequency of "natural" disasters becomes too high. Faced with lower payouts, policyholders are more likely to put pressure on their elected officials to force them to act. This can be done through flood risk prevention plans. These plans, implemented under the leadership authority of the prefectures, make it possible to develop urban space in ways that make it less vulnerable (such as by prohibiting construction in dangerous areas and implementing urban infrastructures for the protection of existing buildings).

It is also possible to act legally on agricultural areas upstream. In this sense, the French rural code (R114-1 to R114-10) authorizes officials to implement environmental constraint zones (*zones soumises à contraintes environnementales* (ZSCEs)) to limit erosion problems, similar to existing protections for catchments. The principle of ZSCEs is to establish boundaries for areas subject to erosion, to propose an action program with assistance for a three-year period, knowing that at the end of these first three years, the official retains the possibility of making some of the measures mandatory if the voluntary action program has not sufficiently advanced. Erosion ZSCEs are relatively rare and most often associated with water quality problems more than with the protection of property and people (such as the ZSCE Seiche in Ille-et-Vilaine).

4.5. The art of avoiding recipes

The materials presented in this chapter are based on the experience of loamy agricultural plains of northern France, which interconnect with highly urbanized areas. This situation is interesting from a pedagogical perspective, because it allows a wide range of erosive processes to be covered (from the formation of slaked crusts to muddy floods). However, it is important to know how to adapt the information taken from this chapter to other contexts. For example, we have shown that in the loamy areas of northern France, winter crops such as wheat lead to high levels of winter runoff arising from the development of slaked crusts. But not all cases of winter wheat lead to high runoff in all regions suffering from water erosion. In the hilly regions of Lauragais, wheat sown on slopes in the winter will not lead to the formation of slaking crusts due to the presence of clay soils that are less sensitive to crusting and which see lower rainfall than in the north of France. The erosion occurring here is essentially "channel–interchannel" type erosion on steeply sloping soils during spring thunderstorms (more common on corn or sunflower crops). Wheat, a crop prone to generating runoff in the north, instead becomes an area subject to higher water infiltration in the south. This simple example illustrates the fact that overly broad recommendations for combating erosion should be avoided, and that it is important to clearly identify the processes at work in each local context, as well as the effects of human activities on these processes.

4.6. References

Agassi, M. and Levy, G.J. (1993). Effect of the dyked furrow technique on potato yield. *Potato Research*, 36, 247–251. doi: 10.1007/BF02360533.

Auzet, A.V., Boiffin, J., Papy, F., Maucorps, J., Ouvry, J.F. (1990). An approach of erosion forms and erosion risk on agricultural land in the Northern Paris Basin, France. In *Soil Erosion on Agricultural Land*, Boardman, J., Foster, I.D.L., Dearing, J.A. (eds). John Wiley and Sons, Hoboken, NJ.

Boardman, J. (1988). Severe erosion on agricultural land in East Sussex, UK October 1987. *Soil Technol.*, 1, 333–348.

Boiffin, J. (1985). Stages and time-dependency of soil crusting in situ. In *International Symposium on the Assessment of Soil Surface Sealing and Crusting*, Ghent.

Boutrais, J. (2000). Elevage et érosion en Adamaoua (Cameroun). *Bulletin du réseau érosion*, 20, 204–217.

Burkhalter, F. (1997). Irrigation et production agricole en Egypte héllinistique et romaine. Histoire, économie et société. *Environnement et développement Economique*, 16(3), 343–352.

Cayot, S. (1993). Etude spatialisée du ruissellement à l'échelle de la parcelle. Mémoire de DEA de Géosciences filière : pédologie "les sols dans les écosystèmes continentaux". Thesis, Université de Nancy I, INAPG, ENSR.

Chenu, C., Le Bissonnais, Y., Arrouays, D. (2000). Organic matter influence on clay wettability and soil aggregate stability. *Soil Sci. Soc. Am. J.*, 64, 1479–1486.

Crawford, C.W. (1991). Estimation of suspended-sediment rating curves and mean suspended sediment loads. *Journal of Hydrology*, 129, 331–348.

Durousset, E. (1994). Contribution à l'étude de l'environnement au niveau local. Pratiques sociales associées à l'érosion des sols viticoles et aux inondations qui en résultent en Saône-et-Loire. PhD Thesis. Université Paris X Nanterre, Paris.

Elyakime, B. and Bruno, J.F. (2000). Gestion de la lutte contre une érosion de versant avec dégâts sur site public. *Économie rurale*, 257, 67–77. doi: 10.3406/ecoru.2000.5182.

Guyomard, H. (2013). Analyse des performances de l'agriculture biologique. Report, Commissariat général à la stratégie et à la prospective [Online]. Available at: https://www.strategie.gouv.fr/sites/strategie.gouv.fr/files/atoms/files/rapport-inra-pour-cgsp-volume-1-web071020131_0.pdf.

Hénin, S. and Gobillot, T. (1950). L'érosion en France. *BTI*, 50, 431–433.

Hjulström, F. (1935). Studies of the morphological activity of rivers as illustrated by the River Fyris. *Bull Geological Institute Upsalsa*, 25, 221–527.

Joannon, A. (2004). Coordination spatiale des systèmes de culture pour la maîtrise de processus écologiques : cas du ruissellement érosif dans les bassisn versants agricoles du Pays de Caux, Haute-Normandie. PhD Thesis, INAPG, Paris.

Lilin, C. (1986). Histoire de la restauration des terrains en montagne au 19ème siècle. *Cah. ORSTOM, Sér. Pédol.*, 22(2), 139–145.

Lilin, C. and Paulet, B. (1987). L'érosion des sols cultivés en France : développement actuel et actions entreprises. *BTI*, 417, 71–94.

Neboit, R. (1984). Érosion des sols et colonisation grecque en Sicile et en Grande Grèce (Soil erosion and greek colonization in Sicily and great Greece). *Bulletin de l'association de géographes français*, 499, 5–13. doi: 10.3406/bagf.1984.5435.

Notebaert, B., Berger, J.-F., Brochier, J.L. (2014). Characterization and quantification of Holocene colluvial and alluvial sediments in the Valdaine Region (southern France). *The Holocene*, 24(10), 1320–1335. doi: 10.1177/0959683614540946.

Schaller, N. (2011). Modélisation des décisions d'assolement des agriculteurs et de l'organisation spatiale des cultures dans les territoires de polyculture-élevage. PhD Thesis, ABIES.

Seghieri, J., Galle, S., Rajot, J.L. (1994). La brousse tigrée dans le Sahel nigérien : étude de la cofluctuation du stock hydrique et de la végétation annuelle. In *Interactions surface continentale/atmosphère : l'expérience HAPEX-Sahel*, Hoepffner, M., Lebel, T., Monteny, B. (eds.). ORSTOM, Paris.

Trimble, S.W. (2010). Donald Worster's "dust bowl". *Aeolian Research*, 2, 1–4.

Voight, B. (1990). The 1985 Nevado del Ruiz volcano catastrophe: Anatomy and retrospection. *J. Volcanol. Geotherm. Res.*, 42, 151–188.

Worster, D. (1979). *Dust Bowl: The Southern Plains in the 1930s*. Oxford University Press, New York.

5

Soils and Agricultural Land Property: Legal Aspects of Sectoral and Functional Protection

Louis DE REDON[1,2]
[1]*SESG, AgroParisTech, INRAE, Université Paris Saclay, Palaiseau, France*
[2]*Institut de Recherche Juridique de la Sorbonne (IRJS), Paris, France*

5.1. Soil, a physical environment like no other?

As an introduction to a work that has been and continues to be a widely used reference in terms of the laws and legal protections of nature – "*L'Écologie et la Loi*" ("Ecology and the law") – Cyrille de Klemm writes that "in nature, universalities are practically all that exists"; "such is the case with ecosystems, where the different biotic and abiotic elements of the environment and the ecological processes that connect them to each other are intimately intertwined".[1] There is no doubt that the physical environments, of which there are three: *soil*, *air* and *water*, participate in these universalities. So much so that the scope of the study of these environments is very difficult to determine: what interest is there, and how effective would it be to protect crops if we do not protect soils, or to protect health if we do not protect air, or to protect fish if we do not protect water?

1 de Klemm, C. (1989). Les éléments de l'environnement en droit positif. In *L'écologie et la loi*. Éditions L'Harmattan, Paris.

Agricultural Soil Science,
coordinated by Yves COQUET and Joël MICHELIN. © ISTE Ltd 2025.

Soil, air and water, common heritage,[2,3] are of a fundamental nature. These physical environments are essential to life, and even to our survival. However, everything that has a fundamental character – the defense of a common value or a common heritage – deserves to be protected by law. In this sense, the responsibility for the protection of physical environments – *terrestrial*, *atmospheric*, or *aquatic* – appears to naturally devolve to the public and collective entity that is the state. It is the duty of the nation to defend and protect the common heritage. And it carries this out, systematically and almost mechanically, by exercising administrative police powers in order to prevent environmental damage (actions taken a priori). While it also does so, in certain instances and in a manner that is more chaotic if not stochastic, by setting in motion public actions within the framework of judicial police missions to fix damages that could not have been avoided, such as pollution (actions taken a posteriori). As evidenced by the content of the French Environmental Code,[4] spanning some 3,651 pages, many texts address prevention (which is therefore the regulation in terms of environmental protection) as some others address repression and reparations (which are therefore objectively, and unfortunately, the exceptions in this area).

Thus, some environments are protected, although they are not protected for themselves but for the habitat that they may constitute or the resource that they may provide, while others are not. The damage caused to them is not suppressed by the harm that an alteration or destruction would cause to the common heritage, but by the reduction of the amenities that they bring to human societies, and even to certain private interests. The legislature is less interested in protecting soil for itself than the functions of it (buildability, extraction of materials, production of crops, etc.), in protecting the atmosphere than the functions of it (breathability, etc.), and in protecting the aquatic environment than the functions of it (potability, navigability, irrigation, etc.). Consequently, it is impossible to identify major general principles of law providing for the protection of physical environments, including the protection of soils. Rather, it is an aggregate of legal rules protecting some of their functions useful for human activities. The law of physical environments is sectoral, and the protection of soil, air and water is functional. In this jumbled body of legal

---

2 "The environment is the common heritage of human beings", French Environmental Charter, recital 3.

3 "Spaces, resources and natural terrestrial and marine environments, the sounds and smells that characterize them, the sites, the daytime and nighttime landscapes, the air quality, living beings and the biodiversity are part of the common heritage of the nation", Article L. 110-1 Environmental Code.

4 Cans, C., Dejean, E., Makowiak, J. (2019). *Code de l'environnement 2019 : annoté & commenté*, 22nd edition. Dalloz, Paris.

regulations, land law is even more so, appearing to suffer an even worse fortune than air law and water law. It is the weakest link in a disparate chain.

Consequently the "land", which legally encompasses the concepts of "soil" and "subsoil" in the scientific sense of these terms, constitutes in the eyes of the law an object which by its nature is immovable and one of property. Any approach that treats it as an element of the environment is incidental, if not nearly nonexistent. It is "a blind spot of the law": "if it is taken into account in some of its aspects, it is not understood in its entirety, as a complex whole with various functions".[5] Therefore, the second book of the French Environmental Code relating to "physical environments" does not have any titles relating to soils at all. It simply covers aqueous environments (aquatic and marine; "Title I") and gaseous environments (air and atmosphere; "Title II"). Solid environments (soils and subsoils) are the most important elements missing from the French Environmental Code.

Consequently, France's laws look rather helpless in providing any protection against the diverse nature and the increasing number of threats to soils, given that there is no legislation or regulations that specifically address the conservation of these fragile environments. But in reality, the law does tackle the issue of soils in a more incidental way: an in-depth study of the legal rules, which are scattered and relate to its conservation, clearly makes it possible to highlight the existence of a diffuse system of legal protections aimed at ensuring soil sustainability: "the scattered nature of the rules relating to soil preservation hides a progressive and underlying adaptation of the law to this problem, which hinges around two axes: the protection and the repair of the soil".

In her thesis on "Soil conservation in French law",[6] Sandrine Staffolani thus demonstrates the existence of a conservation system of soils, established from two main sources of alteration: exploitation and property.

In this legal context, which some may describe as "minimalist", the purpose of this chapter is thus to identify and organize the legal elements of a functional protection of the soil, understood first as a biological environment (section 5.2), that is to say as a substrate covered by the law (section 5.2.1) with the objective of conservation (section 5.2.2), and then of land ownership (section 5.3), that is, as real estate property (section 5.3.1) threatened by the artificialization of territories arising from urban sprawl (section 5.3.2).

---

5 Hermon, C. (2017). La protection du sol en droit. *Droit et ville*, 2, 17–47.

6 Staffolani, S. (2008). La conservation du sol en droit français. PhD Thesis, Université de Limoges, Limoges.

5.2. Soils are covered by the law as a biological environment

5.2.1. *The legal foundations of soil protections*

As Murielle Bertrand correctly notes:

> The images evoked by soil are hardly poetic, except to those who have been able to sense the courage in the calloused hands of those who work the land, the heroism in the daily toil of the seasoned peasants, and deduce from these the richness and the beauty of nature, and the reinforcement of the generous and discreet foundation which nourishes it.[7]

Soils constitute our fundamental substrate: they are at the same time a habitat, a space, a resource and a natural environment. They have both ecological (see Chapters 2 and 7) and economic functions, which are essential for the functioning of ecosystems, for the development of human societies and for the dictionary of nature protection.

> "Soils are a complex and dynamic biological environment formed by the combined action of multiple biotic and abiotic factors. It is composed of mineral and organic elements. The multitude of different components that constitute a soil is the result of a state of equilibrium between climatic factors and vegetation. For the management of spaces and natural resources, soils constitute an element that integrates the ecosystem [...] Neglected in the past, the attacks that have affected the state of soils health are numerous (artificialization, compaction, erosion, pollution, impoverishment) and measures are now tending to reduce them [...].[8]

Thus,

> The ecological functions of soils are understood as the role they play within ecosystems, which can include, for example: serving as a breeding ground for biodiversity, contributing to the storage, filtering, and transformation of nutrients, substances and water, recharging groundwater, sequestering carbon, or regulating carbon.[9]

However, legal experts do not quite see it from this perspective…

---

7 Bertrand, M. (2018). La protection des sols dans le cadre de l'Union européenne. PhD Thesis, Université Lyon 3 Jean Moulin, Lyon.

8 Bioret, F., Sturbois, A., Estève, R. (2009). *Dictionnaire de la protection de la nature*. Presse Universitaire de Rennes, Rennes.

9 Neyret, L. and Martin, G.-J. (2012). *Nomenclature des préjudices environnementaux*. LGDJ, Paris.

As it has come to form a coherent and logical whole, the legal discipline is organized around cardinal distinctions that allow for the proper legal ordering of the subjects and objects of the law. Among these fundamental distinctions, with regard to goods (i.e. objects of law subject to appropriation), the French Civil Code, in Article 516, establishes a classification providing that "all property is movable or immovable". This is the *summa divisio* as concerns all things ("*res*"): their fundamental classification. As a result, there can be no properties that are neither movable nor immovable. All goods fall under one of these two categories, and in order to establish this dichotomy, the law uses two distinguishing criteria: (1) physical (whether or not the good is fixed in place) and (2) economic (the value of the good). It is in this initial legal paradigm that property and land are associated in a way that is as profound as it is ancient, to the detriment of a more all-encompassing understanding of the true nature of soils.

Indeed, from a legal standpoint, there is a fundamental difference between soil compared to water and soil compared to air: unlike these two other physical environments, which are considered commons "res communis" (and as such not subject to appropriation), land is undoubtedly the true primary object of property rights "res propria". Under the terms of Article 518 of the French Civil Code, land is considered to be a "real property by nature": the "land bases", which include the ground surface and the subterranean areas, "are the bases of real property".[10] Legal experts are not interested in the fact that soils are a deep layer at the point where air and bedrock meet, a mobile material that can be removed, transported, exported, decontaminated, etc. Soil, characterized as land, is legally and intangibly a fixed substrate that takes shape under our feet, and which extends to the center of the Earth (including to the very core). And this means to agricultural specialists, that this legal definition of soil is deficient as it does not consider this important biological environment as a natural environment that shall be protected.[11] The French Civil Code, through a scientifically incorrect interpretation, provides for a right *of land* structured by property and is silent on the scientific nature *of soils* (and the wide variety of different soils that can be found): "the Civil Code treats soils less as an object of law and more through the regime of property itself".[12] Despite the fact that it is loose, from a legal standpoint, soils are therefore immovable: their legal quality is literally "absorbed by the property".[13] And this has irrevocable consequences on the legal basis undergirding this natural environment (from a physical, chemical and biological standpoint), one that is so essential and so fragile.

10 Terre, F. and Simler, P. (2014). *Les biens*, 9th edition, Dalloz, Paris.

11 Neyret, L. and Martin, G.J. (2012). *Nomenclature des préjudices environnementaux*, LGDJ, Paris.

12 Bosc, L. (2017). Propriété et protection des sols : réflexions civilistes sur la prise en compte de la qualité des sols. *Droit et villes*, 94, 65–80.

13 Ibid.

At one point, it became part of a legal debate on the reconciliation between polluted soils law and waste law. Essentially, given the absence of laws or regulations relating to the obligations incumbent on the owners of polluted soils, as well as on the operators of polluting installations, the legal case issued in this regard was often tempted to fill the legal void by relying on existing and effective rights in terms of prevention and the repair of goods altered by their use: "any producer or holder of waste" is (1) "obliged to ensure or to arrange for its management, in accordance with the provisions of this chapter", and is (2) "responsible for the management of these wastes until they are definitively reverted to a useful state [...]".[14] This is ultimately what the magistrates of the European Court of Justice did in their historic ruling of September 7, 2004, Public Prosecutor's Office v. Paul Van de Walle (Aff. C-1/03): "for these reasons, the Court rules that: unintentionally spilled hydrocarbons causing the pollution of soils and groundwater are waste [...]", and that "the same applies to all soils polluted by hydrocarbons, including when these soils have not been excavated".[15] Thus, polluted soils, endowed with the status of waste, gave rise to the liability on the part of the producer of the pollution or the holder of soils (owner or tenant/operator) under Title IV "Waste" of Book V "Prevention of pollution, risks, and nuisances" of the French Environmental Code under a strict liability regime establishing principles of prevention, treatment and recovery with the intended outcome of the use or reuse of waste, in this case of the soils from which all pollution has been removed through a restoration process that has become mandatory under the obligations imposed by law regarding the treatment of wastes. This case law was therefore both disruptive and ambitious, and it provoked severe reactions (especially from France).

French jurists heavily criticized the ECJ ruling based on the cardinal distinction between movable and immovable things. In this regard, Article L. 541-1-1 of the French Environmental Code provides that "waste is understood to mean: any substance or any object, or more generally, any movable property, which the holder disposes of or which he intends or is obliged to dispose of". Polluted soils, immovable by nature (even though excavated soils can be mobile), cannot be considered as waste, which is movable by nature (even though some waste can be immobile). Consequently, soils cannot be subjected to any duty of vigilance, prevention of damage and repair of damage under waste law which only concerns "movable" property, which is discarded or can be intended to be abandoned. As a result, in 2008, France was able to obtain a modification of the concept of "waste" under European law through the adoption of a new directive (2008/98), which nullified the ruling of

14 Article L. 541-2, Environmental Code.

15 Maître, M.-P. (2006). Un sol pollué peut être qualifié de "déchet". *Le Moniteur*, 30 November [Online]. Available at: https://www.lemoniteur.fr/article/un-sol-pollue-peut-etre-qualifie-de-dechet.1237749.

the aforementioned legal case in its article 2 in which it held that "the following are excluded from the scope of this directive: [...] b) soils (in situ), including non-excavated polluted soils and buildings permanently connected to the ground".[16] Thus, soils are clearly identified as a legal object independent of any other, and soil conservation can therefore only occur under its own legal regime, provided that the legislatures (French and/or European) and/or the public administration (French and/or European) have the ambition to establish such a legal framework…

5.2.2. *The legal protection of soils*

The damages suffered by soils are numerous, and "the quality of soils is endangered by growing threats that can be defined in a general sense as degradation",[17] that is to say "any form of deterioration of the natural potential of soils that alters the integrity of the ecosystem, either by reducing its ecologically sustainable productivity, or by diminishing its original biological richness and its recovery capacity."[18] These degradations, or pollutions, can be of a physical, chemical or biological origin: compaction, runoff, sealing and erosion, pollution by phytosanitary products or mineral fertilizers, salinization, contamination by manure or slurry, illegal disposal of waste and industrial contamination (local or diffuse), loss of biodiversity, decrease in the content of organic matter, presence of pathogens or invasive species. In order to prevent the negative impacts of these on soil quality, and in particular on the maintenance of soil fertility in agronomic matters, it must first of all be established that the law addresses this subject through an administrative approach. The approach taken by the law favors preventing environmental damage as opposed to sanctioning and remedying it, doing so through the establishment of policing regimes that it endows with significant powers in terms of authorization and control. As a result, for their part, criminal law and civil law are rarely implemented if at all.

Thus, strictly speaking, there are no measures in the French Environmental Code or elsewhere that establish criminalization, seeking to establish action that would damage soils as objectionable and punishable behavior, with the exception of Article L. 173-3 of the French Environmental Code which, in a very broad sense, considers damage to soils as similar to that done to the quality of air and water. Thus,

16 European Union (2008). Directive 2008/98/ec of the European Parliament and of the Council of November 19, 2008 on waste and repealing certain directives. Document, JOUE L 312, 22/11/2008, p. 3.

17 Bertrand, M. (2018). op. cit., p. 55.

18 Scientific and Technical Advisory Panel of the Global Environmental Facility (1999). Report of the STAP Expert Group on Land Degradation. Report, GEF/C.14/Inf [Online]. Available at: https://www.thegef.org.

(1) after the cessation of activities of an operation, an installation or a work, the fact that the restoration obligations or the oversight measures prescribed by the administrative authority have not been met", that is, the decontamination of the site and/or the monitoring of residual pollution; as well as (2) "the fact of the continuation of certain operations or activities, the operation of an installation or a work, or the execution of activities subject to declaration, authorization, or derogation, without complying with a formal notice issued by the competent administrative authority", are subject to punishment in the form of penalties ranging from one to five years of imprisonment and from €75,000 to €300,000 in fines, which only occurs when they have seriously harmed the health or safety of persons or when they have provoked "a substantial degradation" of the fauna and flora, or of the "quality of soils, air or water".[19]

It is therefore quite clear that "most of the behaviors sanctioned by criminal law regarding the environment are not described as attacks on nature [...], but as acts of disobedience against administrative prescriptions", for which the consequence is "that accused parties will invariably fail in their defense if such defense is based on the fact (even if it can be proven) that the acts of which they are accused have not had a damaging effect on the environment: it will suffice for the administration to have prohibited it".[20] Conversely, in a similar sense, an environmental protection association will also fail to sanction a polluter if it bases its complaint on the fact (even if it can be proven) that the pollution caused has had harmful effects on the environment. It is sufficient for the administration to have authorized it. We must therefore note that "the tight grip that administrative law holds on criminal law distorts the function of such law: polluters are no longer considered offenders like others, whom we seek to intimidate or resocialize, but the subject of a kind of very rigorous disciplinary law"; "they are guilty because they disobey the orders of the administration."[21]

Finally, the conservation measures established within a diffuse and sectorized legal structure in favor of maintaining good ecological conditions of soils, that is, their being considered as a natural environment and a fragile habitat, often remain as incentives. In law, nothing obliges a farmer to subscribe to the agricultural/environmental and climatic measures (MAEC) incorporating obligations to protect and restore soils (i.e. limiting their tillage as much as possible, setting up a canopy throughout the year, diversifying crop rotations on field crop lands, replacing mechanical tillage with work by soil organisms and practicing direct sowing under plant cover). Nor is there

19 Article L. 173-3 Environmental Code.

20 Robert, J.-H. (2005). Le contentieux pénal. In *Conférence des présidents des cours suprêmes des États francophones d'Afrique sur la contribution du droit au développement durable* [Online]. Available at: www.courdecassation.fr/venements_23/colloques_4/2005_2033/intervention_m._robert_8131.html.

21 Ibid.

anything that would require farmers to respect their agricultural or environmental commitments in favor of soil protection and restoration, provided that they have subscribed to them, except of course their interests in receiving the full payment of European aid from the Common Agricultural Policy (PAC).[22] Therefore, soils, their protection as well as their restoration are left in a kind of legal *wasteland*.

As agriculture and environmental protection are shared within the competences of the European Union,[23] the European Commission has tried, in vain, to integrate an ecological dimension into its soil conservation policies. In 2008, its "thematic soil strategy"[24] resulted in a draft European directive establishing a legal framework for the protection of soils and "amending the directive on environmental responsibility with regard to the prevention and repair of environmental damage".[25] The recitals of the directive were ambitious, which considered soils as "a natural resource" that is (1) "nonrenewable in the sense that its degradation rates can be fast while its formation and regeneration processes are extremely slow"; forming (2) "a very dynamic system that fulfills many functions and plays a crucial role for human activity and the survival of ecosystems"; and (3) is "of general interest, which is suffering the effects of the increasing pressure exerted on the environment and which, as such, must be protected from degradation". Unfortunately, due to the lack of a majority in the Council, the draft directive was abandoned, and the European Union remains powerless to confront the ecological and agricultural challenges of soils conservation.

The Commission, which does not seem to lose interest on soils, issued a communication on November 17, 2021, regarding the establishment of an EU strategy for soil protection by 2030, with the aim of "reaping the benefits of healthy soils for people, food, nature, and climate".[26] To this end, by 2050, all soil

22 Hermon, C. (2014). Agriculture et environnement : un nouveau projet pour la PAC ? *Revue de l'Union Européenne*, 574, 52.

23 European Union (2022). Répartition des compétences au sein de l'Union européenne. Document, TFUE [Online]. Available at: https://eur-lex.europa.eu/legal-content/FR/TXT/?uri=legissum:ai0020.

24 European Union (2006). Implementation of the thematic strategy for soil protection and ongoing activities. Document, 52012DC0046 [Online]. Available at: https://eur-lex.europa.eu/legal-content/EN/TXT/HTML/?uri=CELEX:52012DC0046.

25 European Union (2004). Directive 2004/35/CE du Parlement européen et du Conseil du 21 avril 2004 sur la responsabilité environnementale en ce qui concerne la prévention et la réparation des dommages environnementaux. Document, 32004L0035 [Online]. Available at: https://eur-lex.europa.eu/legal-content/FR/TXT/?uri=celex%3A32004L0035.

26 European Commission (2021). EU Soil Strategy for 2030. Reaping the benefits of healthy soils for people, food, nature and climate. Document, COM 2021 699 final [Online]. Available at: https://eur-lex.europa.eu/legal-content/EN/TXT/PDF/?uri=CELEX:52021DC0699&from=EN.

ecosystems throughout the EU would have to be healthy, and therefore more resilient, which would require decisive changes during the decade of 2020–2030. Thus, certain objectives are included in accordance with the United Nations SDGs, and in particular:

– **Medium-term objectives by 2030:**

- combat desertification, restore degraded lands and soils, in particular lands that are affected by drought and floods, and seek to achieve a world without soils degradation;

- restore large portions of degraded and carbon-rich ecosystems, including soils;

- achieve 310 million metric tons of CO_2 equivalent of net greenhouse gas removals per year across the European Union in the land use, land-use change and forestry sectors;

- achieve the proper ecological and chemical state of surface water and a good chemical and quantitative state of groundwater by 2027;

- reduce nutrient losses by at least 50%, reduce the overall use of chemical pesticides and the risks associated with them by 50% and reduce the use of more dangerous pesticides by 50%;

- make significant progress in the remediation of contaminated sites.

– **Long-term goals by 2050:**

- achieve zero net artificialization (ZNA) of soils;

- reduce soil pollution to levels that are no longer considered harmful to human health and the health of natural ecosystems, and that respect the limits of our planet, thus creating an environment free of toxic substances;

- achieve a climate-neutral Europe and, as an initial step, seek to obtain terrestrial climate neutrality within the EU by 2035;

- achieve a climate-resilient society within the EU, fully adapted to the inevitable repercussions of these changes.

As progress is made on the next legislative and regulatory developments at European level, which will make it possible to implement the public policies necessary to carry out these objectives – *as necessary as they are ambitious* – the scope of action of the justice system in terms of soil protection appears to be very limited at this time. And the well-intentioned goals of the European Commission appear to be more wishful thinking than anything resembling a system of coercive law.

Thus, the action of polluting soils is not punishable as such. The simple presence of polluted soils commits the operators to carry out a restoration determined by the future use of the site, rather than on its return to a satisfactory ecological condition[27]. The scope of such a commitment is fixed unilaterally by the competent administrative authority, which is not usually an environmental authority. A French parliamentary report from 2020, "Industrial and mine-related soil pollution: Taking responsibility, repairing the mistakes of the past, and thinking sustainably for the future",[28] has highlighted this lack of legislative framework in terms of soil protection and restoration by proposing to "add to the environmental code a legislative definition of soil pollution, as exists for water and air" in order for more effective soil conservation.

5.3. Soils are covered by the law as a property right

5.3.1. *The legal foundations of agricultural land ownership*

All too often, agricultural land is considered to be a reserve of land property by many public authorities and economic actors: this definition is used for the implementation of major urban planning projects, making it possible to meet the need for housing, as well as being used to create new business areas. Agricultural land may thus appear to be a simple variable to add to the adjustments of urban planning policies without any real consideration for the essential functions provided by these spaces, which still represent more than half of France's national territory (totaling to 51% of French land in 2020).[29] However, the rise in the artificialization of soils in France is increasing at a staggering pace: by 0.8% per year (60,000 ha/year – the equivalent of one of the country's administrative departments every 10 years).[30] The changes made to the intended use of the land, which in addition to the rampant destruction of agricultural surfaces and natural habitats, are inevitably

27 Article L. 556-3 Environmental Code.

28 Jourda, G. (2020). Pollutions industrielles et minières des sols : assumer ses responsabilités, réparer les erreurs du passé et penser durablement l'avenir. Report 700 (2019–2020) [Online]. Available at: www.senat.fr/rap/r19-700-1/r19-700-1_mono.html#toc162.

29 République française (2014). Moindres pertes de terres agricoles depuis 2008, après le pic de 2006–2008. *Pri313* [Online]. Available at: https://agreste.agriculture.gouv.fr/agreste-web/disaron/Pri313/detail/.

30 Planetoscope (2012). L'artificialisation des sols en France [Online]. Available at: www.planetoscope.com/sols/2024-l-artificialisation-des-sols-en-france.html.

accompanied by an increase in noise, light and aqueous, chemical, and plastic waste products, in addition to which new logistical systems and transport infrastructures must also be added, are all just as detrimental for nature as they are for agriculture.[31,32] Excavated or impermeable soils are no longer able to perform their functions in the natural cycle. And yet, every day, 25 ha of agricultural land disappear and due to the increase in population, there is no objective reason to predict any naturally occurring decrease in anthropogenic pressure on soils around urbanized centers and the areas surrounding them.

To prevent this dramatic issue, the French Citizens' Climate Convention proposed in its 2019 report to set an objective of ZNA by 2050,[33] consistent with the announcements of the European Commission detailed above. It is a very ambitious principle that will have a lasting impact on France's relationship with land, provided that it survives the many attacks and criticisms levied against it since its inscription in the climate-resilience law.[34,35] This "ZNA" objective will be carried out by putting a halt to the removal of agricultural and natural areas by 2050, as well as through the use of compensation mechanisms[36,37]: the artificialization of new spaces

31 Girardet, X. (2013). Paysage et infrastructures de transport : modélisation des impacts des infrastructures sur les réseaux écologiques. PhD Thesis, Université de Besançon, Besançon.

32 Chéry, P., Lee, A., Commagnac, L., Thomas-Chery, A.L., Jalabert, S., Slak, M.F. (2014). Impact de l'artificialisation sur les ressources en sol et les milieux en France métropolitaine : Evaluation selon trois sources d'informations indépendantes. *Cybergeo: European Journal of Geography*, 668.

33 Convention Citoyenne pour le Climat (2021). Les propositions de la Convention Citoyenne pour le Climat. Report [Online]. Available at: https://propositions.conventioncitoyennepourleclimat.fr/pdf/ccc-rapport-final.pdf.

34 French Law no. 2021-1104 of August 22, 2021 on the fight against climate change and the strengthening of resilience to address its effects. Available at: https://www.legifrance.gouv.fr/jorf/id/JORFTEXT000043956924.

35 Rule 191: "In order to achieve the nation's objective of zero net artificialization of soils by 2050, the pace of soil artificialization in the ten years following this law's entry into effect must be such that, over this period, the total space consumption observed nationwide is less than half of that observed over the 10 years preceding this date. [...] These objectives shall be applied in unique and territory-specific ways, under the conditions established by law".

36 Article 192 amends Article L. 101-2 of the French Urban Planning Code, which now provides that "the fight against the artificialization of soils" will include "an objective of the absence of net artificialization in the long term".

37 Article 192 also creates a new article, L. 101-2-1, which is added into the French Urban Planning code and provides that the ZNA objective "results from the balance" between, in particular: "(1) The limiting of urban sprawl"; "(5) the preservation and restoration of biodiversity and nature in cities"; "(6) the protection of the soils in natural, agricultural, and forested areas" and "(7) the re-naturalization of artificial soils" (pts 5 and 6) (al. 1).

can thus be compensated by the renaturation of old artificialized spaces,[38] such as by the ecological restoration of abandoned areas like urban wastelands. Finally, it should be noted that the law has provided a legal definition of the "artificialization of soils",[39] as well as of the "net artificialization of soils".[40]

In this context, maintaining land-use diversity in the face of the pressure for greater urbanization now appears to be of general interest, as the French Urban Planning Code has long enshrined a collective understanding of national land management in its first article: "The French territory is the common heritage of the nation"[41]. The law establishes public authorities as "managers" and "guarantors" of this said heritage, given that "a fund, even if appropriate, nevertheless represents a portion of the territory: the territory is not an object that is possessed by individuals, but a place to which they belong"[42]. And it is heritage that is threatened with disappearance, given that building permits must be understood and considered, above all else, as permits to destroy, to carry out the destruction of natural spaces and the destruction of agricultural land. Therefore, a balanced and measured use of the natural resource that is "soil" is essential to maintain the major balances of development, and consequently the fight against urbanization and land artificialization, has more recently been tepidly identified as a priority policy of the French government, with the objective of stabilizing the overall rate of artificialization of the national territory at below 10%.[43]

In terms of land management as conducted by the public authority, local urban plans (*Plans Locaux d'Urbanisme*) – whether intercommunal (PLUIs) or otherwise (simple PLUs) with municipal maps – are clearly the best legal tools available to

---

38 Article L. 101-2-1-1 French Urban Code, paragraph 3: "The re-naturalizationtion or de-artificialization of a soil consists of actions or operations taken to restore or improve the functionality of such soil, having the effect of transforming an artificial soil into a non-artificial soil".

39 Article L. 101-2-1-1 French Urban Code, paragraph 2: "Artificialization is defined as the sustainable alteration of all or part of the ecological functions of a soil, in particular its biological, hydrological, and climate-related functions, as well as its agricultural potential by its occupation or use".

40 Article L. 101-2-1-1 of the French Urban Code paragraph 4: "The net artificialization of soils is defined as the balance between the artificialization and the re-naturation of soils observed on a given perimeter and over a given period".

41 Article L. 101-1, Urban Code.

42 Grimonprez, B. (2015). La fonction environnementale de la propriété. *Revue trimestrielle de droit civil*, 3, 539–550.

43 See: https://artificialisation.developpement-durable.gouv.fr/.

local authorities under the urban planning code.[44] In practice, the drafting of a PLU(I) must allow for the establishment of a territorial regulation that ensures the balance between urban renewal, controlled urban development, the economical use of natural spaces and the preservation of spaces allocated to agricultural and forestry activities. The authors of the urban planning documents must therefore ensure that they determine the nature and purpose of the natural and agricultural areas by providing the motivations for their choices in the presentation reports of the urban planning documents.[45] Thus, PLU(I)s approved by the deliberative bodies of local and regional authorities after public consultation[46,47] will have five parts (the presentation report, the planning and sustainable development project, the planning and scheduling guidelines, the town planning regulations and the annexes),[48] and they delineate four main zones for the territory (urban areas "U", areas to be urbanized "AU", natural areas "N" and agricultural areas "A").[49,50,51,52,53]

Agricultural zones are therefore areas specific to the PLU(I) in the sense that the legislature has expressly defined the methods of land use reserved for agricultural activities, particularly authorized activities that can be conducted and structures that can be built (and those that cannot). Of course, the general principles of land use set out in L. 110 and L. 121-1 of the French Urban Planning Code remain applicable: the land must be managed economically, and agricultural land must be protected. Thus, the ability to build in agricultural areas is strictly regulated. The regulations specify the conditions of the height, location and density of the constructions, making it possible to ensure that they are suited to their environment and that they allow the agricultural nature of that environment to endure. In particular, the regulations may designate buildings which, because of their architectural or heritage interest, may be subject to changes of intended uses or limited extensions, provided that the projects do not compromise agricultural operations. The change of intended uses and the work authorizations are subject to the assent of the departmental commission for the consumption of agricultural spaces[54], with the exception of the

44 Article L. 121-1 to 121-15 and Article L. 151-1 to 154-4, Urban Code.

45 Article L. 151-4, Urban Code.

46 Article L. 153-11, Urban Code.

47 Article L. 153-19 and 20, Urban Code.

48 Article L. 151-2, Urban Code.

49 Article L. 123-1, Urban Code.

50 Article R. 123-5, Urban Code.

51 Article R. 123-6, Urban Code.

52 Article R. 123-8, Urban Code.

53 Article R. 123-7, Urban Code.

54 Article L. 112-1-1, Rural Code.

installations necessary for the agricultural operation itself and the constructions necessary for collective equipment or public services. However, such protection of agricultural land, if it is real and effective, is relegated to the short and medium term, since the urban planning documents are simply the expression of a political will of the present time and can therefore be revised at the whims of local elected officials, particularly following elections and changes in governance. Thus, agricultural land is much more likely to be subject to urbanization after a revision of a PLU(I).

To avoid this risk, safeguards can be set at higher regulatory levels by integrating the conservation of agricultural land within a Territorial Coherence Scheme (*Schéma de Cohérence Territorial* (SCoT)).[55] Indeed, PLUs, PLU(I)s and community charters are required to be compatible with the SCoT.[56] The choices made by local-elected officials in terms of urban development must necessarily fall within the framework of the principles of land management established within the territorial coherence schemes, which "present the diagnosis established with regard to economic and demographic forecasts and the needs identified in terms of economic development, spatial planning, the environment, the social balance of housing, transport, equipment, and services".[57] As such, the SCoTs can, in particular, "implement the reform and sustainable development projects" involving the territory, and therefore "determine the major balances between spaces that are already or will soon be urbanized, and natural and agricultural or forest areas", while enjoying "the foreseeable impacts of these guidelines on the environment".[58]

This allows local elected officials to include the conservation of agricultural land on the scale of a large territory when adopting a SCoT, a goal which is then set in motion over the long term within all municipalities and intercommunity areas covered by it. They are then required to ensure their urban planning documents are compatible with the conservation issues of agricultural land established in the SCoT (provided that one exists). However, what is required in this case is simply compatibility and not compliance: the authors of the PLU(I) are allowed a certain margin of appreciation. Indeed, "with the exception of the limited cases provided for by law in which the territorial coherence schemes may contain prescriptive norms, these must be limited to setting orientations and objectives".[59] Consequently, the planning documents do not have to be strictly in line with each provision or

55 Article L. 122-1 to 122-19, Urban Code.

56 Article L. 131-1, Urban Code.

57 Article L. 122-1, Urban Code.

58 Ibid.

59 République française (2017). 6ème-1ère chambres réunies. Report, 395216 [Online]. Available at: https://www.legifrance.gouv.fr/ceta/id/CETATEXT000036233084.

particular objective of the SCoT. Rather, they should provide more of a "global analysis" that seeks "to be placed on the scale of the entire territory covered" in such a way that "the plan does not contradict the objectives imposed by the scheme" taking into account "the orientations adopted and their degree of precision".[60]

5.3.2. *The legal protection of agricultural land*

Though agricultural land is considered by law as land, it is nevertheless a very special type of land, given the general interest issues that in which it is implied, as well as its fragile nature. As such, it is given the full attention of the legislature and is the subject of special legal protection. At the center of this protection system are entities known as SAFERs (*Sociétés d'Aménagement Foncier et d'Établissement Rural* (Land Development and Rural Settlement Companies)). Established in Article 15 of French law no. 60-808 of August 5, 1960 regarding agricultural orientation, the SAFERs were initially responsible for "acquiring land or agricultural holdings freely put up for sale by their owners, as well as uncultivated land, intended to be transferred for the purpose of subsequent development", seeking, in particular, "to improve agrarian structures, to increase the area of certain agricultural holdings, and to facilitate the cultivation of the soil and the installation of farmers". As of that time, its missions have been listed in Article L. 141-1 of the French Rural Code, which provides that the SAFERs are constituted (1) "to work primarily for the protection of agricultural, natural, and forested areas", (2) "to contribute to the diversity of landscapes, the protection of natural resources, and the maintenance of biological diversity", (3) to contribute "to the sustainable development of rural territories" and (4) "to ensure the transparency of the rural land market".

Though the word "soil" does not appear, the conservation of soils undergirds the text. After all, first, how can the sustainability of agricultural farms be guaranteed without the protection of agricultural land? And second, how can the commercial exploitation of this agricultural land be carried out without the protection of the ecological and agronomic quality of the soils? Obviously, the quantitative as well as qualitative maintenance of agricultural land is essential to carrying out the missions assigned to the SAFERs. To this end, they are given exorbitant prerogatives under common law, such as the ability "to acquire rural property, land, agricultural, or forested land holdings with the goal of transferring them", "to substitute one or more assignees to carry out the transfer [...]", "to acquire shares or units of companies having as their main objective the commercial exploitation or the agricultural property [...]" and "to engage or lend their assistance in real estate transactions under a written mandate [...]".

60 Ibid.

Finally, the SAFERs have a right of preemption established in Article L. 143-1 of the French Rural Code: "it is established that SAFERs shall have a right of preemption in the event of the transfer for valuable consideration of real estate for agricultural use [...]". This right of preemption is therefore exercised on undeveloped immovable property for agricultural use or barren land to be used for agricultural purposes. The latter is understood as land located either in a protected agricultural area, or within a protected perimeter within agricultural and natural peri-urban areas, or in an agricultural area or a natural or forested area. In the absence of an urban planning document, lands located in sectors yet to be urbanized in municipalities (excluding woods and forests) are also considered as agricultural land. Finally, properties without structures must have a minimum area defined by the decree which gives the SAFERs the right of preemption, which varies according to each department (a threshold that is lowered to 0 in zones A "agricultural" or N "natural" in urban planning documents). The exercise of this right of preemption has, among other things, the purpose of "the fight against land speculation" (Article L. 143-2, 5th Rural Code), "environmental protection" (Article L. 143-2, 8th Rural Code) and "the protection and enhancement of agricultural and natural peri-urban areas" (Article L. 143-2, 9th Rural Code).

Thus, the SAFERs are central, if not essential (or even mandatory) operators with regard to the transfer of agricultural land: they are the first to make contact with future buyers, and they play a central role in the vast majority of sales/acquisitions (providing advice to sellers and buyers, setting sales prices, publicizing sales, selecting buyers with a given agricultural activity, purchasing and reselling land as intermediaries). However, their activity has not been able to significantly contain the progressive artificialization of soils considered as an object of property, especially in peri-urban areas, and the issue of the protection of soils considered as biological objects has seen little to no consideration despite the changes in law. However, a number of regulatory tools exist that make it possible to establish stronger protection of agricultural land in order to ensure its preservation over the long term. They can be adopted by communities who can utilize them to create, for example, "protected agricultural zones" (*Zones Agricoles Protégées* (ZAPs)) and/or "natural and agricultural peri-urban areas" (previously PAENs).

Accordingly, the first tool for protecting agricultural land is ZAP. Established by the agricultural guidance law of July 8, 1999,[61] ZAPs are zones with reinforced protections for agriculture, which have a status of a public utility easement: "agricultural areas whose preservation is of general interest due either to the quality

61 République française (1999). Loi no. 99-574 du 9 juillet 1999 d'orientation agricole. Document, AGRX9800053L [Online]. Available at: https://www.legifrance.gouv.fr/eli/loi/1999/7/9/AGRX9800053L/jo/texte..

of their production, their geographical location, or their agricultural quality can be classified as protected agricultural areas".[62] It should be noted that the existence of small wooded plots within a ZAP does not prevent its establishment.

The ZAP is established through a prefectural decree at the request of the municipalities, making it possible to protect agricultural areas whose preservation is in the public interest (either due to the quality of the production there, or their geographical location). The creation of a ZAP on a given site implies that any change in assignment or mode of land use that could "permanently alter the agricultural, biological, or economic potential of the area", is subject to the opinions of the Chamber of Agriculture and the Agricultural Policy Committee.[63] In the event that any of these entities issue an unfavorable opinion, the change can only be authorized on a reasoned decision by the Prefect.[64] Consequently, ZAPs offer many advantages for the conservation of agricultural land: they protect their agricultural uses, they avoid land speculation and limit the price of land, they protect land from sprawl, they secure and promote the revival of agricultural activities (particularly recultivation), and finally they express a true recognition of the agricultural identity of a territory by also encouraging the implementation of long-term agricultural development strategies. Despite this, ZAPs have some weaknesses, such as (1) the fact that they are strictly a municipal initiative and thus cannot be created if there is a lack of political will, and (2) the fact that a zone being classified as a ZAP does not definitively prohibit changes in land use, despite the important safeguards established by the law.

The second tool for the protection of agricultural land is the natural and agricultural peri-urban areas (*espaces naturels et agricoles périurbains* (ENAPs)), formerly the "perimeters of protection and enhancement of agricultural and natural peri-urban areas" (*périmètres de protection et de mise en valeur des espaces agricoles et naturels périurbains* (PAENs)). To preserve unbuilt peri-urban spaces, the law of February 23, 2005 relating to the development of rural territories gave the departments and public institutions of intercommunal cooperation (*établissements publics de coopération intercommunale* (EPCIs)) competence in the area of the protection and development of agricultural and natural peri-urban spaces.[65] This

---

62 Article L. 112-2, Rural Code.

63 Article R. 425-20, paragraph 1, Urban Code.

64 Article R. 425-20, paragraph 2, Urban Code.

65 République française (2005). Loi no. 2005-157 du 23 février 2005 relative au développement des territoires ruraux. Document, AGRX0300111L [Online]. Available at: https://www.legifrance.gouv.fr/loda/id/JORFTEXT000000257340.

possibility has been replaced without altering existing laws through the "agricultural and peri-urban natural areas"[66] by the recodification order of September 23, 2015[67]:

> The department or a public institution mentioned in Article L. 143-16 may delimit intervention areas relating to action programs with the agreement of the municipality or municipalities concerned or public institutions competent in matters of local urban planning, after advice from the departmental chamber of agriculture and public inquiry.[68]

These *natural and agricultural peri-urban areas* constitute reinforced protection perimeters, which applies to zones A and N of the PLU: within the same legal tool, they bring together the possibility for land management and a project for the development and rearrangement of land by consolidating their agricultural and natural vocation beyond the revisions and modifications of urban planning documents (including territorial coherence schemes). Thus, the ENAPs must form part of a logic of compatibility with the SCoTs, the PLU, and potentially the charters of Regional Natural Parks if necessary.[69] The perimeter of these zones cannot include plots located in an urban area (or an area to be urbanized), or in a deferred development zone perimeter.[70] They are established by the department with the agreement of the municipality or municipalities involved, and after receiving guidance from the Chamber of Agriculture (and possibly the National Forest Agency if the perimeter includes plots that are subject to the forest regime).[71]

Moreover, in addition to the zoning of the ENAPs, the departmental councils must develop:

> An action program, in agreement with the competent municipality (or municipalities) or EPCIs, that specifies the developments and management guidelines intended to promote agricultural exploitation, forest management, preservation, and enhancement of natural spaces and landscapes within the perimeter.[72]

---

66 Article L. 113-15, Urban Code.

67 République française (2015). Ordonnance no. 2015-1174 du 23 septembre 2015 relative à la partie législative du livre Ier du code de l'urbanisme. Document, ETLL1511519R [Online]. Available at: https://www.legifrance.gouv.fr/loda/id/JORFTEXT000031204500.

68 Article L. 113-16, Urban Code.

69 Article L. 113-22, Urban Code.

70 Article L. 113-17, Urban Code.

71 Article L. 113-21, Urban Code.

72 Article L. 143-2, Urban Code.

Finally, via the SAFERs, the departments can acquire agricultural land through a right of expropriation and a right of pre-emption granted to them by law on these peri-urban areas.[73,74] Consequently, the ENAPs are an interesting tool for the conservation of agricultural land insofar as this zoning can only be modified through an inter-ministerial decree,[75] while the right of expropriation and the right of preemption provide local authorities with the option to carry out land control operations in agricultural matters on their territory. However, for this policy to be implemented, the choices made by the departments and the EPCIs remain subject to approval by all of the municipalities implicated in the zoning (which constitutes a significant obstacle to their establishment).

5.4. Toward a global protection of physical environments?

While these new urban planning policies have allowed for a slowdown in the dynamics of land artificialization in recent years, the French agricultural space has nevertheless decreased by more than 20% in 50 years with a loss of more than five million ha of agricultural land.[76] Thus, the ambitions shown in the zero net artificialization (ZNA) established by the climate-resilience law are objectively a welcome development. However, translating this legislative expectation into effective and efficient public policy in the field will require close monitoring given that, despite the fact that 2050 remains far off, there are already growing calls in favor of a relaxation of the ZNA through waivers and exceptions, or even the outright removal of the objective (which would imply a significant risk of impropriety in French law with regard to European law).

Objectively, agricultural land has become a source of tension. As agricultural land is slowly eaten up by urban extension and the construction of infrastructures, it is put into permanent competition with other uses with which it has a very difficult time competing from an economic point of view. Thus, it would appear to be more urgent and important than ever to recall "that it is the dissociation of private (economic) utilities and collective utilities (amenities) that gives rise to environmental aspect of property"[77]; an aspect undergirding the current moment, which deserves to be affirmed as a fundamental principle in the exercise of the legal possession.

73 Article L. 113-24, Urban Code.

74 Article L. 113-25, Urban Code.

75 Article L. 113-19, Urban Code.

76 Pointereau, P. and Coulon, F. (2009). Abandon et artificialisation des terres agricoles. *Courrier de l'environnement de l'INRA*, 57, 109–120.

77 Grimonprez, B. (2015), op. cit.

Thus, it would be necessary to first of all strengthen and perpetuate the existing legal tools: to maintain ZNAs over the long term by introducing *ratchet effects* through public policies put in place in favor of soil conservation (to avoid the recurring questioning of these and to apply a "non-regression" policy regarding the regulations).[78] This would also involve integrating the qualitative issues of soils (i.e. the consideration of soils as a fragile natural environment to be protected, and not as land considered as a simple good to be owned, governed under the logic of property rights). Thus, the inclusion of a principle of renaturalization in the French Urban Planning Code through the climate-resilience law is also good news, subject to its maintenance and implementation in public policies within the territories.

Finally, it would also be worth adding a Title III to Book Two of the French Environmental Code on "physical environments" that covers "soils". This is a title that would address, and (*if necessary*) would establish, the main principles of soil management: its quantitative conservation (an ownership approach) and its qualitative protection (ecological approach). Thus, the special provisions of the French Rural Code, Forestry Code and Urban Planning Code with regard to land uses would be subject to compliance with the principles of conservation of land, which would be considered as a natural resource under the Environment Code.

Finally, while the work of the French Citizens' Climate Convention has undoubtedly gone in the right direction, it is regrettable that the climate resilience law did not create a wider definition for the crime of ecocide as it excludes soils from its scope for now.[79] Indeed, these could have ultimately been protected by criminal law in the same way as water and air. No such developments have happened yet, despite the proposals that had been made.

78 Article L. 110-1 French Environmental Code: "[...] 9: The principle of non-regression, according to which the protection of the environment, ensured by the legislative and regulatory provisions relating to the environment, must only be subject to constant improvement, taking into account the scientific and technical knowledge of the moment".

79 New article L. 231-1 of the French Environmental Code: "The action, carried out as a manifestly deliberate violation of a particular obligation of prudence or safety provided for by law or regulation, of emitting into the air, throwing, dumping or allowing to flow into surface or underground waters or into the waters of the sea within the limits of territorial waters, directly or indirectly, one or more substances whose actions or reactions would cause serious and lasting harmful effects to public health, flora, fauna, [...] or serious changes to the normal water supply regime is punishable by five years of imprisonment, and a fine of one million euros, which may be increased up to five times the amounts deemed to have arisen from the committing of the offense [...]".

New Article L. 231-3 of the French Environmental Code: "The offense provided for in Article L. 231-1 constitutes an ecocide when the acts are committed intentionally".

The subject of soils protection is therefore a pressing topic, but there is no reason to think that in the near future, the law will not move in the direction of an effective protection of a natural environment that has been neglected by the legislature. Perhaps the necessary, wide-reaching and broadly encompassing legal solution to soils protection can be found in a place outside the Palais Bourbon, such as the Strasbourg European Parliament? Before the seeds for a few additional (and inevitably insufficient) sectoral regulations are planted, maybe it is time to plow the field of public awareness.

6

Methods for Estimating the Agronomic Value of Soils

Joël MICHELIN
UMR ECOSYS, AgroParisTech, INRAE, Université Paris Saclay, Palaiseau, France

6.1. Why evaluate the agronomic value or the cultural suitability of soils?

According to an estimate of the soil resource worldwide (INRA 2009), the vast majority of landmasses (78%) contain soils that are "non-productive" due to various causes (they are covered with ice, too cold, too dry, too steep, too thin, too wet, or too poor in nutrients). Only 22% of the land surface (3,300 Mha) have soils considered "productive", of which about 45% are actually cultivated soils or "arable land" (1,500 Mha). The rest (55%) are considered "in reserve", and mainly located in developing countries (southern regions/Global South). Among the "productive" soils, 15% are "highly productive" soils, compared to 58% of "low productive" soils and 27% of "moderately productive" soils. These differences are largely linked to the diversity of soils around the world, which are the result of various genesis and evolution mechanisms (Ruellan 2010; Jamagne 2011). This broad variety of soils translates into a great diversity of functioning and variable potential opportunities and constraints for agricultural use. In many situations, peoples' considerations for the cultural aptitudes of the soils form the basis of the current uses of soils. Throughout history, soils that have been unsuitable for cultivation or too difficult to cultivate for various reasons (slope, rocks, too much or not enough water, etc.) have been left as forests, permanent meadows, or natural vegetation. In some cases, improvements have been made to make the soils suitable for cultivation (i.e. by creating terraces, drainage and irrigation) (Ruellan 2010). Since the formation of

Agricultural Soil Science,
coordinated by Yves COQUET and Joël MICHELIN. © ISTE Ltd 2025.

soils is a very slow process, they constitute a limited and non-renewable resource on a human scale. "Productive" soils used for cultivation that possess good agricultural production capacities are valuable, but they are in danger around the world. The annual loss of "productive soil" capital averaged 0.5% from 1975 to 2000 (INRA 2009). The summary report on the state of land and water resources (FAO 2021) indicates that resources have never been under more pressure, as a result of the increase in the world population and the effects of climate change. The report shows that the area of agricultural land has changed little since 2000, but the increase in urban areas has led to a shift in land use, with a decrease in forest cover and permanent meadows and pastures and an increase in permanent and irrigated crops from 2000 to 2019. The report estimates that 31% of cultivated lands are threatened, generally including lands that have been put into production recently, and with the majority in irrigated areas.

Evaluating the agricultural value or the agricultural suitability of soils is therefore an important issue for managing agricultural production at different scales, and also for providing advice for planning documents. The soils most suitable for production must be identified and reserved for use in agricultural production, and not destined for other uses such as urbanization. The assessment of soil potential and constraints must also lead to adapted practices to limit soil degradation and maintain their suitability for cultivation.

The agricultural value of soils is based on the possibilities that soils offer farmers to cultivate plants with optimal growth to meet the needs for food production, even as current agricultural production moves toward other outlets. To assess the agronomic value of soils, we must first look at the functions of soils for crop production (physical support, air, water and nutrient resources), and then see the characteristics and intrinsic properties of soils that allow soils to perform production functions. From a practical point of view, the soil characteristics that are easily observable are prioritized. Extrinsic factors are also considered, along with the factors of the physical environment, such as the climate and relief, and technical and socioeconomic factors, such as land use, production systems and facilities.

The evaluation of agronomic values can be carried out using different levels of approaches, from regions to plots. The principles and methods of evaluation come in the form of indices, the recording of constraints or potential opportunities, and can create rankings and maps.

Some examples of observation methods carried out in situ and diagnostic procedures will be given at the end of the chapter.

6.2. Soil functions for agricultural production

For the agricultural production of plants, the soil must offer favorable conditions for the proper development of the roots to meet the needs of the plants.

6.2.1. *The soil provides physical support for the plants*

For plants, soil is first and foremost a physical base for anchoring itself via the root system and stabilizes the parts of the plant above ground. However, the soil is not limited to this physical support function; it must also meet the needs for the plant's growth via the root system. The needs of plants can be summarized simply as follows: air, or more precisely oxygen (root respiration), water and requirements in nutrient elements, mainly nitrogen, phosphorus, potassium and also various other minerals and trace elements.

The soil must therefore allow the roots to develop laterally and vertically to most effectively ensure the functions of anchoring and nourishing the plant.

Roots grow in the spaces between the solid particles of the soil. These empty spaces, or soil porosity (see section 1.2), come in several forms: they are spaces between elementary particles (textural porosity), or they are spaces between aggregates (structural porosity), or they are spaces created through biological activity (porosity of biological origin). Soil fauna (see Chapter 2), including earthworms, can dig galleries of different sizes, which are beneficial to the roots. The roots can also penetrate non-aggregated soil or aggregates, if the material is loose enough to allow the roots to grow between the elementary particles. The roots can thus contribute to the porosity generated through biological processes as they decompose after their death. These spaces will be conducive to the development of new roots.

Finally, to promote root growth, empty spaces can be created by tilling the soil with tools (plows, tined or disc tools). The tools fragment the soil and create aggregates (or clods) of various sizes, with spaces between these clods. The organization of the clods can be categorized through the "cultivation profile" method (see section 1.2.3).

6.2.2. *Soil, a reservoir for air and water*

For proper root development, soil porosity is essential. However, for roots to survive and grow, air and water must be present in the voids. If all the pores are filled with water (in soils saturated with water), no more air, and therefore no more oxygen, is left for the roots to respire. The roots of cultivated plants do not develop, or the existing roots undergo necrosis (with a few exceptions, such as rice). Excess water in the soil is a strong constraint to the development of cultivated plants.

Conversely, the lack of water in the soil is also a limiting factor for plant growth. When soil moisture is lower than the so-called wilting point level for moisture, the water that remains is no longer accessible, since the suction capacity of the roots is lower than the water retention forces on the solid phase of the soil (Calvet 2003a).

Plants will absorb water from the soil via the roots, but two conditions are necessary, as shown above: the roots must have air to develop, and the roots must be able to extract water from the soil.

6.2.3. *The soil: a source of nutrients*

The roots will absorb the nutrients dissolved in the soil water. The elements in this solution come from the mineralization of the organic matter of the soil through biological activity, from the dissolution of elements from the mineral phase of the soil, or from the release of elements adsorbed on the adsorbent complex of the soil (clays and organic materials) (Calvet 2003a). The nutrients can be supplied by the farmers in mineral or organic forms and thus feed into the soil solution directly, but most often through their integration into the biogeochemical cycle of these elements in the soil.

The assessment of the capacities of soils to fulfill physical support resources, and act as resources for the needs of cultivated plants, can be made with regard to the intrinsic characteristics and properties of soils.

6.3. Intrinsic properties and characteristics of soils

Soil properties are analyzed using soil characteristics observed in situ, and/or determined by laboratory analyses on soil samples, or by measurements in situ.

6.3.1. *Characteristics that are relatively easy to obtain*

Characteristics that require large investments of technical resources and time cannot be systematically determined during soil evaluations due to the availability of resources, whether human or financial. Also, in practice, the evaluations favor soil characteristics that can be easily observed in the field or through routine laboratory analyses. Soil properties will be deduced from the characteristics that are easy to obtain, supplemented by the results of research work, and possibly the use of pedotransfer functions – in other words, statistical relationships that make it possible to relate characteristics that are easy to obtain to properties that are more difficult to measure (Baize 2016).

In the field, it is easy to observe and take soil samples by drilling with a hand auger up to 120 cm (standard auger length). The easily observable characteristics in situ with an auger are the texture, soil thickness, soil color and stains and presence of limestone (through an effervescence test with hydrochloric acid). The texture can be estimated very quickly in the field by tactile tests, but doing so requires a person who is specially trained. It can also be deduced from the particle size analyses of samples (through routine laboratory analyses on samples taken with an auger).

However, not all soil characteristics can be described using auger probes (Baize and Jabiol 2011). The abundance of coarse elements (CE; elements larger than 2 mm in size) can be assessed with an auger, but only for gravel. Larger elements (pebbles, stones, blocks) will only be observable on the surface or in the upper part of the soil profile using a spade.

The structure cannot be observed with an auger drill, as it will destroy the aggregates. It must be observed using a soil pit. However, for the upper part of the soil, the structure can be observed using a spade test (Turillon et al. 2018) or a cultivation profile (see section 1.2.3). With the help of a telescopic loader, a machine that many farms have, it is possible to create a miniature 3D profile up to about 70 cm deep. The miniature 3D profile, which is minimally destructive and quick to produce, makes it possible to observe the structure, the rooting and the effects of biological activity and technical pathways (Tomis 2017).

Table 6.1 proposes a comparison of soil functions with properties and characteristics. This list of soil properties and characteristics is not exhaustive: these properties are relatively easy to estimate from physical or chemical characteristics of soils observable in the field or data obtained by routine analyses carried out in the laboratory, or deduced from pedotranfer functions. In certain contexts, particularly in the Mediterranean, salinity and sodicity must be taken into account, as they have consequences on agriculture in terms of toxicity, water supply and plant nutrition (see section 8.5). The biological properties and characteristics are not developed. The routine use of bioindicators is still limited (see section 2.5.1).

Functions	Properties	Features
Water resource	Maximum available water content (Available water)	Water retention
		Abundance of coarse elements
		Rooting depth
Air resource	Natural drainage	Depth at which signs of hydromorphy appear
	Surface aeration	Structural stability of the surface horizon
Nutrient resource	Stock	Rooting depth
		Abundance of coarse elements (weighted average)
		CEC
	Availability	pH
Physical medium	Ease of tilling (seed bed)	Surface texture
		Concentration of coarse elements (surface horizon)
	Practicability/compaction	Slope
		Depth at which signs of hydromorphy appear
		Structural stability
	"Aggressiveness of the soil"	Abundance of coarse elements (surface horizon)

Table 6.1. *Relationships between the functions, properties and characteristics of the soil*

6.3.2. *The water reserve of the soil*

For the water resources function, the corresponding soil property is the maximum available water content (MAWC), more commonly known as the "available water" (AW), which is the amount of water stored in the soil and accessible to plants. The MAWC is conventionally defined as the amount of water in the soil between the soil moisture level at field capacity (M_{FC}), also called "retention capacity" (RC), and the soil moisture at the wilting point (M_{WP}) (Baize and Jabiol 2011; Baize 2016; Bouthier et al. 2022). The soil moisture at field capacity corresponds to the volume of water retained in the soil's pores after the soil has been drained, with the free water having flowed away by gravity. Soil moisture at the wilting point is the volume of water that the roots cannot extract because it is firmly held in place through the fine porosity with forces greater than the suction forces of the roots. Plants can no longer extract water and will wilt, hence the term "wilting point" for soil moisture.

Determining the MAWC of the soil involves calculating the difference in water volume between the M_{FC} and the M_{WP}. Measuring moisture volume requires measurements taken in situ or in a laboratory on non-modified soils, so as to preserve their structural and biological porosity. Gravimetric moisture content can be measured more easily, but this implies determining the soil's bulk density (section 1.2.1). These measurements require specific equipment, and laboratories equipped with such equipment are rare. Therefore, they are too complex and expensive to be carried out on a regular basis. Instead, simpler methods are used (Baize and Jabiol 2011; Bouthier et al. 2022) to arrive at quantitative estimates of soil MAWC without complex laboratory measurements or measurements in situ. The easiest of these methods is an estimation by pedotransfer functions using textural classes. A series of specially prepared tables will give an AW content for each texture class. These values were obtained from numerous measurements on soil samples following different methods (Jamagne et al. 1977; Bruand et al. 2002; Baize 2018). The values obtained by the Soil Mapping Service of Aisne (Jamagne et al. 1977) have been used for a long time in France. They result from measurements done on samples previously dried and sieved to 2 mm, and only for the soils from the department of Aisne. More recent works, with measurements on undried clumps measuring from 5 to 10 cm^3, allow for a clearer accounting of soil structure. In addition, the soil samples come from about 20 French departments; therefore, the results are more relevant (Bruand et al. 2002). Therefore, on the basis of the textures assessed in the field by an experienced operator or from simple granulometric analyses, it is possible to locate the soil on the texture classification diagram of the Soil Mapping Services of the Aisne Chamber of Agriculture, known as the "Aisne soil map triangle" or "Jamagne Triangle" (Figure 6.1), and to directly obtain an order of magnitude of the MAWC, assigned according to the classes of textures (Figure 6.2).

The MAWC is then estimated for fine earth (fine earth MAWC) of the soil (particles of sizes <2 mm). If the soil contains CE (of sizes >2 mm), how should these be included? If the concentration of CE is less than 10%, they can be neglected for the calculation of the MAWC (Bouthier et al. 2022). If the soil is stony (with a concentration greater than 10%), then the volume of CE is left out of the calculation, which will be made by horizon (MAWC horizon) according to its thickness (E), using the following formula:

$$\text{MAWC of horizon} = \text{MAWC of fine earth} \cdot (100 - \text{EG})/100 \cdot \text{E}$$

where the MAWC of the horizon is in mm, fine earth MAWC is in mm/cm, E is in cm and EG is abundance (%) of CE.

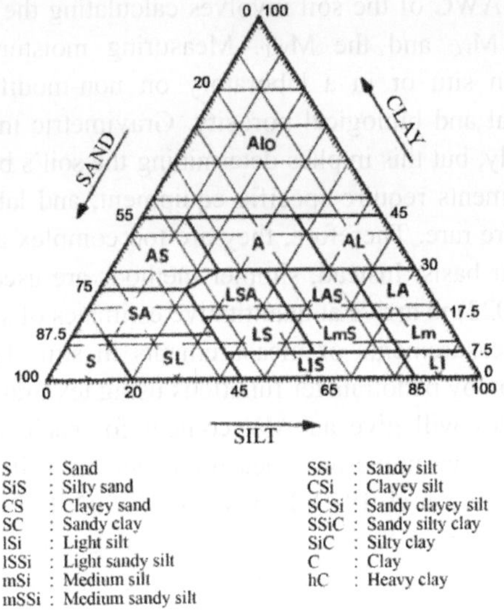

Figure 6.1. *Texture class triangle (Aisne soil map triangle or Jamagne triangle)*

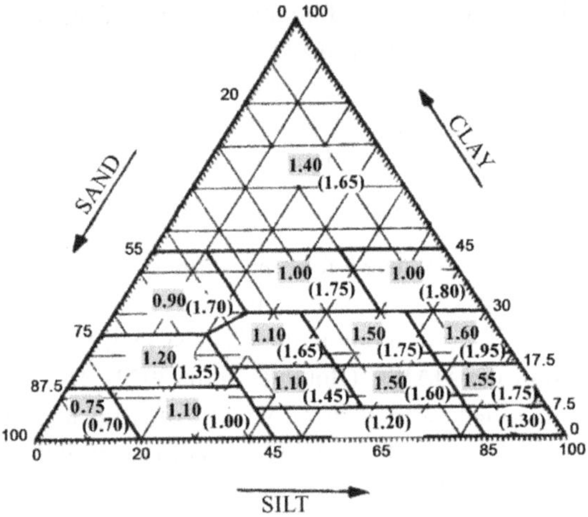

Figure 6.2. *Values of maximum available water content (MAWC) of fine earth in mm per cm of soil depth, placed on the texture class triangle of the Aisne soil map of Figure 6.1. Values in parentheses taken from Jamagne et al. (1977); values highlighted in yellow from Baize and Jabiol (2011), calculated from data from Bruand et al. (2002) for water retention and average bulk densities from Jamagne et al. (1977). For a color version of this figure, see www.iste.co.uk/coquet/agricultural.zip*

In this case, the AW content for the CE is neglected, due the fact that it has no porosity or a water reserve inaccessible to the roots. Nevertheless, coarse porous elements, such as chalk (Ballif et al. 1995) or Beauce limestone (Coutadeur et al. 2000), can contain AW. The data available for evaluating AW content for CE are quite hard to come by (Bouthier et al. 2022). This makes including them difficult.

The MAWC of the soil (MAWCsoil) will be the sum of the MAWC of the soil horizons, which are prospected by the roots of the plants. It is necessary to define the maximum root depth. The water stored in the soil can only be used if the roots have access to it, except in very special cases where it is possible for capillary rises to occur, as is the case of chalk in Champagne (Ballif et al. 1995). Rooting can be stopped by a physical obstacle (such as an impenetrable compact material), a chemical obstacle (such as elements toxic to the plant) or a water obstacle (water saturation with anoxic conditions). In practice, observing the maximum rooting depth is not easy and requires observations from soil pits. From observations made using an auger, the maximum rooting depth can be estimated either by the maximum depth that can be analyzed using the auger, or by the observation of waterlogging, or by noting the presence of toxic materials. In deep soils without obstacles, the maximum rooting depth varies according to the cultivated species. It can exceed 2 m for winter wheat and barley (Bouthier et al. 2022). For annual crops, the MAWC of the soil is frequently estimated for a depth limited to 120 cm (the length of conventional augers). The rooting limit depth, defined as the depth beyond which the density of the roots is no longer sufficient to ensure a usage rate of the water reserve of 100%, is less than 120 cm for most species in field crops (Bouthier et al. 2022). The value of 120 cm of the auger is quite common, making it possible to compare soils for a given maximum rooting depth. This limitation of 120 cm does not make sense in the case of multi-annual crops (i.e. alfalfa) or perennial crops (i.e. tree crops and vineyards).

Given the different approximations and estimates, evaluating the MAWC without taking complex measurements in the field and in the laboratory remains at estimates down to levels of at least 5 or 10 mm (Baize and Jabiol 2011).

6.3.3. *Natural soil drainage*

The air resource function of the soil will depend on the soil's natural drainage property, and therefore on the ability of the soil to dry out and thus obtain a porosity containing water and air. When the natural drainage of the soil is limited by an impermeable obstacle in the soil or the subsoil, then the water can occupy all of the pores. The soil is then saturated with water – this is known as being "waterlogged" or having "an excess of water" (Trouche and Morlon 2017). Soil can be diagnosed as waterlogged by observing the signs (or traces) of hydromorphy. These are morphological traits that result from the dynamics of iron and manganese according to the conditions of excess water. Rusty spots of reoxidized iron and/or ferromanganic black concretions (iron and manganese) indicate temporary waterlogging. In the absence of air, biological activity will quickly consume all oxygen, including oxygen from water. The soil then begins to have anoxic and reductive conditions. The iron and manganese will be reduced and take soluble forms. Then, once the soil is no longer waterlogged (through drainage, evaporation or consumption by vegetation), some of the pores find air. Under oxidizing conditions (with the presence of oxygen), iron and manganese will be reoxidized and precipitate in the form of rusty spots and black concretions.

A greenish-gray or bluish-gray color (the "Gley" pages from the Munsell chart[1]) results from the presence of ferrous iron under permanent reductive conditions due to permanent waterlogging.

Determining the depth at which signs of hydromorphy appear, which is observable using an auger, makes it possible to identify the thickness of the soil where the roots will have air to develop. However, a diagnosis of waterlogging requires a good deal expertise on soil (MEDDE, GIS Sol 2013). Colors occurring due to the alteration of iron-rich materials should not be confused with signs of hydromorphy. In the absence of iron or in the presence of a circulating or oscillating table of highly oxygenated water, there will be no signs of a reduction in iron. The conditions of observation should also be considered: in a dry and powdery soil, the abundance and the contrast of rust spots is difficult to estimate with an auger.

6.3.4. *Aeration of the soil*

A second important property for air resources is surface aeration. The macroporosity on the surface must be durable to ensure the entry of air into the soil. The quality of the soil structure will be important for the exchange of air between the soil and the atmosphere (Girard et al. 2011a). The characteristic used is the structural stability of the surface horizon. A structured horizon will have a high porosity. The degradation of the structure on the surface will lead to a closure of the pores, with clumping or the formation of a slaking crust after rains

1 Munsell soil color charts: a color chart used in the field to identify and objectively code the colors of soils.

(see sections 1.2.2.3 and 4.2.1). The structural stability can be estimated by various measurement methods on samples in a laboratory or in situ (Le Bissonnais and Le Souder 1995). These measures are difficult to implement systematically. Also, in practice, structural stability can be estimated from texture (Monnier and Stengel 1982; Baize 2018), and thus related to primary consequences (Figure 6.3). This is only an approximation, because the structural stability also depends on organic matter (Chenu et al. 2000) and other soil constituents such as limestone, iron, or aluminum oxides, as well as pH (Girard et al. 2011).

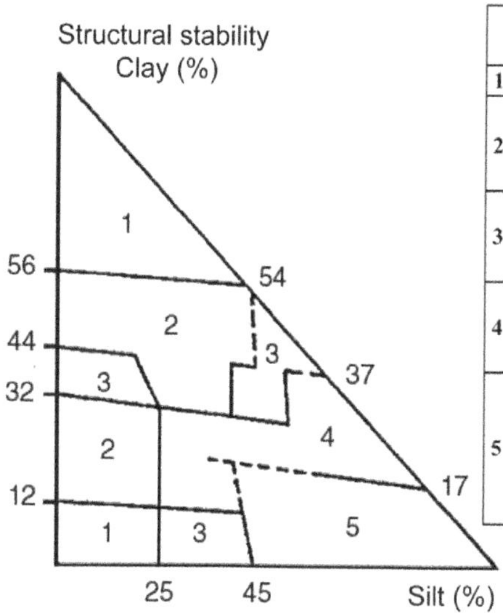

Structural stability	Primary consequences
1 - Very stable	Durable structure
2 - Stable	Slaking unlikely or insignificant Clumping rare
3 - Average	Frequent slaking Clumping possible Erosion possible on steep slopes
4 - Unstable	Very frequent slaking Frequent clumping Frequent erosion
5 - Very unstable	Significant slaking and clumping Soil impermeable to water in the winter Erosion on slight slopes

Figure 6.3. *Relationship between texture, structural stability, and primary consequences (according to Monnier and Stengel 1982; Baize 2018)*

6.3.5. *Stock and availability of nutrients*

The nutrient resource function will depend on the stock of nutrients accessible to the roots and the availability of these elements to be absorbed. The minerals present in the soil will mainly be conditioned by the nature of the constituent materials of the soil, following alteration and pedogenesis processes (AFES 2008). The plants will have access to the elements located across the depth of the soil that is accessible to the roots. Rooting depth is therefore a characteristic to take into consideration. CE are not a source of nutrients without prior alteration. Also, the abundance of CE will be estimated and subtracted from the volume of soil that is prospected by the roots.

The cation exchange capacity (CEC), the ability of the soil to retain cations (Ca^{2+}, K^+, Mg^{2+}, etc.) while keeping them readily available for plants to feed on, is a function of the abundance and nature of clays and organic materials. The CEC can be determined through a routine laboratory analysis. In the field, the observation of constituents, the estimation of texture and the abundance of organic matter can indicate a CEC range (i.e. low, medium and high).

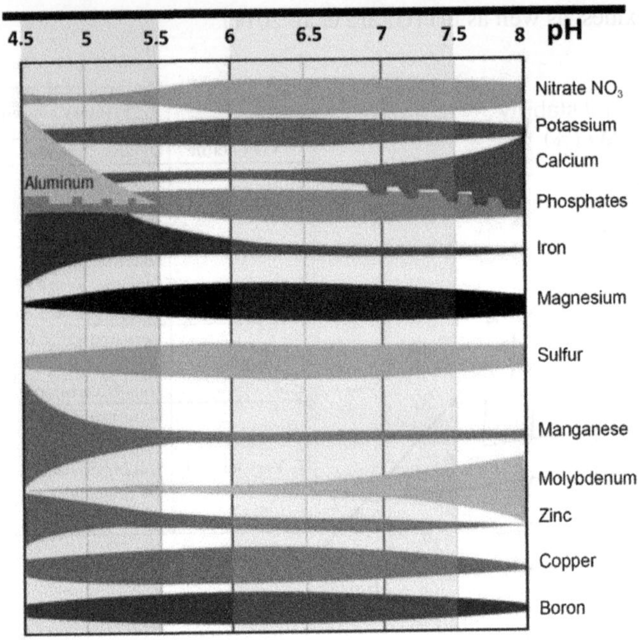

Figure 6.4. *Assimilation diagram of mineral elements as a function of pH (from Truog 1948; UNIFA 2024). For a color version of this figure, see www.iste.co.uk/coquet/agricultural.zip*

The elements stored in the soil will only be able to contribute to the nutrition of the plants if they are available, and therefore present in the solution to be absorbed by the roots. Soil pH plays an important role in the availability of mineral elements (Figure 6.4). The assimilation of the primary nutrients and trace elements is optimal between pH 6 and pH 7.5 for non-calcareous soils. An acidic pH, at 5.5 or lower, generates risks of toxicity by metallic elements; in particular, there is a risk of aluminum toxicity for cereal crops (Baize 2018).

pH level is routinely determined in the laboratory. In the field, it is possible to use a portable pH meter or to estimate it quickly with a "pH tester" consisting of a special reagent and a color code, or even more simply, by using a piece of "pH test paper". The HCl effervescence test on fine earth makes it possible to identify the presence of limestone and to conclude that the soil has a high pH, generally between 7.9 and 8.5 (Baize and Jabiol 2011). In calcareous soils (pH > 8), the availability of certain elements is limited (Fe, Mn, P, etc.).

6.3.6. *Workability of the soil*

The soil's physical support function can be translated into properties that allow for planting or sowing in good conditions. In the context of mechanized agriculture, the workability of the soil with agricultural machinery will be limited when the soil is waterlogged. Farm equipment will not be able to move through the plots without forming ruts or becoming stuck. The depth of the appearance of signs of hydromorphy makes it possible to identify problems of excess water and to deduce the workability of the soil. Wet soil is more susceptible to compaction. Good structural stability will also limit soil clumping and settlement (see section 1.2.2.5).

Finally, the ease of performing work with equipment will also depend on the slope, a factor that is extrinsic to soil (see section 6.4.1.2).

6.3.7. *Soil tillage*

The soil's support function can also be determined by the ability of the soil to be worked, more or less easily by the tools, in particular through the preparation of a structured seedbed that allows the germination of the seeds and later rooting (section 1.2.2.6). The ease with which the soil can be tilled will depend on the texture of the soil and the abundance of CE. A "balanced" texture is considered to be best suited for crops.

The CE in the surface horizon are a source of greater soil "aggressiveness" toward agricultural equipment. For example, a soil that is rich in hard and angular flint will cause the equipment (i.e. plow blades and tines) to wear rapidly and tires to degrade.

To complete the diagnosis of the cultural aptitudes of soils, in addition to the intrinsic properties and characteristics of soils, it is necessary to consider the extrinsic factors.

6.4. Extrinsic factors

The extrinsic factors at play can be divided into two categories: factors of the physical environment (climate and topography) and technical and socioeconomic factors (land use, production systems, practices and development).

6.4.1. *Physical environment factors*

6.4.1.1. *Climate*

The climate will have an impact on the functioning of the soil and its ability to be cultivated.

Each cultivated plant will have its own corresponding "base temperature", that is, the temperature below which the plant will not be able to grow, and a "sum of temperatures", the amount of heat necessary for the development of the plant. If these thermal conditions cannot be reached within the local climate (in climates such as in mountainous areas), then the constraints linked to the soil will be secondary: the main factor limiting the production of the plant will be the climate, and not the suitability of the soil. Extreme temperatures (such as temperatures that are very cold in the winter or very hot in the spring or summer, or frost in the spring) will have consequences in terms of plant development and the expression of production capacity. Climate accidents during the growth cycle may be the main factor limiting production, regardless of the suitability of the soils to crops.

The amount and distribution of precipitation can be major factors in relation to the MAWC of the soil. If the effective amount of rain received is less than the amount necessary to replenish the MAWC, the amount of water in the "soil reservoir" will not be at its maximum. For crop management, it will be necessary to evaluate the AW, which can be lower than the MAWC. If the effective rains are evenly distributed throughout the year and they ensure sufficient amounts of water to provide the water supply for the plants, then the MAWC of the soil will not be a factor to consider. On the other hand, if the rains are unevenly distributed with dry periods when the plant needs water, then the MAWC of the soil will be a primary factor. The intensity of rainfall must be considered: a high intensity of rains may lead to runoff and erosion, with consequences for crop production and on the territory (see Chapter 4). The suitability of soils for certain plant productions will also have to be examined according to the risks generated in relation to the climatic context.

Therefore, the agricultural value of the soil cannot be dissociated from local climatic conditions in order to optimize plant production.

6.4.1.2. *Land relief*

Land relief is characterized by its altitude and its slope. In the mountains, the higher altitudes generate lower temperature conditions, which is the main factor limiting agricultural production.

For low altitude areas, at either the low points when considering a watershed zone, or the coastal areas on the scale of a country or a continent, the low points may be areas that can be flooded by rivers or areas of submersion under sea water. This factor of soil location in the low points of topographical reliefs will have to be considered in crop suitability evaluations.

Slope intensity is a limiting factor in the possibilities of cultivation and mechanized tillage. The synthesis work carried out by the GEPPA (Group for the Study of Applied Pedology Problems) (Baize and Jabiol 2011) indicates that soil cannot be cultivated at a slope of 30% or greater, and the soil is difficult to cultivate at a slope of 15–30%. The grazing limit for cattle is a slope of 75%.

6.4.2. *Technical and socioeconomic factors*

6.4.2.1. *Cultivation systems*

Depending on the specific needs of each cultivation system for production in quantity and quality, farmers will look for different soil properties. Higher quality viticulture favors healthy soils (without excess water) that are not very thick and are rockier, which ideally obtain water stress to promote quality and not quantity. Vegetable farming favors healthy, sandy, non-stony soils that can be saturated rapidly to facilitate tillage and irrigation. Grain farming favors healthy, thick soils with a balanced texture, with the objective of maximizing the water reserve and the nutrient potential. In addition, crop successions can combine cereal crops and industrial crops (such as sugar beets and potatoes). The needs and constraints for these different productions can vary. The growth of potatoes requires a non-stony soil with a light texture to facilitate mechanical harvesting. Sugar beets, which have small seeds, require a seedbed with small aggregates for good germination and good structural stability to limit slaking before the sprouting of the seedlings. To satisfy multiple types of crop successions, the evaluation criteria must all add up; such criteria will not necessarily all be compatible. The suitability of the soils to the crops will be partly determined on the basis of the technical and socioeconomic contexts. We must note that the criteria will be different in a production system that uses non-mechanized potato harvesting; pebbles and stones are not an obstacle to manual harvesting.

The evaluation of the agricultural value of soils will frequently be defined on the basis of the cultivation systems.

6.4.2.2. *Layout and practices*

To develop agriculture, humans have always sought to reduce environmental constraints through adjustments and corrective practices. Steep sloping grades have been converted into terraces. Excess water in certain soils has been remediated through ditches, channels and buried drains. Soils with a water deficit have been irrigated (Ruellan 2010). In the 20th century, with the rise in mechanization, technological progress and the development of mineral fertilizers, more constraints could be overcome and more improvements were made (Figure 6.5). Soils formerly left in wastelands or pastures were cultivated, with high production capacities. In France, the chalky Champagne region is a classic example of this. After World War Two, this region, once known as "Champagne pouilleuse" (poor Champagne), with its shallow limestone soils, was mostly wooded and only saw crop development near towns. Mechanization, the consolidation of plots and the use of mineral fertilizers have transformed the region, which now contains more than 80% of cultivated land and high yields of grains, beets and industrial crops (Garnotel 1985; Ballif et al. 1995).

Figure 6.5. *Examples of agricultural layouts. Left image: vineyard on a very steep slope arranged into terraces (2018, Slovenia); right image: drainage site in Brie in 2002 (deep silt soils). For a color version of this figure, see www.iste.co.uk/coquet/agricultural.zip*

Another example is Brie, a region with thick loamy soils subjected to excess water due to its clay substrate with low permeability. The region was primarily used for livestock grazing with pastures and some cultivated plots, drained locally with pottery drains. The development of mechanization, machines and drainage techniques using buried PVC drain pipes has allowed the almost systematic drainage of all deep silt soils (Figure 6.5, in 2002: creation of a second drainage 30 years after the first). With the problems of waterlogged soils solved, Brie is currently occupied primarily by field crops.

Layout is thus an important consideration when evaluating the agricultural value of soils. Certain practices must also be taken into account, such as the practice of liming, which has the goal of maintaining a pH level close to 7 for non-calcareous soils that tend to acidify (Groupe Chaulage du COMIFER 2009).

6.5. Evaluation of agricultural values

6.5.1. *Different levels of approach*

The evaluation of the agronomic value of the soils can be done at different levels of investigation, from the regional level to the plot level (Girard et al. 2011b). A regional approach, with low resolution, makes it possible to provide information to local authorities in charge of regional or departmental planning. An example is the departmental map of the Oise soils at 1/100,000. This inventory map of agricultural soils is complemented by a map of cultural aptitudes (Begon et al. 1977). The resolution does not allow for a detailed diagnosis, but provides information on the potential of the soils for the different areas of the department.

An approach at the level of the small natural region offers a better resolution, but it is not yet sufficient to evaluate the lands for an agricultural operation. It can provide knowledge for the dynamics of a group of farmers, a cooperative or a chamber of agriculture to guide collective choices for agricultural development or to guide local development and urbanization policies. An example is the production in the 1980s in France of departmental maps of agricultural land at 1/50,000 (Thiebaut and Nicaud 1986). For almost all departments, a map has been drawn up on the area of a topographic sheet at 1/50,000 of the IGN, using six productivity classes (1: very high productivity; 2: good productivity; 3: average productivity; 4: poor productivity; 5: low productivity; 6: very low or zero productivity). For each sheet, the general definition of the six classes is completed by taking into account the local factors of the physical environment (soil, climate and relief) and the production systems in place (agrosystems). The classification also includes developments that improve productivity (land development, drainage and irrigation).

The plot level approach makes it possible to provide advice to the farmer for the management of each plot within the farm (such as the choice of crops, planting, crop rotation, cultivation practices and tools). However, the notion of plot as the level of the approach needs to be carefully applied. A single plot does not necessarily contain just one type of soil. For large plots (spanning several hectares, or even tens of hectares), it is common to have several types of soil with different constraints and potential opportunities within the same plot. This means that an intra-plot approach will be necessary. This is the case for precision agriculture: each homogeneous area of the plot is managed differently (see section 3.5.1).

6.5.2. *Principles of evaluation and classification*

6.5.2.1. *Scoring of potential opportunities or constraints*

The favorable or unfavorable nature of the intrinsic properties and characteristics, and of the extrinsic factors, is defined in relation to the needs for optimal production. It is easier to classify constraints than potential opportunities. The classification is established according to the intensity of the constraint: from zero constraint, representing optimal conditions, to major constraints, where the conditions are the most unfavorable. Between these two extremes, intermediate classes are established. The classes of constraints are generally denoted from 1 (zero constraints) to X (major constraints).

For example, for slopes, the GEPPA (Baize and Jabiol 2011) proposes 10 classes of slope intensity and identifies the related constraints for land use, tillage, irrigation, drainage using buried pipes and the risk of erosion by runoff. If we only consider the ability to plow the soil, then the classification is reduced to five classes. For a slope intensity of up to 5%, that is, a very low to moderate slope, the soil can be cultivated without constraint for tillage (class 1); from 5% to 10%, the slope is moderate and the tillage constraints are minor (class 2); from 10% to 15%, the slope is moderate to strong and the tillage constraints are specific (class 3). From 15% to 30%, the slope is steep, the soil is difficult to cultivate, tillage requires suitable equipment and techniques, and the constraint is strong (class 4). Beyond 30%, the slope is very steep, the soil cannot be cultivated and the constraint is major (class 5).

There is no universal evaluation grid for denoting the intensity of the constraints according to the different criteria. Each classification defines a number of classes of constraints, and ranges of values for the different variables of each class. The threshold values for defining the classes of constraints result from experimental results or expertise, depending on the contexts and the productions. The classes are defined according to the objectives of the evaluation, the desired accuracy and the accuracy of the data used.

6.5.2.2. *Example of constraint scoring*

To illustrate this approach, we will now look at an example of the scoring of constraints as used for field crops in the Parisian Basin. Table 6.2 provides an example of a grid to be used for evaluating and scoring constraints in five classes on the basis of a series of characteristics or properties of soils and the slope factor.

Characteristics or properties	Scoring of class/intensity of constraint				
	1/none	2/low	3/moderate	4/strong	5/major
Depth at which signs of hydromorphy appear (cm)	>120	80–120	50–80	30–50	<30
Abundance of coarse elements (%)	<5	5–15	15–30	30–60	>60
Rooting depth (cm)	>120	80–120	50–80	30–50	<30
MAW (mm)	>250	150–250	150–100	50–100	<50
CEC (cmol+/kg)	>30	20–30	10–20	5–10	<5
pH	6.5–7.4	7.5–8.4	5–6.4	<5	>8.5
Slope intensity (%)	<5	5–10	10–15	15–30	>30

Table 6.2. *Example of a five-class grid for the evaluation and scoring of constraints to be used for field crops in the Parisian Basin*

Depending on the available data, the criteria can be refined to a greater or lesser extent. For example, for the abundance of CE, a five-class scoring system can be done across the board (Table 6.2) or based on the different dimensions of the CE (Table 6.3).

To complete the scoring of the constraints of the properties and characteristics given in Table 6.2, a five-class scoring of the texture and the structural stability of the surface is shown in Figure 6.6. The scores are placed on the Jamagne texture triangle shown in Figure 6.1.

Classes of concentration of coarse elements by dimensions					
Concentration (%)	<5	5–15	15–30	30–60	>60
Gravel (0.2 to 2 cm)	1	1	2	3	4
Pebbles (2 to 7.5 cm)	1	1	3	4	5
Stones (7.5 to 20 cm)	1	2	3	5	5
Boulders (>20 cm)	1	2	4	5	5
Dimensions taken from the Pedological reference base 2008 (AFES 2008)					

Table 6.3. *Example of a scoring grid for the constraint of the concentration coarse elements by their dimensions (ratings: 1: zero constraint; 2: low constraint, 3: moderate constraint, 4: strong constraint, 5: major constraint)*

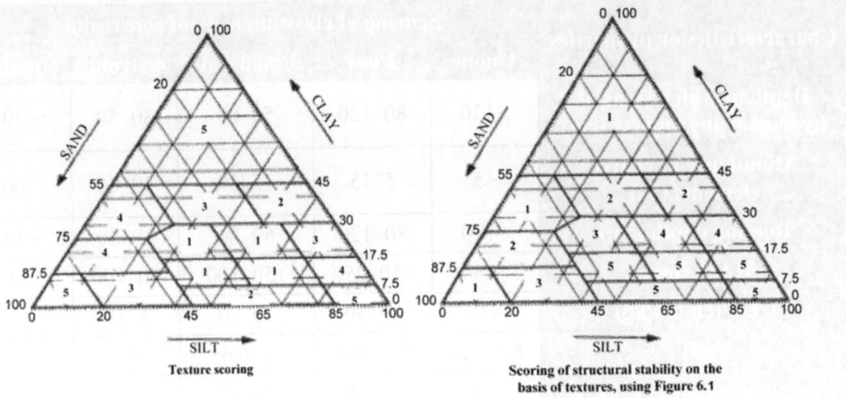

Figure 6.6. *Five-class scoring system for texture and the structural stability using the Jamagne texture triangle from Figure 6.1*

6.5.2.3. *Classification of soils by additive model of constraint scores*

A total constraint score for each soil is obtained by adding up the constraint scores of the different criteria. Then, classification of the soils is organized into aptitude classes according to the range of scores obtained. The number of classes is not standardized. However, it should be limited to less than a dozen so as not to become too complex. In addition, data obtained through field estimates do not necessarily allow the classes to be multiplied.

The additive model can incorporate weightings for the scores so as to take into account the relative importance of the different criteria. An example of an additive model with weighting is the method developed for improving the soils of Provence by the soil scientists of the Société du Canal de Provence (Duclos 1971). After defining the factors and characteristics to be taken into consideration for the context to be studied, the constraints for the production are separated based on important factors, secondary factors and penalizing constraints. The three important factors are the slope, soil depth and the AW. They are scored according to five ratings from the strongest to the weakest constraint levels using the following ratings: 0, 5, 10, 15 and 20 (if the slope is greater than 30%, the soil is considered unsuitable, as noted above). The two secondary factors are the texture (a score of 5: balanced textures without slaking; score 3: sandy textures that are easy to work with; score 2: loamy textures prone to slaking; score 1: clay-rich textures, heavy soils) and the potential fertility (a score of 5: favorable; score 3: somewhat favorable; score 1: not very favorable). The penalizing constraints are noted negatively, grouped into four classes for excess water (a score of –2 for low, –5 for medium, –10 for intense and

–20 for marshes) and salinity (–2 for low, –5 for average, –10 for notable and –20 for significant), and three classes for the abundance of CE (a score of –1 for an abundance of 15%–30%, –3 for 30%–60%, –5 for >60%), active limestone (very fine limestone) (a score of –1 for between 7% and 15%, –3 between 15% and 25%, and –5 for > 25%) and for the nature of the substrate (a score of –1 if it is loose or crumbly, –3 if soft or cracked, and –5 if hard or compact). These scores are added up for each soil. The scores obtained range from 0 to 70. Also, seven soil classes are established with a classification scale from 1 to 10, broken down into 10 points. The aptitudes for the enhancement of the soils of Provence are then defined, ranging from class I for highly suitable soils (score >60 points) to class VII for unsuitable soils (score <10 points). By adapting the criteria, the ratings and the prioritization, the same method has been applied to evaluate the soils of the Mediterranean region (Duclos 1994).

The weightings of the criteria are variable and complex. Certain criteria are used to define multiple soil properties (such as the depth of the rooting, the depth at which signs of hydromorphy appear; see Table 6.2). This is an initial weighting of the criteria.

6.5.2.4. *Classification of soils using a hierarchical model of constraints*

The constraints are prioritized according to their impacts on the suitability of the soils, grouped into absolute constraints, major constraints and minor constraints.

Absolute limiting factors make cultivation impossible, which can include factors such as a slope of more than 30%, or permanent waterlogging of the soil on the surface with frequent submersions. These absolute constraints cannot be overcome by simple adjustments.

The factors that considerably limit cultivation capabilities will form a list of major constraints. The different classes are defined according to the intensity of the constraints, such as the abundance of CE, slope intensity, soil depth or the intensity of excess water. The classification will prioritize these constraints and give threshold values for the declassification. The classification can incorporate the possibilities of reasonable technical and/or economic investments to remove the constraints.

Minor constraints do not limit the soils' ability to be cultivated. The constraints are very low or non-existent; no development is necessary.

An example of a hierarchical model of constraints is presented in Table 6.4, according to the map of agricultural soil suitability for agricultural soils in the department of Oise (Begon et al. 1977). The classification gives six classes of possible crop ranges from class I in which farmers' choices are not limited in any way to V in which their choices are limited to meadows without prior drainage. Class VI represents the bottom sections of non-farmable wet valleys, occupied by marshes or poplar groves. The major constraints are excess water, soil depth (the main factor of the MAWC), the stoniness of the soil and the slope. The reduction of the classification is done when one of these major constraints is considered to apply. Excess water is the main constraint.

Class/Range of possible crops	Constraints (at least one constraint from the list triggers a reclassification)
I/ Broad range	No constraints or minor constraints (low depth of hydromorphy, low slope <5%)
II/ Low range limitation	Surface stoniness <10–15%; moderate hydromorphy from 40cm; obstacle more than 80cm; moderate slope <10%
III/ Average limitation of the range	Surface stoniness of 15–20%; medium hydromorphy in the plowing zone and moderate below 40cm; impermeable at more than 80cm; obstacle between 40 and 80cm; slope <15%
IV/ Strong limitation of the range	Surface stoniness >20–25%; medium hydromorphy in the plowing zone and moderate below 40cm and sandy texture; obstacles less than 30cm; slope of 15 to 20%
V/ Very strong limitation of the range	Fairly strong hydromorphy at less than 40cm
VI/ No crops	Strong surface hydromorphy (permanent waterlogging) and frequent submersion

Table 6.4. *Example of a hierarchical classification model. Classification into six classes using a range of possible crops on the basis of the constraints according to the map of cultural aptitudes of agricultural soils of the Department of Oise (Begon et al. 1977)*

Another example of a hierarchical classification model is that used in the departmental maps of agricultural land (*cartes départementales des terres agricoles*, CDTA). The classification into six productivity classes (see section 6.5.1) is carried out by a hierarchy of constraints which are not the same from one sheet to another, depending on the physical context and the agrosystems. For example, the sheet for Clermont-Ferrand (SCET AGRI 1983a) initially distinguishes two large natural regions, "mountains" and "plains". The mountain area is a location for dairy production and livestock farming. The plains area is subdivided into an area of mixed farming and permanent crops (Limagne viticulture) and an area for large wheat crops (Limagne agriculture). For each zone, the declassification criteria vary. For the mountains and the plains with mixed farming and permanent crops, no soil is in class 1 (in the national classification of CDTA). The absolute constraints are the slope and the quantity of CE (class 6). The major constraints are the depth, the water reserve, the amount of CE and the drainage. For the plains area with large cereal crops, the majority of the soils are in class 1. The first major constraint for the classification breakdown is salinity (class 4), followed by the AW (class 2). Another example, the sheet for the Macon region (SCET AGRI 1983b), also distinguishes the natural regions between the Bresse and the Mâconnais. In the Mâconnais region, the vineyards with protected origin designations (*appellations d'origine contrôlée* (AOC)) are mapped according to their designation – they are not classified according to the CDTA nomenclature. The map does not indicate any class 1 land. For the Bresse part, the first major constraint is the poor drainage. Depending on the intensity of the excess water and the risk of flooding, the land is broken down into classifications from class 3 to class 5. Then, the slaking characteristics are used, giving a categorization from class 3 to class 4. For the Mâconnais part, the main constraints are the depth, the slope and the coarse element content (for a classification of class 5). For the classification of class 3 and 4, the major constraints are excess water, depth and slope. The absolute constraint is the outcroppings of hard rocks (class 6). For the Bresse region, the sandy regions of the Saône Valley, used for market gardening and vegetable crops in open fields with irrigation, are classified higher, as class 2 instead of classes 3 or 4. In this case, the operating systems and the arrangements are considered to allow the land to have good productivity levels (Class 2).

These different examples serve as a reminder that the evaluation of soils is done according to soil constraints, as well as in relation to their agricultural use within a given technical and socioeconomic context. A classification done on identical bases on the national level translates on the local level into specific variations, adapted to the different contexts of the small natural regions and the agro-systems that can be found. The CDTA map for Chinon illustrates this perfectly. Across a varying territory, the classification is divided into three agro-systems with a sub-classification specific to each of them (SCET AGRI 1985). If the context is homogeneous on a sheet with a scale of 1/50,000, the classification does not vary according to the agro-systems. For example, for the Saint-Just-en-Chaussée sheet (BDPA 1985), located entirely on the Picardy Plateau, the approach is based first on the characteristics of the soils according to their geomorphology: the soils of the loamy plateaus, the loamy soils of the borders of the plateaus, the soils on altered chalk clay on the edges of the plateaus, the soils on slopes and the soils of dry or wet valleys.

6.6. Examples of observation and classification methods

6.6.1. *Approach using field pedological surveys*

Soil evaluation can be carried out through field pedological surveys. If it is deemed more advantageous to use easily observable characteristics to establish the diagnosis and classification, soil evaluation will be carried out on the basis of descriptions and simple measurements in the field (see section 6.3.1). Observations using boreholes made with an auger may be sufficient. The number of observation points and the density of these points will depend on the resolution scale and the budget allotted for the study. The location of the observation points is determined according to the characteristics of the environment, topography, geomorphology, land use and vegetation (using indicator plants). An analysis of the different existing documents, topographic maps, geological maps, soil maps, aerial photographs and satellite images makes it possible to characterize the area to be surveyed and to prepare the field phase (Baize and Jabiol 2011). An analysis of the immediately surrounding landscape on the ground makes it possible to refine the location of the observation points. If necessary, samples can be taken for laboratory analyses to complete the observations made on site (Baize and Jabiol 2011; Baize 2018).

For example, the Muencheberg Soil Quality Rating (M-SQR) is a relatively simple and fast field method for assessing soil quality (Mueller et al. 2007). The method is semi-quantitative with empirical notation using a field manual. The characteristics are evaluated mainly by visual methods, with a minimum amount of equipment in the field (which includes an auger, spade, knife, meter, a pH probe or pH test strips, an electrical conductivity probe, a photographic camera and a GPS tracker). The assessment is based on the analysis of easily observable soil properties that limit production. The overall scores characterize the yield potentials of cultivated lands or prairies. Depending on the limitations, they range from 0 (the worst) to 100 (the best). In the field classification, five classes of soil quality are used: very poor (<20), poor (20–40), moderate (40–60), good (60–80) and very good (>80).

6.6.2. *Approach using soil maps and databases*

When a pedological map exists at the scale desired for the evaluation, then this soil map can be exploited to produce a map of the soil's suitability for crops.

One example of this is the map of the suitability for crops of agricultural soils from the department of Oise, made from the 1/100,000 scale soil map (Begon et al. 1977).

Another example is the soil map of the department of Indre, at 1/50,000 scale and in a digital format. The soil database associated with this pedological map allows for the production of various thematic maps that are derived from this map, including the agricultural potential map (INRA and CA36 2006; Moulin and Eimberck 2011).

However, this production faces a problem: there is no soil map with this level of precision that spans France's national territory. Even the national soil mapping program, at 1/100,000 scale, could not be completed for all of France (see Chapter 3). For a local study, it is necessary to identify the existing soil maps. Refersols, the national directory of soil mapping studies, can be used (see section 3.3.2.1). On the other hand, the soil maps of the "Regional Pedological Reference Frame" (*Référentiels Régionaux Pédologiques* (RRP)), covering the entire territory of France at 1/250,000 scale, are available on the IGN Geoportal (Messant et al. 2021). This makes it possible to have access to soil characterization sheets for the different typological units of soils. In principle, this makes it possible to attach the soils observed in the field to typological soil units using a simple and fast expert survey.

Since 2015, the TypTerres project (RMT Sols et territoires 2019; Sauter et al. 2019; Laroche et al. 2020) has been developing an agronomic typology of soils based on the RRP. The goal of this project is to provide free online access to a unique and easy to use typology that covers all of France. The project was first developed in two test-case territories, West-Central France and Alsace. It currently remains under development, a sample of which can be seen in the TypTerres progress map given in Figure 6.7. This online tool has already been finalized for several regions. The TypeSol Burgundy project, marked in blue on the map, is an earlier project, more similar to TypTerres but developed exclusively in Burgundy. The data are available on the RMT Soils website and on the websites of the territories. The principle of the TypTerres system is to group together the soil unit types of the RRPs (UTS) into agronomic TypTerres types (UTT = TypTerres Typological Unit). The main objectives of the system are to provide a correspondence between the soil types from the RPP (UTS) and the TypTerres agronomic types (UTT) and to record the most readily observable characteristics of the soils that can be seen directly in the field, determining their agricultural behavior and therefore their suitability for crops (texture, depth, natural drainage, CE, etc.). Another of its goals is not to exceed roughly 50 category types (± 10 types) per department. For example, for the four departments of Vendée, Deux Sèvres, Vienne and Indre, which form the territory of the West-Central France, the 913 typological units of soils (UTS) are grouped into 65 Typeterres typological units (UTT). For each UTT, a determination key is defined. This makes it possible to associate a soil from a given agricultural plot to a UTT based on simple observations in the field. Eventually, each user will be able to use a mobile application on a tablet or cellphone to assign soils to a UTT in the field.

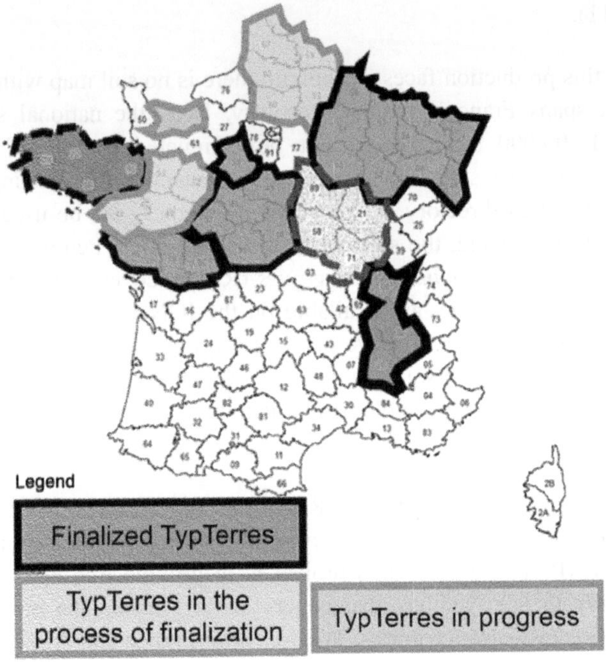

Figure 6.7. *Progress made by TypTerres (November 2022) (RMT Sols et territoires); blue: the TypeSol Burgundy project. For a color version of this figure, see www.iste.co.uk/coquet/agricultural.zip*

6.7. Conclusion

The methods for evaluating the agronomic value of soils or their suitability for crops are complex and still relatively empirical. They include both the needs of cultivated plants to develop optimally, as well as the technical and socioeconomic aspects of local agricultural production, in addition to the climate and geomorphological context. There is no universal method for evaluation. Depending on the context, different intrinsic and extrinsic soil factors are taken into account, and the rankings and prioritizations of the constraints limiting the choices and the potentialities of the productions may vary. For field surveying, methods are based on easily and quickly observable soil characteristics. Then, certain properties can be estimated from existing data and by using pedotransfer functions. Maps in soil suitability classes are produced at different scales and according to different classification methods. Software programs, such as TypTerres, are currently being developed to use maps and soil inventory databases in order to streamline the assessment processes at the national level, and to promote the use of these data for field diagnostics based on a minimum of observations made in the field. The evaluation methods give little if any consideration to the biological characteristics and functioning of the soils. Research work to develop the routine use of bioindicators should make it easier to integrate the consideration of biological activity into the evaluations of the agronomic value of soils (see Chapter 2).

6.8. References

AFES (2008). *Référentiel pédologique 2008*. Éditions Quæ, Versailles.

Baize, D. (2016). *Petit lexique de Pédologie. Nouvelle édition augmentée*. Éditions Quæ, Versailles.

Baize, D. (2018). *Guide des analyses en pédologie : 3e édition revue et augmentée*. Éditions Quæ, Versailles.

Baize, D. and Jabiol, B. (2011). *Guide pour la description des sols : nouvelle édition*. Éditions Quæ, Versailles.

Ballif, J.L., Guérin, H., Muller, J.C. (1995). *Eléments d'agronomie champenoise. Connaissance des sols et de leur fonctionnement. Rendzines sur craie et sols associés. Esquisse géomorphopédologique*. INRA, Paris.

BDPA (1985). Carte départementale des terres agricoles – St Just en Chaussée. Map and Leaflet, Ministère de l'agriculture, Direction départementale de l'agriculture de l'Oise.

Begon, J.C., Hardy, R., Mori, A., Roque, J., Jamagne, M. (1977). *Les sols du département de l'Oise : cartes et notices d'utilisation. Services d'etudes des sols et de la carte pedologique de France*. Centre National de Recherches Agronomiques, Versailles.

Bouthier, A., Scheurer, O., Seger, M., Lagacherie, P., Beaudoin, N., Deschamps, T., Sauter, J., Fort, J.L., Cousin, I. (2022). *Réservoir en eau du sol utilisable par les cultures*. Editions Arvalis, Paris.

Bruand, A., Pérez Fernández, P., Duval, O., Quétin, P., Nicoullaud, B., Gaillard, H., Raison, L., Pessaud, J.-F., Prud'Homme, L. (2002). Estimation des propriétés de rétention en eau des sols : utilisation de classes de pédotransfert après stratifications texturale et texturo-structurale. *Etude et gestion des sols*, 9(2), 105–125.

Calvet, R. (2003a). *Le sol, propriétés et fonctions. Tome 1 : constitution et structure, phénomènes aux interfaces*. Editions France Agricole, Paris.

Calvet, R. (2003b). *Le sol, propriétés et fonctions. Tome 2 : phénomènes physiques et chimiques, applications agronomiques et environnementales*. Editions France Agricole, Paris.

Chenu, C., Le Bissonnais, Y., Arrouays, D. (2000). Organic matter influence on clay wettability and soil aggregate stability. *Soil Science Society of America Journal*, 64, 1479–1486.

Coutadeur, C., Cousin, I., Nicoullaud, B. (2000). Influence de la phase caillouteuse sur le réserve en eau des sols. Cas des sols de petite Beauce du Loiret. *Etude et gestion des sols*, 7(3), 191–205.

Duclos, G. (1971). Appréciation de l'aptitude à la mise en valeur des sols de Provence. *Bulletin de l'Association française pour l'etude du sol*, 6, 33–46.

Duclos, G. (1994). *Atlas des sols de la region Provence – Alpes – Côte d'Azur*. Société du canal de Provence et d'aménagement de la région provençale.

FAO (2021). L'état des ressources en terres et en eau pour l'alimentation et l'agriculture dans le monde – Des systèmes au bord de la rupture. Summary Report, Rome.

Garnotel, J. (1985). *L'ascension d'une grande agriculture – Champagne pouilleuse – Champagne crayeuse*, volume 1. Economica, Paris.

Girard, M.C., Schvartz, C., Jabiol, B. (2011a). *Etude des sols : description, cartographie, utilisation*. Dunod, Paris.

Girard, M.C., Walter, C., Rémy, J.C., Berthelin, J., Morel, J.L. (2011b). *Sols et environnement*, 2nd edition. Dunod, Paris.

Groupe Chaulage du COMIFER (2009). *Le chaulage. Des bases pour le raisonner*. Comifer, Puteaux.

INRA (2009). *Le sol. Dossier INRA. INRA mensuel*. Éditions Quæ, Versailles.

INRA and CA36 (2006). Département de l'Indre : cartographie des potentiels agronomiques. Document [Online]. Available at: https://centre-valdeloire.chambres-agriculture.fr/agroenvironnement/sols/cartographie-des-sols/indre/ [Accessed 31 July 2023].

Jamagne, M. (2011). *Grands paysages pédologiques de France*. Éditions Quæ, Versailles.

Jamagne, M., Bétrémieux, R., Bégon, J.-C., Mori, A. (1977). Quelques données sur la variabilité dans le milieu naturel de la réserve en eau des sols. *Bulletin technique d'information*, 324–325, 627–641.

Laroche, B., Degan, F., Koller, R., Scheurer, O., Bouthier, A., Moulin, J., Sauter, J., Ducommun, C., Fort, J.-L., Maillant, S. et al. (2020). TYPTERRES. Vers une typologie agronomique partagée. *Etude et gestion des sols*, 27, 241–256.

Le Bissonnais, Y. and Le Souder, C. (1995). Mesurer la stabilité structurale des sols pour évaluer leur sensibilité à la battance et à l'érosion. *Etude et gestion des sols*, 2(1), 43–56.

MEDDE, GIS Sol (2013). Guide pour l'identification et la délimitation des sols de zones humides. Guide, Ministère de l'Écologie, du Développement Durable et de l'Énergie, Groupement d'Intérêt Scientifique Sol.

Messant, A., Lehmann, S., Moulin, J., Lagacherie, P., Jalabert, S., Noraz, A., Lemercier, B., Chafchafi, A. (2021). Diffusion des référentiels régionaux pédologiques sous la forme d'une carte des sols dominants (France métropolitaine Hors - Corse) accessible sur le Géoportail. *Etude et gestion des sols*, 28, 57–69.

Monnier, G. and Stengel, P. (1982). La composition granulométrique des sols : un moyen de prévoir leur fertilité physique. *Bull. Tech. Inf.*, 370–372, 503–512.

Moulin, J. and Eimberck, M. (2011). La cartographie des sols de l'indre pour des applications thématiques diversifiées. *Etude et gestion des sols*, 18(2), 75–89.

Mueller, L., Schindler, U., Behrendt, A., Eulenstein, F., Dannowski, R. (2007). Muencheberg Soil Quality Rating (SQR). Field manual for detecting and assessing properties and limitations of soils for cropping and grazing. *Leibniz Germeinschaft,* 102.

RMT Sols et territoires (2019). Les typologies agronomiques de sols TypTerres [Online]. Available at: https://sols-et-territoires.org/produits-du-rmt/typterres [Accessed 31 July 2023].

Ruellan, A. (2010). *Des sols et des hommes. Un lien menacé*. Editions IRD, Bondy.

Sauter, J., Moulin, J., Fort, J.-L. (2019). Les typologies agronomiques des sols, indispensables pour valoriser les référentiels régionaux en pédologie. *Agronomie environnement & sociétés*, 9–2, 195–197.

SCET AGRI (1983a). Carte départementale des terres agricoles – Clermont-Ferrand XXV-31. Map and Instructions, Ministère de l'agriculture, direction départementale de l'agriculture du Puy-de-Dôme.

SCET AGRI (1983b). Carte départementale des terres agricoles – Macon XXX-28. Map and Instructions, Ministère de l'agriculture, direction départementale de l'agriculture de Saône-et-Loire.

SCET AGRI (1985). Carte départementale des terres agricoles – Chinon XVII-23. Map and Instructions, Ministère de l'agriculture, direction départementale de l'agriculture de l'Indre-et-Loire.

Thiebaut, L. and Nicaud, A. (1986). Les cartes départementales des terres agricoles et l'environnement : un outil pour quel débat social ? *Revue juridique de l'environnement*, 2–3, 210–230.

Tomis V. (2017). Guide méthodique du mini-profil 3D. Sol-D'Phy – Agro-Transfert RT. Guide [Online]. Available at: http://www.agro-transfert-rt.org/wp-content/uploads/2017/04/Guide-m%C3%A9thodique-du-mini-profil-3D-version-web-6M.pdf [Accessed 31 July 2023].

Trouche, G. and Morlon, P. (2017). *Excès d'eau : conséquences. Les mots de l'agronomie. Histoire et critique*, INRA [Online]. Available at: https://lorexplor.istex.fr/Wicri/Europe/France/InraMotsAgro/fr/index.php?title=Exc%C3%A8s_d%27eau_:_cons%C3%A9quences&oldid=2293 [Accessed 31 July 2023].

Truog, E. (1948). Lime in relation to availability of plant nutrients. *Soil Science*, 65, 1.

Turillon, C., Crétin, V, Tomis, V., Duparque, A. (2018). Guide méthodique du test bêche structure et action des vers de terre et fiche de notation de la structure par horizon. Sol-D'Phy – Agro-Transfert RT. Guide [Online] Available at: http://www.agro-transfert-rt.org/wp-content/uploads/2018/08/Guide-m%C3%A9thode-beche-web.pdf [Accessed 31 July 2023].

UNIFA (2024). Parlons fertilisation [Online]. Available at: https://fertilisation-edu.fr/cultures-fiches-pratiques.html [Accessed 9 November 2024].

7

The Concept of Ecosystem Services and Its Application to Soil: Between Promises and Reality

David MONTAGNE[1], Ottone SCAMMACCA[1,2], Christian WALTER[3] and Isabelle COUSIN[4]

[1]*UMR ECOSYS, AgroParisTech, INRAE, Université Paris-Saclay, Palaiseau, France*
[2]*UMR Prodig, AgroParisTech, IRD, CNRS, Université Paris 1 Panthéon-Sorbonne, Aubervilliers, France*
[3]*UMR SAS, INRAE, Institut Agro Rennes-Angers, France*
[4]*Info&Sol, INRAE, Orléans, France*

7.1. Introduction

The concept of "Ecosystem Services" emerged in the 1980s (Gómez-Baggethun et al. 2010; Baveye et al. 2016), a hybridized form of ecology and environmental economy. It emerged as a possible response to the increasingly evident biodiversity loss and environmental pollution. It involved (re)creating how humans related to their ecosystem in an explicitly multidimensional, functional and utilitarian manner. Through its explicitly multi-dimensional nature, the concept of Ecosystem Services intends to go beyond the traditional view of ecosystems as a source for primary materials such as minerals, plants and animals, shifting toward a recognition of the multiple benefits humans can directly and indirectly derive from properly functioning ecosystems. This is especially in terms of regulating the quality of

Agricultural Soil Science,
coordinated by Yves COQUET and Joël MICHELIN. © ISTE Ltd 2025.

different parts of the environment (air, water, soil), regulating natural risks (flooding, landslides etc.), and also in terms of cultural benefits (leisure, spirituality, artistic inspiration, etc.). The functional dimension aims to shift our view from what an ecosystem *is* to what it *does*. Finally, by emphasizing how humans benefit, directly or indirectly, from properly functioning ecosystems and their biodiversity, the utilitarian dimension of Ecosystem Services aims to motivate more people to preserve ecosystems, when their interest may not be aroused by moral arguments alone.

As with all ecosystems, human interest in soil has, for a long time, focused on assessing and improving soil fertility. This is defined as the soil's capacity to support agricultural and forest production (Bunemann et al. 2018). It is indisputable today that focusing interest on agricultural and forest production alone has been accompanied by many perverse effects ranging from the degradation of various soil properties and functions (loss of organic matter, compaction, various forms of contamination, etc.) to total or partial destruction through erosion or due to urbanization pressures. In this context, adapting the concept of Ecosystem Services to soils opens up the possibility of radically renewing the relationship between human societies and their soils. Soils will no longer be simply considered as factors in agricultural or forest production (or as infrastructure support), but will be seen as an essential part of the environment, contributing to multiple functions and services that are indispensable to the well-being of human societies (Robinson et al. 2009; Dominati et al. 2010; Bunemann et al. 2018). By allowing us to consider the challenges of agricultural or forest production alongside concerns related to regulating the quality of the environment or biodiversity conservation, the concept of Ecosystem Services could be a major tool for an agro-ecological transition. More specifically, it is a tool for designing, guiding and assessing sustainable soil management modalities.

This chapter presents a synthesis of the current knowledge of Ecosystem Services. The following topics will be approached, in this order: (i) the definition of the concept of Ecosystem Services; (ii) their classification; (iii) their quantification using biophysical approaches; and (iv) the mapping of these services. Finally, the last section focuses specifically on how to apply this concept to the "soil" part of ecosystems.

7.2. A short history of Ecosystem Services: from inception to the international institutionalization of this concept

The concept of Ecosystem Services emerged in the United States in the early 1980s. The goal was to promote interest in the conservation of biodiversity and in the role played by biodiversity in ensuring the well-being of human societies (Gómez-Baggethun et al. 2010; Chaudhary et al. 2015; Baveye et al. 2016). Over four decades, the concept has slowly built up around a new scientific discipline: *Ecological Economics*. It has dedicated scientific journals, including *Ecosystem Services*. It has a global network, the Ecosystem Services Partnership, and in English alone, over 10,000 scientific articles have been published on the subject. As the concept spread through the scientific community, it gradually became institutionalized in politics and economics. The creation of tools, especially market and payment provisions, made it possible to transform the concept into an operational ecosystem management tool, implemented in different places and at different levels around the world (Gómez-Baggethun et al. 2010; Chaudhary et al. 2015). The success story of this scientific concept, which went from being an emerging idea to an ecosystem management tool within a few decades, can be divided into three successive phases: (i) the emergence of the idea in the 1970s and 1980s; (ii) the development and spread of the concept over the 1990s; and finally, (iii) the institutionalization of the concept and its market entry in the 2000s (Gómez-Baggethun et al. 2010).

7.2.1. *An emerging concept*

The concept of Ecosystem Services was born in the early 1980s from the coming together of sections of ecological sciences and economic sciences (Braat and de Groot 2012; Costanza et al. 2017). This coming together was motivated by two observations. First, in the field of ecological sciences, the objectivization of how human activities that impacted the environment did not result in any change in political or economic choices, nor did they halt the damage done to ecosystems. The second, in the field of economics, given that there was no way to measure how ecosystems contributed to human activities, it was not possible to take into account environmental issues when making political and economic decisions (Gómez-Baggethun et al. 2010; Costanza et al. 2017). Consequently, the promoters of the Ecosystem Services concept hoped to get through to public and private leaders by expressing the links between biodiversity, the functioning of ecosystems and the well-being of human societies in terms of dominant political and economic structures, thus formalizing a utilitarian and monetized vision of ecosystems (Gómez-Baggethun et al. 2010). This is how the term *ecosystem function*, defined in the 1960s as "the set of processes operating in an ecosystem, independent of their utility for humans", was gradually replaced by the terms *nature's services* (Westman 1977) and finally *Ecosystem Services* (Ehrlich and Ehrlich 1981), which explicitly refers to the sub-set of functions that contribute, directly or indirectly, to the well-being of human societies (Gómez-Baggethun et al. 2010; Braat and de Groot 2012).

7.2.2. *Development and spread of the concept*

In the 1980s and 1990s, the concept of Ecosystem Services was largely restricted to the academic domain. The creation of Ecological Economics, which sought to quantify relations between nature and societies using economic approaches (Costanza et al. 1997a), was accompanied by a boom in publications on these concepts, evaluation methods, especially monetary evaluation, and case studies (Braat and de Groot 2012; Costanza et al. 2017). This second phase was marked by publications by Daily (1997) (which offered the first documented description of many services provided by the main types of ecosystem, thereby consolidating the scientific bases of the concept) and Costanza et al. (1997b) (which provided the first global monetary assessment of Ecosystem Services). Both these publications helped heighten the attractiveness of Ecosystem Services in the academic world, and provided the arguments that would help its subsequent entry into the political world (Gómez-Baggethun et al. 2010; Chaudhary et al. 2015).

7.2.3. *Institutionalization and market entry*

The launch of the Millennium Ecosystem Assessment in 2005 was a key factor in the explosion of research and publications on Ecosystem Services. These expanded into new disciplines such as agro-ecology, hydrology and marine ecology in the world of natural sciences, and political ecology, human geography, and land use planning in the political and social sciences (Gómez-Baggethun et al. 2010; Chaudhary et al. 2015). This rapid expansion into new scientific disciplines reinforced the Ecosystem Services approach's ability to inform political leaders and orient decision-making processes. Furthermore, the coming together of scientists, political decision-makers, and many other actors to participate in the same process was clearly conducive not only to the creation and large-scale dissemination of knowledge on Ecosystem Services but also to the adoption of these concepts by political leaders and their institutionalization across multiple transnational and international projects (Ruhl and Salzman 2007; Chaudhary et al. 2015). This is notably the case with The Economics of Ecosystem and Biodiversity (TEEB) project launched in 2007 by the G8+5 environment ministers in order to establish a global assessment of the economic benefits derived from biodiversity and ecosystems, as well as the costs of biodiversity and ecosystem losses (Chaudhary et al. 2015). In 2012, these initiatives, and others that followed, finally resulted in the creation of the Intergovernmental Science-Policy Platform on Biodiversity and Ecosystem Services (IPBES), where scientific data are analyzed and synthesized to inform and support the decisions taken by public and private actors. The Mapping and Assessment of Ecosystems and their Services (MAES), developed for Europe within the framework of the European Strategy for Biodiversity (European Commission 2014) or the French Assessment of Ecosystems and Ecosystem Services (EFESE) in France (Monnoyer-Smith 2016; Theron et al. 2017), share the same objectives. Following the institutionalization of the concept of Ecosystem Services and the development of research into how they could be monetized, political decision-makers gradually started focusing on designing market tools that could lead to economic motivations for conserving ecosystems. In the early-2000s, this led to the creation of the European carbon market and the development of payments for Ecosystem Services (Gómez-Baggethun et al. 2010).

7.2.4. *What is the future of Ecosystem Services?*

The strategic and pragmatic option chosen by ecologists and economists in the United States was to present the relationships between biodiversity, the functioning of ecosystems and the well-being of human societies in utilitarian and monetary terms compatible with the dominant political and economic structures in place. This proved to be very effective in popularizing the concept of Ecosystem Services. In just about four decades, this concept has spread through many countries, across multiple scientific disciplines, and among a very wide variety of actors in the academic, political and economic spheres. While it was initially conceived of as a communication tool to sensitize people to the role that biodiversity and ecosystems play in the well-being of civilizations, Ecosystem Services today form the basis for tools that use market logic in the field of nature conservation. The question now is whether current scientific knowledge is well established enough for the concept of Ecosystem Services to be fully operational. At a more fundamental level, many authors have highlighted the risk that when this concept is appropriated by, and integrated into, the same economic and political frameworks that are the cause of ecological degradation, it can accelerate the commodification of nature and, in fine, lead to counterproductive effects that are far-removed from the hopes of those who first conceived of this concept. This is why, in recent years, a new concept has emerged under the aegis of the IPBES. This is Nature's Contribution to People (NCP) and aims to reinforce the position of social sciences in the approaches used to assess ecosystems and to provide a better understanding of the multiple conceptualizations of human–nature ties, going beyond the Western conceptualization that currently predominates in Ecosystem Services (Pascual et al. 2017; Díaz et al. 2018).

7.3. Definition and conceptualization of Ecosystem Services

7.3.1. *Ecosystem Services: a product derived from ecological systems or a co-product of ecological and socioeconomic systems*

Despite being the subject of intense discussion for over two decades, there is still no unanimously accepted definition for the concept of Ecosystem Services, instead we have a large variety of definitions, which are often inaccurate and sometimes compete with each other (Nahlik et al. 2012). We differentiate between two broad families of definitions (Nahlik et al. 2012). The first group, largely taken from economics, considers that Ecosystem Services correspond to the benefits that human beings derive from the combined functioning of ecological and socioeconomic systems (Figure 7.1(a)). In this case, Ecosystem Services come under the socio-economic system (Nahlik et al. 2012). The second group of definitions, largely derived from the ecological sciences, considers that Ecosystem Services are an integral part of ecological systems. Here, Ecosystem Services correspond to the contributions from ecosystems that generate benefits for human beings, with or without any additional contributions from socioeconomic systems (Figure 7.1(b)).

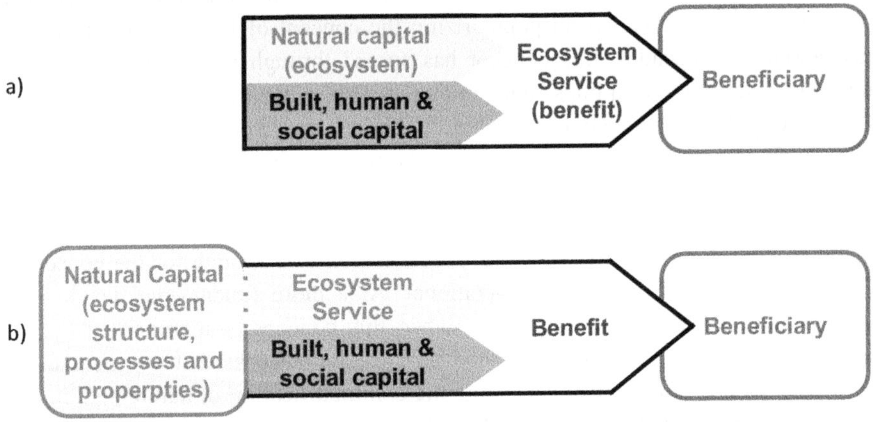

Figure 7.1. *Representation of the two main families of definitions of Ecosystem Services: (a) as a part of the socioeconomic system and a co-product of the natural capital and anthropic capital, and (b) as a part of the ecosystem and the contribution of the ecosystem to the well-being of human societies (adapted from Theron et al. 2017). For a color version of this figure, see www.iste.co.uk/coquet/agricultural.zip*

While, at first glance, this distinction may seem subtle or anecdotal, in reality it is fundamental, especially in agroecosystems. The example of the biomass production service is particularly illustrative of this. According to the first, socioeconomic definition, the biomass production service corresponds to the quantity of biomass produced, as a result of the combined functioning of the ecosystem and multiple human interventions linked to crop practices. All other things being equal, the level of this service would generally tend to increase with the level of human intervention. In the second, ecological sciences definition, the biomass production service corresponds to the fraction of biomass production that is related to the functioning of the ecosystem, independent of any human intervention. By this definition, the biomass production service may therefore be potentially higher in an agroecosystem that has low human capital, than in a very intensive agroecosystem where human capital (selecting enhanced seeds, tillage, fertilization, irrigation, crop protection, etc.) complements or supplements the contribution of the ecosystem, including when the final production levels are higher in the intensive system. For the rest of the chapter, we will focus more on the second definition, even if a non-negligible part of the literature that is cited may itself prefer the first definition.

7.3.2. *Conceptualization using a cascade model*

Condition	The physical, chemical and biological characteristics that describe the structure, composition and processes in an ecosystem.
Function	A function is a process, or combination of processes, which depends on, as well as underpins, the dynamic of the structure and composition of an ecosystem.
Services	Ecosystem services are the direct and indirect contributions made by ecosystems to the well-being of human societies.
Services related to the soil	Ecosystem Services related to the soil are the sub-set of Ecosystem Services that are directly and quantitively governed by soil characteristics, processes and functions.
Advantage	This is a tangible product (goods) or intangible product (benefit) derived from Ecosystem Services, which enhances the well-being of one or more beneficiaries, with or without being combined with any anthropic capital.
Value	A translation or conversion, often in monetary terms, of the advantages derived from Ecosystem Services by a particular group of beneficiaries in a given context.
Supply (potential)	A (potential) service supply refers to the capacity of a given ecosystem to generate Ecosystem Services, independent of whether any demand or beneficiaries exist.
Flow	Service flow refers to the fraction of service offerings directly or indirectly used or acknowledged by the totality of the beneficiaries.
Demand	Demand for a service denotes the total quantity of the service desired by all potential beneficiaries.
Bundle	A bundle of services is the subset of Ecosystem Services that occur together in time and space.
Indicator	An indicator is a variable, or combination of variables, chosen to represent a phenomenon of interest, often in a simplified manner.

Box 7.1. *Definition of the principal terms related to the concept of Ecosystem Services*

Potschin and Haines-Young (2011) developed an "Ecosystem Services Cascade Model" to try and precisely describe the concept of Ecosystem Services (Figure 7.2). This model is based on the distinction made between the two sub-systems: the ecosystem, on the left, and the socioeconomic system, on the right. Ecosystem Services are thus the final products of the ecosystem in the chain that links the ecosystem conditions (i.e. the set of characteristics that describe the structure and the functioning of the ecosystem) to the creation of economic value. In this model, Ecosystem Services (Box 7.1) are defined as the contributions made by these ecosystems to human well-being. They are distinct from upstream functions and downstream goods and benefits.

Functions refer to all processes at work in ecosystems, independent of their utility for human beings, while *services* are defined by the existence of a beneficiary (Box 7.1). While the existence of functions can be considered to be an intrinsic quality of ecosystems, this is not the case with Ecosystem Services, whose existence varies over time and space, depending on human beneficiaries and their varying needs. The *advantages* (Box 7.1) refer to objects or parts, tangible (goods) and intangible (benefits), respectively, of the ecosystem that will lead to an increase in well-being through consumption or usage. The realization of this gain most often involves contributions from elements of the economic and social system. For example, advantages from a forest would require the gathering of wood or, more simply, require signposting for easy exploration of a natural site. This helps in crossing the *production boundary* (Figure 7.2), which separates the ecosystem from the socioeconomic system. For example, primary production is a function of ecosystems, creating a volume of usable wood (biomass production service), all or part of which can be gathered (Goods) and can finally gain economic value in the form of firewood (Value). Similarly, the water purification function generates a volume of purified water (water quality regulating service), part of which can be used (Goods) and finally gain value in the form of potable water, generating economic value. While this model may prove useful in identifying and formalizing the various steps required to transform ecosystem characteristics into value (monetary or otherwise) for human beings, it remains a contentious issue today (Potschin and Haines-Young 2011; Costanza et al. 2017). For example, Costanza et al. (2017) consider that maintaining a boundary between the natural system and the socioeconomic system is a continuation of the conventional opposition between nature and humanity, which is largely at odds with the objectives that were sought to be achieved through the creation of the concept of Ecosystem Services. Furthermore, Potschin and Haines-Young (2011) recognize that it is difficult to establish boundaries between the concepts of "condition", "function", "service" and "benefit" without reference to a specific situation and that we are yet to arrive at a global consensus on the conceptual framework and terminology used.

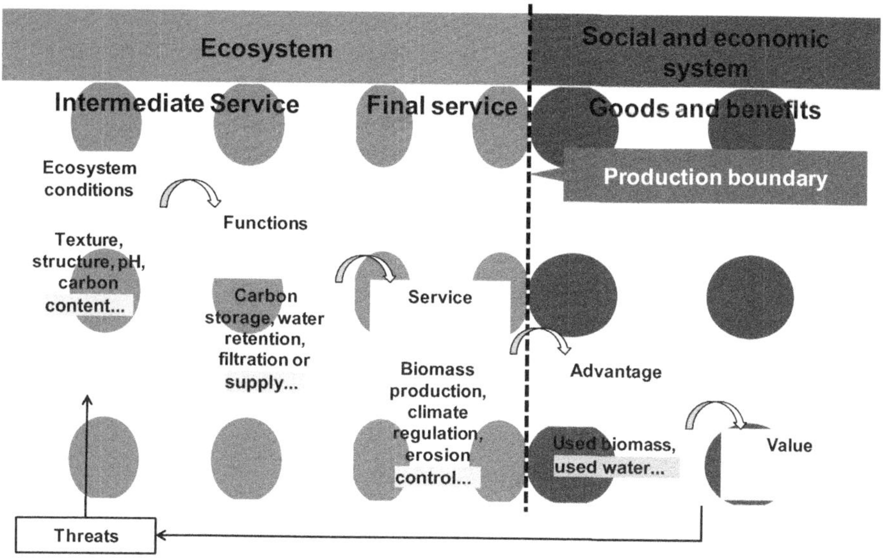

Figure 7.2. *Conceptualizing Ecosystem Services through a "cascade model". Examples of properties, functions, services and advantages that are specifically related to soils (adapted from Potschin and Haines-Young 2011). For a color version of this figure, see www.iste.co.uk/coquet/agricultural.zip*

7.3.3. *Supply, flow and demand*

In recent times, more authors (Villamagna et al. 2013; Burkhard et al. 2014; Schröter et al. 2014; Hein et al. 2016) have started distinguishing between the different facets of the provision of Ecosystem Services, especially (i) supply, (ii) flow and (iii) demand. "Supply" refers to the maximum quantity of Ecosystem Services that a particular ecosystem is able to generate under current use and management conditions (Box 7.1). "Service flow" refers to the quantity of Ecosystem Services that are actually consumed or acknowledged by beneficiaries under actual usage and management conditions for the ecosystem under consideration (Box 7.1). Finally, "demand" refers to the total quantity of services that society would like (Box 7.1). The distinction between supply, flow and demand enables us to clarify the shift from the ecosystem from which the supply is drawn toward the socioeconomic system that benefits from the flow and is also the source of the demand. Furthermore, this distinction also makes it possible to assess the sustainability of these uses or the degree of satisfaction of the demand. These are especially useful for managing ecosystems and supporting public or private policies (Baró et al. 2016). Therefore, comparing supply with flow makes it possible to identify potentially sustainable uses (the flow is smaller than the supply) and unsustainable uses (the flow is greater than the supply). Comparing the flow and demand enables us to identify situations in which society's demand is met (the flow is greater than demand) or unfulfilled (the flow is smaller than the demand). Nonetheless, as with the concept of Ecosystem Services, the definition of these different facets of Ecosystem Services, as well as the vocabulary used for them, is far from universal. Other terms used for "supply" include *capacity* (Villamagna et al. 2013; Schröter et al. 2014) or *potential supply* (Burkhard et al. 2014; Hein et al. 2016). Similarly, *use* and *current supply* are two terms used for service flows (Geijzendorffer et al. 2017). At the same time, other authors use some of these concepts in distinctly different ways. For example, Hein et al. (2016) use *capacity* to denote the maximum level of service use that does not affect future supply across the services provided by the ecosystem.

7.4. Classification of Ecosystem Services

7.4.1. *Classification from the Millennium Ecosystem Assessment*

There are multiple and varied relations between the characteristics of an ecosystem and the well-being of human beings. Several classifications have been proposed, since the early 2000s, that would help provide an exhaustive and precise quantification of services (Costanza et al. 1997b; Daily 1997; Corvalan et al. 2005; Haines-Young and Potschin 2018). One of the best-known classifications is the one developed for the Millennium Ecosystem Assessment framework (Table 7.1). It identifies 21 main types of Ecosystem Services, grouped into four broad categories.

1) *Provisioning services*: these refer to all tangible products from ecosystems that are directly consumed by human societies. In practice, this is biomass (cultivated or otherwise), water, mineral substances, genetic material or energy.

2) *Regulating services*: these are all of the processes and mechanisms that help maintain favorable living conditions and, consequently, regulate the health, safety and comfort of human beings. In concrete terms, these involve regulating the flow of matter in the environment (gas, water, particles and organisms), the quality of different parts of the ecosystem (air, water and soil) and all processes and mechanisms that allow living systems to reproduce, including pollination.

3) *Cultural services*: these refer to all characteristics of the ecosystem that have an effect on the physical and psychological state of human beings. These are organisms, locations (e.g. a lake), or situations (e.g. a sunset) that lead to a physical interaction (e.g. sports), an intellectual interaction (e.g. an esthetic or pedagogic

experience), or a spiritual interaction (e.g. a religious practice) between a human being and their ecosystem.

4) *Supporting services*: these cover all processes required to ensure the provision of the other services, such as soil formation processes and nutrient cycling.

While this classification popularized the concept of Ecosystem Services in academic, political and economic spheres as well as among the lay public, and although it is still widely used, it is also highly criticized. For one, considered as intermediary services that generated other service categories, supporting services were associated with a double-accounting risk and were slowly abandoned in favor of the other three service categories, which are sometimes called "Final Services" (Boyd and Banzhaf 2007; Potschin and Haines-Young 2011; Nahlik et al. 2012; Potschin-Young et al. 2018). The definitions used for cultural services were also criticized, chiefly because of how imprecise they were when compared to the definitions of other service categories (Boyd and Banzhaf 2007; Costanza et al. 2017). Finally, the definitions for the different services are not completely consistent with each other or even with the definition of the concept of services. Consequently, although the definition of provisioning services refers most often to tangible products from ecosystems, such as cultivated biomass, water or energy, the definition of regulating services refers to ecological processes such as water purification or climate regulation (Boyd and Banzhaf 2007; Nahlik et al. 2012).

CICES v5.1		MEA (2005)
Code	**Class**	
Biotic provisioning services		
1.1.1.1	Cultivated terrestrial plants grown for nutritional purposes	Fibers, biochemicals and ornamental resources
1.1.1.2	Materials from cultivated plants for direct use or processing	Fibers and biochemicals
1.1.1.3	Cultivated plants grown as a source of energy	Fuel
1.1.2.1	Cultivated aquatic plants grown for nutritional purposes	Food
1.1.2.2	Materials from cultivated aquatic plants for direct use or processing	Fibers and biochemicals
1.1.2.3	Cultivated aquatic plants grown as a source of energy	Fuel
1.1.3.1	Animals reared for nutritional purposes	Food
1.1.3.2	Materials from reared animals for direct use or processing	Fibers and biochemicals

1.1.3.3	Animals reared to provide energy	Fuel
1.1.4.1	Animals reared by in situ aquaculture for nutritional purposes	Food
1.1.4.2	Materials from animals reared by in situ aquaculture for direct use or processing	Fibers, biochemicals and ornamental resources
1.1.4.3	Animals reared by in situ aquaculture as an energy source	Fuel
1.1.5.1	Wild plants used for nutritional purposes	Food
1.1.5.2	Materials from wild plants for direct use or processing	Fibers, biochemicals and ornamental resources
1.1.5.3	Wild plants used as a source of energy	Fuel
1.1.6.1	Wild animals used for nutritional purposes	Food
1.1.6.2	Materials from wild animals for direct use or processing	Fibers, biochemicals and ornamental resources
1.1.6.3	Wild animals used as a source of energy	Fuel
1.2.1.1	Seeds and other plant materials collected for maintaining or establishing populations	Genetic resources
1.2.1.2	Whole higher or lower plants used to breed new strains or varieties	Genetic resources
1.2.1.3	Individual genes extracted from plants	Genetic resources
1.2.2.1	Animals collected for the purposes of maintaining or establishing a population	Genetic resources
1.2.2.2	Wild animals used to breed new strains or varieties	Genetic resources
1.2.2.3	Individual genes extracted from organisms	Genetic resources
Biotic regulation and maintenance services		
2.1.1.1	Bio-remediation of wastes or toxic substances	Air and water purification
2.1.1.2	Filtration/sequestration/storage/accumulation of wastes or toxic substances	Air and water purification
2.1.2.1	Smell reduction	Air and water purification
2.1.2.2	Noise reduction	
2.1.2.3	Visual screening	
2.2.1.1	Control of erosion rates	Erosion control
2.2.1.2	Buffering and attenuation of mass movement	Storm protection
2.2.1.3	Hydrological cycle and water flow regulation	Water regulation
2.2.1.4	Wind protection	Storm protection
2.2.1.5	Fire protection	

2.2.2.1	Pollination	Pollination
2.2.2.2	Seed dispersal	
2.2.2.3	Maintaining nursery populations and habitats	
2.2.3.1	Pest control	Biological control
2.2.3.2	Disease control	Regulation of human disease
2.2.4.1	Weathering processes	Soil formation
2.2.4.2	Decomposition and fixing processes	Soil formation
2.2.5.1	Regulation of the chemical condition of freshwaters	Water purification
2.2.5.2	Regulation of the chemical condition of salt waters	Water purification
2.2.6.1	Regulation of the chemical condition of atmosphere and oceans	Water purification
2.2.6.2	Regulation of temperature and humidity	(Local) climate regulation
Biotic cultural services		
3.1.1.1	Active or immersive interactions with natural environment	Recreation and ecotourism
3.1.1.2	Passive or observational interactions with natural environment	Recreation and ecotourism
3.1.2.1	Creation of traditional ecological knowledge	Knowledge systems
3.1.2.2	Education and training	Educational values
3.1.2.3	Cultural resonance	Cultural heritage values and cultural diversity
3.1.2.4	Aesthetic experiences	Aesthetic values
3.2.1.1	Symbolic meaning	Sense of place
3.2.1.2	Sacred or religious meaning	Spiritual and religious values
3.2.1.3	Entertainment and representation	Inspiration
3.2.2.1	Existence value	
3.2.2.2	Bequest value	

Table 7.1. *Classification of biological production services, regulating services and cultural services according to the Common International Classification of Ecosystem Services (CICES v5.1) and the proposed correspondence with the classification from the Millennium Ecosystem Assessment (MEA). The names of the services in the CICES and MEA classifications have been simplified*

7.4.2. *The CICES classification*

The Common International Classification of Ecosystem Services was developed by the European Environmental Agency in the context of the revision of the integrated economic and environmental accounting system (SEEA), led by the statistics division of the United Nations. The objective was to construct a standardized classification of Ecosystem Services (Table 7.1). Initially published in 2009 and first revised in 2013 (V4.3), followed by a second revision in 2018 (V5.1), the classification now includes almost a hundred Ecosystem Services that have been systematically defined on the basis of (i) an *ecological clause*, which describes what is produced by the ecosystem, and (ii) a *usage clause*, which specifies the benefit that is derived from this (Haines-Young and Potschin 2018). For example, the production service "plants grown for food", is defined as the contribution of ecosystems to the growth of cultivated plants (ecological clause) that can be harvested and used as a primary material to produce food (usage clause). Similarly, the erosion regulating service is defined as the decrease in loss of matter thanks to the stabilizing effect of plants and animals (ecological cause), which mitigates or prevents damage to human utilization of the environment or to the health and safety of humans (usage clause). These services are classified into sections, divisions, groups and classes of services (Haines-Young and Potschin 2018). Sections correspond to the three broad categories of services (production, regulating and maintenance and cultural services) with the category of supporting services being partially integrated into the category of regulating and maintenance services, and partially abandoned, to avoid the problem of double-accounting (Haines-Young and Potschin 2018). *Divisions* differentiate between products and types of interactions from, or with, ecosystems, whether these are biomass, water, or other biotic or abiotic substances (Haines-Young and Potschin 2018). *Groups* detail the types of products or interactions that are used or acknowledged by human beings, such as the biomass from cultivated terrestrial plants, the biomass from livestock or the biomass from wild animals (Haines-Young and Potschin 2018). Finally, *classes* specify the use of these different resources, such as using cultivated, terrestrial plants for food, production of energy or production of fibers (Table 7.1). Compared to most existing classifications, which largely focus on the contribution made by biodiversity to the well-being of human populations (Baveye et al. 2018), the CICES classification is notably characterized by the recognition and classification of services supplied by the abiotic components of ecosystems. This initial acknowledgment of abiotic services led to criticism, both from the promoters of abiotic services as well as those who were dismissive of the concept. The promoters considered that this acknowledgement was still insufficient, as seen in the asymmetries in status, and in the number and accuracy of the definitions of biotic and abiotic services (Gray 2018). The sceptics dismissed the recognition of abiotic services because they thought that it was an implicit challenge to the concept of Ecosystem Services, which were essentially related to biodiversity and, therefore, to living things.

While most existing classifications are largely similar, especially since they all recognize provisioning, regulating and cultural services (Costanza et al. 2017; Englund et al. 2017), there are many differences in the finer details (Table 7.1). Thus, considering its hierarchical and highly detailed structure, several services given in the CICES classification have no equivalent in earlier classifications (Englund et al. 2017; Haines-Young and Potschin 2018). When there are equivalents, they are often complex, multiple and crossed (Englund et al. 2017; Haines-Young and Potschin 2018). Finally, using identical vocabulary may mask the fact that there are very different realities, for instance, the difference highlighted in the definitions of the biomass production service. In an agroecosystem, this may be considered to be the total biomass co-produced by the ecosystem's own functioning and human interventions, or it may be considered to be the contribution of the ecosystem's functioning to the growth of a plant that may be harvested. Although the development of the early classifications of Ecosystem Services was a key factor in popularizing the concept, a consistent and harmonized classification is yet to be created, despite the major advances already made. This is a major challenge, especially as concerns the classification of cultural services (Costanza et al. 2017).

7.5. The biophysical quantification of Ecosystem Services

The cascade model of Ecosystem Services (Figure 7.2) made it possible to showcase the production chain connecting ecosystem characteristics to the well-being of human populations. Biophysical quantification approaches can, more or less easily, enable the measurement of characteristics of ecosystems, their functions, the supply or the flow of services, that is, all of the factors that are highly dependent on the ecosystem. On the other hand, the quantification of all factors arising from the socioeconomic system, including benefits, value or demand, require socioeconomic and monetary quantification approaches (Gómez-Baggethun et al. 2016), which will not be discussed here. The quantification of any Ecosystem Service comes down to two essential questions: (1) what is to be quantified and (2) how can it be quantified?

7.5.1. *What is to be quantified?*

7.5.1.1. *Indicators for Ecosystem Services*

The first step in the quantification of an Ecosystem Service consists of choosing a variable that is assumed to represent the service under consideration and to indicate its level. This variable is called the indicator variable, or simply "indicator" (Box 7.1). The choice of indicator depends on multiple criteria that have recently been summarized and classified into four broad groups: (i) credibility, (ii) salience, (iii) legitimacy and (iv) feasibility (van Oudenhoven et al. 2018).

Credibility corresponds to the user's confidence in the indicator and the information that this indicator aims to provide. The credibility of an indicator is first assessed by its ability to accurately represent the target object. In the context of the biophysical quantification of Ecosystem Services, the question of what is to be assessed arises: is it the ecosystem condition, its functions, its supply or the flow of services? The conditions for the credibility of an indicator are as follows: it is embedded in a conceptual framework such as the cascade model (Figure 7.2); scientific backing from experts or in literature; and an assessment of the indicator based on data from reliable and verified sources (van Oudenhoven et al. 2018; Rendon et al. 2022).

Salience refers to the relevance of the information conveyed by the indicator with respect to the user's needs. Therefore, the indicator must be closely associated with specific objectives, while also being easily accessible to most stakeholders (e.g. public and private decision-makers, scientists and beneficiaries), which may prove to be contradictory conditions. The indicators must also be sensitive to variations over time and space, to allow us to track across time and to carry out comparisons between different areas. Finally, in an ideal situation, the indicators must be usable across multiple scales of time and space, for instance, allowing comparisons between local and national assessments.

The *legitimacy* of an indicator allows it to be accepted by different stakeholders and allows it to be integrated into a decision-making process. Selecting the indicators through a participative and inclusive process that involves all stakeholders makes it possible to give further legitimacy to the indicators (van Oudenhoven et al. 2018).

Finally, *feasibility* refers to the possibility of tracking the indicator across time and space. In practice, this criterion refers to whether the indicator or the base variables required to compute the indicator are available in existing databases. More generally, the human, material or financial resources required to measure the selected indicator must remain sustainable.

Unfortunately, it is not always easy, or even possible, to satisfy all of these criteria. Thus, choosing indicators that are simple, comprehensible even to the uninitiated, and which already exist in databases, most often results in choosing indicators with reduced scientific or operational pertinence (Schröter et al. 2015; van

Oudenhoven et al. 2018). Conversely, scientifically proven indicators, specific to a particular local context are most often unavailable in existing databases, are expensive to estimate, and are difficult to understand (Schröter et al. 2015; van Oudenhoven et al. 2018). In theory, the antagonism between credibility and feasibility does not really pose a problem as long as simple indicators are limited to situations that are not critical, such as raising awareness, and as long as critical uses, such as operational management of natural resources, are based on complex indicators (Schröter et al. 2015). Unfortunately, there is limited availability of data in many parts of the world, especially data that have the appropriate spatial and temporal resolution (Andrew et al. 2015). Furthermore, feasibility is also the main criterion that governs the selection of indicators, so that the selected indicators are more indicative of what has already been measured, or what can be measured, than they are of the factors to be measured in the absolute (Remme et al. 2014; Heink et al. 2016; Czucz et al. 2020) This limitation therefore brings up two difficulties: (i) a very large diversity in the indicators, which restricts comparisons between the different quantifications of Ecosystem Services, and (ii) the selection of simple, or even simplistic, indicators despite these having debatable or even dubious scientific pertinence (Baveye et al. 2016). In their analysis of 427 indicators representing 33 Ecosystem Services, Czucz et al. (2020) observed that around 20% of the indicators corresponded to the quantification of the condition of the ecosystems, 20% corresponded to the quantification of the potential supply of Ecosystem Services, 20% corresponded to the quantification of the flow of services, 15% corresponded to the quantification of benefits derived from the functioning of the ecosystems, while 25% remained ambiguous. Furthermore, provisioning services and cultural services are usually assessed using the benefit indicators, obtained from various statistics available for agricultural or forest productions, water use, or on infrastructure and tourism numbers, while regulating services are assessed based on indicators of conditions, which are better known and easier to measure (Czucz et al. 2020).

7.5.1.2. *Quantification: single or multiple services?*

When we consider assessing Ecosystem Services, whether biophysical or monetary evaluation, the question that comes up is which of these services are to be quantified. Scientific literature proposes assessing individual services most often. However, it also proposes the joint assessment of several services at the same time, creating *service bundles* (Mouchet et al. 2017; Hölting et al. 2019; Obiang Ndong et al. 2021). This kind of simultaneous assessment of multiple Ecosystem Services thus aims to identify synergies (when the level of one service rises, the level of another also rises) or possible antagonisms (when the level of one service increases the level of another drops) (Raudsepp-Hearne et al. 2010; Lee and Lautenbach 2016). The synergies or antagonisms between services may be direct, when the level of one service directly influences the level of another, or indirect, when the level of multiple services are simultaneously influenced by the same underlying ecological process (Bennett et al. 2009). The analysis and understanding of these relations or interactions are likely to be the most useful type of information for decision-makers to make their decisions (Obiang Ndong et al. 2020). Consequently, it is probably most useful to steer quantification efforts further towards multi-service assessment.

7.5.2. *How to quantify services?*

The emergence and dissemination of the concept of Ecosystem Services has been accompanied by the development of any number of approaches for their quantification. These approaches may be classified under five main categories: (i) approaches using direct measurements, (ii) expert-based approaches, (iii) approaches using *proxy variables*, (iv) approaches using modeling and (v) integrated approaches.

7.5.2.1. *Approaches using direct measurements*

Direct measurement approaches consist of directly measuring the levels of one or more Ecosystem Services in the field. These approaches are practically nonexistent in the literature on Ecosystem Services (Baveye et al. 2016) and the rare existing quantifications are most often related to only a few services in highly specific contexts (Chalhoub et al. 2020). Indeed, whether services can even be measured is still very much an open question. While goods, such as the quantity of biomass, of water, or of materials taken from ecosystems, are frequently measured (for example by the direct usage of these resources) and then used as indicators for provisioning services, the measurement of regulating services quickly proved to be a complex task, and cultural services proved virtually impossible to measure in this way. Let us look at an example: while the quantity of eroded soil is frequently measured, we do not directly measure the erosion regulating service, which is more related to the quantity of soil that has not been eroded thanks to the functioning of the ecosystem. Similarly, the measuring regulating services of the quantity of different parts of an ecosystem (air, water, soil, etc.) would involve, in an absolute sense, the measurement of a very large variety of organic contaminants, which is never taken into consideration for one single site. Consequently, Chalhoub et al. (2020) deliberately restricted their measurement of the “filtration of contaminants” to two herbicides, which were chosen based on their behavior in the environment and differentiated application modalities. The quantification of cultural services is even more difficult since it depends not only on the characteristics of the environment but also on local cultural practices (Plieninger et al. 2015; Fish et al. 2016).

Furthermore, when it is possible to directly measure an Ecosystem Service, the question of the spatial and temporal pertinence of this measurement comes up immediately. Indeed, the measurement is most often limited to the specific measurement conditions, which becomes even more problematic when the measurements are complicated to carry out and the supply of the services are themselves variables across time and space. Thus, while the erosion regulating service provided by a buffer strip may easily be measured in particular pedo-climatic and growing conditions, it is basically unrealistic to think of systematically measuring the same service across all of the filter strips within a small, natural area. Despite this, such direct measurements for Ecosystem Services must be developed and generalized across as many services as possible (Baveye et al. 2016). These measurements are essential tools in quantifying the uncertainties associated with the various approaches used to indirectly assess Ecosystem Services (Choquet et al. 2021).

7.5.2.2. *Expert-based approaches*

Expert-based approaches to assess service levels were among the earliest to have been developed and the most widely used (Jacobs et al. 2015; Roche and Campagne 2019). These emerged from a dual problem: (i) the diversity of Ecosystem Services and the complexity in assessing them and (ii) the non-availability of data required to implement biophysical modeling approaches and the difficulty in obtaining these data within a reasonable cost and timeframe (Jacobs et al. 2015). In this context, turning to multidisciplinary panels of experts made it possible to bring together multiple points of view (ecological, socioeconomic, political, etc.) to arrive at the first assessments of many Ecosystem Services, including those that were more complicated to assess, such as cultural services (Jacobs et al. 2015; Roche and Campagne 2019). However, the approaches were criticized for their subjectivity, lack of reproducibility and their semi-quantitative dimension (Baveye et al. 2016). "Expert opinions" depend not only on each expert's own knowledge, but also include biases that are not only linked to the implementation of the method (composition of the panel of experts, scoring scale etc.) but also the individual situation of each expert (function, personal goals, proximity to the situation being studied, behavior) and the external conditions during the exercise (e.g. current political or socioeconomic affairs). For example, it is often more difficult for an expert to modify an assessment that has been proposed previously than it is to carry out their own assessment, even if this study takes longer to implement (Campagne and Roche 2018). Experts may also overestimate levels for services that they know well, or from which they benefit directly, compared to services that are less known or less used (Jacobs et al. 2015). How the experts arrive at a consensus is particularly dependent on the behavior of the experts, some of whom may be exceptionally charismatic, persuasive, voluble or may even play up their position, especially in panels made up of scientific and technical experts as well as “lay” users (Campagne and Roche 2018). Finally, the importance given to one or more services in scientific or media reporting may also influence the results of the expert consultation process (Rutgers et al. 2012). However, a meticulous construction of expert-based approaches can limit some of these biases (Jacobs et al. 2015; Campagne and Roche 2018) and, under some conditions at least, it is possible to obtain assessments that are comparable to those obtained through biophysical modeling, which is often deemed to be more relevant and accurate (Roche and Campagne 2019). An “ideal” panel of experts would consist of 10–20 members, bringing together scientific and technical experts, service users and policy-makers involved in managing these services, and the panel would also focus on diversifying profiles rather than redundancy (Campagne and Roche 2018). In any case, it is useful to complement the estimation of service levels with an assessment of their level of uncertainty and the variability between experts (Campagne and Roche 2018), and then compare expert-based estimations with direct measurements or biophysical modeling approaches (Jacobs et al. 2015; Campagne and Roche 2018).

7.5.2.3. *Proxy variable approaches*

Proxy variable approaches are based on the quantification of one, or sometimes multiple, variables that are assumed to control the level of a given Ecosystem Service (Baveye et al. 2016, Greiner et al. 2017). These approaches involve three main steps: (i) selecting the proxy variables; (ii) choosing a reference situation or defining (empirically) the relationship between the value of the proxy variable and the level of the service being examined and (iii) allocating a weight to each proxy variable used, if there are multiple variables (Hewitt et al. 2015).

The variables used as proxy variables are most often selected by experts (Rutgers et al. 2012; Hewitt et al. 2015). These are simple variables, descriptive of the chief physical, chemical and biological characteristics of ecosystems. The major advantages they offer are that they are available in existing databases (Rutgers et al. 2012) or can be calculated through the combination of existing data, or they are (relatively) inexpensive to measure (Le Clec’h et al. 2016). Considering these selection criteria, proxy variables and Ecosystem Services are generally related through not one, but multiple relationships: several proxy variables can influence the same Ecosystem Service, and several Ecosystem Services may be influenced partially or completely by the same proxy variable. For example, Rutgers et al. (2012) used five variables to assess the service of nutrient retention and release. These included organic carbon concentration, the abundance of earthworms and the pH. In parallel, they used organic carbon concentrations in the assessment of nine Ecosystem Services, including hydrological regulation services.

Given the absence of any quantitative relationships between the proxy variable and the “target” variable (here, the level of the Ecosystem Service), such approaches only allow us to quantify service levels relative to a particular state (Rutgers et al. 2012; van Wijnen et al. 2012) or relative to reference levels (Hewitt et al. 2015). Choosing a relationship between proxy variables and the “target” variable (most often a positive or negative linear correlation) makes it possible to translate the variations in the proxy variable into variations in the level of the Ecosystem Services (Rutgers et al. 2012; van Wijnen et al. 2012; Obriot et al. 2016). Finally, when different proxy variables are used in parallel, a weighting is attributed to each of these variables, either based on expert recommendations (Hewitt et al. 2015), or based on statistical approaches (Obriot et al. 2016).

Approaches of this kind can allow a quick assessment, using little data, of the levels of an Ecosystem Service and its variations over time or space, with respect to a reference state. However, the risk, and amplitude, of the errors associated with these approaches become larger when the relationship between the proxy variable and the service levels are indirect, non-monotonic or nonlinear (Eigenbrod et al. 2010). For example, the concentration of clay particles in soil horizons plays a crucial role in the production of biomass (Adhikari and Hartemink 2016). While moderate clay concentrations are generally conducive to biomass production, very small or very high concentration are generally not favorable. Furthermore, the level of the biomass production service depends not only on the concentration of clay particles, but on various other soil parameters that cannot be captured through the quantification of the clay particle fraction alone.

7.5.2.4. *Modeling approaches*

Models are simplified representations of existing relationships between one or more predictive variables (here, the biophysical or socioeconomic characteristics of ecosystems) and one or more variables that we wish to predict (here, the levels of Ecosystem Services). There is a wide variety of models depending on the relations between the predictive variables and the predicted variables, how the temporal variability between the predictive and predicted variables is taken into account, the existence of a random component in the predictive and/or predicted variables, etc. Of course, this chapter will not provide an exhaustive overview of environmental modeling approaches but will be restricted to two broad families of models that are used to assess Ecosystem Services: empirical models and mechanistic models.

7.5.2.4.1. Approaches using empirical modeling

In empirical modeling approaches, the relationships between the predictive variable and the predicted variables are either <<IF THEN>>, most often established by experts and represented as a decision-tree, or they are quantitative relationships measured experimentally, but without knowing exactly what the causal links are. Empirical modeling approaches have been widely used for a long time in environmental sciences to estimate a large number of variables, such as whether certain soils are suited to agricultural production; the risk of erosion, infiltration and water retention; retention and degradation of pollutants; carbon storage; etc. (Greiner et al. 2017; Choquet et al. 2021). For historical reasons, the vast majority of these models were developed to quantify functions or services related to production and the regulation in cultivated ecosystems. They were more rarely used for functions and services in forest or prairie ecosystems, and almost never used for cultural services. Indeed, the modeling of cultural services is a very recent phenomenon (Greiner et al. 2017; Grêt-Regamey et al. 2017a). Additionally, the vast majority of these models is specific to one function or service. Quantifying multiple Ecosystem Services thus requires the identification and selection of a series of models (Choquet et al. 2021). The risk of incompatibility between models becomes greater with increased diversity in philosophy, architecture or complexity between the different models.

Most of these models were also developed before the emergence of the concept of Ecosystem Services. For example, the Storie Index was used to evaluate the potential for agricultural production (Storie 1978), while the USLE (Universal Soil Loss Equation) model was used to assess the risk of erosion (Wischmeier and Smith 1978). Consequently, the variables being modeled are most often associated with threats to the environment, or some impact on the environment, rather than with services, strictly speaking. This is notably the case with models that predict erosion rather than erosion control by the ecosystem, or which predict the leaching of nitrates rather than their retention. The outputs from these models must therefore be processed (often through inversion) before being interpreted in terms of Ecosystem Services (Guerra et al. 2016; Choquet et al. 2021). Finally, most of these models are also static models (Greiner et al. 2017) in which the main objective is to estimate the predicted variable either at a given instant, or averaged over an annual period. Therefore, climatic conditions, usage modalities or the management of the ecosystem are represented in these models in a highly simplified manner (Greiner et al. 2017). It is therefore not possible to use this type of modeling to model the *current supply* of a service, but only the average supply in average conditions, which is a situation very rarely encountered in reality (Greiner et al. 2017).

7.5.2.4.2. Mechanistic modeling approaches

Mechanistic modeling approaches are based on the mathematical representations of physical, chemical and biological processes that are at work in ecosystems, unlike empirical models, which only describe the cause–and–effect relationships, independent of the underlying mechanisms. These approaches, which came into use recently, are becoming more and more widespread in the assessment of Ecosystem Services (Dominati et al. 2016; Francesconi et al. 2016; Demestihas et al. 2018; Choquet et al. 2021; Ellili-Bargaoui et al. 2021). They are often considered the most reliable and accurate assessment approaches (Bagstad et al. 2013; Greiner et al. 2017). Furthermore, inasmuch as they can explicitly take into account the temporal variability in environmental conditions (climate, soil, etc.), in land use or agricultural or forestry practices, these approaches are particularly popular when it comes to assessing the impact of climate change or land use on the supply of Ecosystem Services.

There are different types of models. The first type includes models that are specific to the various disciplines within environmental sciences. For examples, dynamic carbon models that are frequently used to assess carbon storage and, in fine, the climate regulating service or, again, hydrological models used to quantify water provisioning or regulating services (Francesconi et al. 2016). While models of this type can be used in natural ecosystems as well as in cultivated ecosystems, they can only quantify a single Ecosystem Service.

Crop models are the second type of mechanistic models used to quantify Ecosystem Services. Initially designed to model the flows of water, carbon and nitrogen in the soil-plant-atmosphere continuum, crop models, unlike more specific models, are able to quantify several Ecosystem Services, especially the biomass production service, the climate regulation service through the modeling of carbon storage, the regulation of water flows through the modeling of vertical drainage or, again, the regulating of water quality through the modeling of the leaching of nitrate (Dominati et al. 2016; Choquet et al. 2021; Ellili-Bargaoui et al. 2021). However, by construction, crop models cannot model forest ecosystems, nor can the majority of them model permanent crops, prairies, orchards or market gardening. Within the single domain of field crops alone, they are only useful for pure key crops, almost never for mixed crops, and above all are best-suited to thick, homogeneous and well-drained soils, which represent only a small fraction of the diversity of cultivated soils (Choquet et al. 2021).

The last type of mechanistic model used to model Ecosystem Services are "global" models (*dynamic vegetation global models*, *earth system models*, etc.). As with models for specific disciplines, by construction these are adapted to the modeling of natural and forest ecosystems, and as with crops models, they enable the quantification, in parallel, of all services related to water, carbon and nitrogen cycles (Rabin et al. 2020). On the other hand, when used as tools for modeling the terrestrial system in its entirety, the representation of the different parts of the ecosystem (soil, plants, etc.) is highly simplified. Similarly, models of this type very rarely integrate an explicit representation of agricultural or forestry practices.

Thus, there is a large diversity in mechanistic modeling approaches (and an even larger number of models), and it is possible to make use of and capitalize on the complementary nature of these. Therefore, while an approach that is only limited to cultivated ecosystems will benefit from using a crop model to quantify the largest possible number of Ecosystem Services, quantifying climate regulation using a dynamic carbon model can make it possible to work with complex landscapes. It must be noted that at present there is no single mechanistic modeling tool that can be used for assessments, even if we only consider the principal Ecosystem Services in complex landscapes (agricultural and forest). Furthermore, mechanistic modeling approaches are by far the most data intensive in a data-poor world, especially on a planetary scale. These approaches are also highly complicated to implement without a specific support. It is therefore not always easy to know which assessment approach is ultimately the best performing: expert-based approaches and proxy variable approaches are considered to be subjective and imprecise but can be implemented based on generally available data and skills; mechanistic models, believed to be objective and accurate, are often lacking data and the necessary competencies (Bagstad et al. 2013).

7.5.2.5. *Integrated approaches*

As research on Ecosystem Services progressed, many assessment tools and platforms were developed (Bagstad et al. 2013; Turner et al. 2016; Grêt-Regamey et al. 2017a). It is possible to distinguish between existing tools – around 20 tools identified by Bagstad et al. (2013) and around 60 tools identified by Grêt-Regamey et al. (2017a) – based on several criteria. First of all, although some tools allow for the assessment of a more or less large range of Ecosystem Services in parallel, especially provisioning and regulating services, others are specific to a single service (Grêt-Regamey et al. 2017a). Similarly, some tools enable the assessment of the biophysical value (e.g. EnviroAtlas (Pickard et al. 2015)), the monetary value (e.g. Benefit Transfer Toolkit (Loomis et al. 2008)), the social and cultural value (e.g. SolVES (Sherrouse et al. 2011)), or a combination of the different values attributed to services (e.g. MIMES (Boumans et al. 2015)). While certain tools do not include an explicit spatial dimension (e.g. Soil Navigator (Debeljak et al. 2019)), others allow users to not only assess but also map Ecosystem Services over different local or global scales (e.g. LUCI (Jackson et al. 2013)). The assessment of Ecosystem Services by these tools may rely on a wide variety of approaches (Turner et al. 2016) including participative approaches (e.g. SolVES (Sherrouse et al. 2011)), approaches that generalize data from literature (e.g. Benefit Transfer Toolkit (Loomis et al. 2008)), empirical modeling approaches (e.g. Artificial Intelligence for

Ecosystem Services (ARIES) (Villa et al. 2014)) or mechanistic modeling approaches (e.g. Integrated Valuation of Ecosystem Services (InVEST) (Natural Capital Project 2022)). Depending on the approaches that are used, some tools may be very simple (spreadsheet, a small GIS application, etc.), while others contain complex informatic models. Finally, while many of these tools are free to use and often well-documented, a certain number of them rely on the use of proprietary software that is often less well-documented (Bagstad et al. 2013). InVEST (Natural Capital Project 2022) and, to a lesser degree, ARIES (Villa et al. 2014) are among the best-known and most widely used multi-service tools today (Grêt-Regamey et al. 2017a).

One advantage of the large variety of models that are currently available is that for any specific situation it is possible to offer a turnkey solution that is adapted in terms of the range of services, output type (biophysical, economic and multiple types), spatial cover and availability of technical skills or data. However, these different options are not equitably represented. As a result, regulating and provisioning services are more often assessed than cultural services (Grêt-Regamey et al. 2017a). Within cultural services, physical interactions in the context of recreational physical activities are integrated into more tools than esthetic or spiritual interactions. The biophysical and monetary values of services are given much greater importance compared to social or cultural value (Bagstad et al. 2013; Grêt-Regamey et al. 2017b). Although the majority of tools is adapted for the regional scale, only a few allow global assessments. Finally, at best, half of currently available tools indicate the uncertainties associated with the assessment and mapping results, and very few of these have gone through a validation procedure (Bagstad et al. 2013; Grêt-Regamey et al. 2017a). Therefore, the consequences of the choice of a particular tool must be anticipated. These consequences simultaneously impact the number and type of services considered, their conceptualization, the type and form of assessment and the confidence that can be attributed to the assessments that are produced.

7.5.2.6. *Conclusion*

There is a large variety of approaches for assessing Ecosystem Services (and multiple methods and tools within each family of approaches) from expert-based approaches to the use of integrated platforms with tiered approaches designed to make the most of their complementary nature (Grêt-Regamey et al. 2017b). Thus, in an initial approach, a quick expert-based approach can identify the services provided and give a rough assessment of these. Assuming that data and resources (technical and human) are not limiting factors, biophysical modeling can then refine the levels of the services provided, improve the spatial precision of their mapping and assess the impact of different management scenarios on the supply of all or part of the Ecosystem Services identified in the preceding step (Bagstad et al. 2013; Grêt-Regamey et al. 2017b). However, while these different approaches and methodologies are capable of generating results that are relatively consistent with each other, they are often difficult to compare (Bagstad et al. 2013; Vorstius and Spray 2015; Roche and Campagne 2019; Choquet et al. 2021). In effect, these different methods are based on variable conceptualizations, indicators or units. Thus, when it comes to arbitrating between services or beneficiaries, the diversity of approaches (and in the obtained results) can become a handicap, reducing stakeholder confidence. Disputes between actors become even more problematic when there is any information on the uncertainties in different approaches and when many of them have little or no scientific validation. This is why the direct measurements of Ecosystem Services (Chalhoub et al. 2020) are so appealing: they are essential for the construction, calibration and validation of indirect approaches (Baveye et al. 2016; Baveye 2017).

7.6. Mapping Ecosystem Services

Research on Ecosystem Services recognized its explicitly spatial dimension early on, whether it involved (i) recognizing that the demand, production and direct or indirect use of Ecosystem Services is not localized to any point in space, but present specific spatial distributions, involving possible spatial shifts between these different dimensions; (ii) identifying and characterizing spatial relations (synergy, compromise, independence) that exist between the different services; or (iii) locating *hotspots* and *cold spots* in demand, supply, or use of Ecosystem Services (Crossman et al. 2013; Schulp et al. 2014; Englund et al. 2017; Fossey et al. 2020). This explicitly spatial dimension of Ecosystem Services and the associated challenges has led to many projects that aimed at mapping Ecosystem Services and analyzing their spatial distribution. These maps serve multiple purposes, from the representation and visualization of information in order to raise awareness to operational use in land use planning (Andrew et al. 2015; Englund et al. 2017).

Nevertheless, it is not an easy task to spatialize Ecosystem Services. The ecological processes underlying the supply of services, as well as the socioeconomic and political dynamics underlying the demand for, and use of, these services, take place on (and across) different spatial scales (Raudsepp-Hearne and Peterson 2016; Lavorel et al. 2017; Xu et al. 2017). Furthermore, the data required to map Ecosystem Services is not always available at this scale or with a spatial resolution that is adequate to fulfil the objective of the mapping. Thus, there are constraints and compromises in the choice of mapping methods and products, the impacts of which are not yet well studied (Grêt-Regamey et al. 2014; Bagstad et al. 2018; Scammacca et al. 2023). Mapping often complements the quantification process, and like quantification, two main questions arise: (1) what is to be mapped and (2) how is the mapping to be carried out?

7.6.1. *What is to be mapped?*

As in the assessment phase, the first question leads to the selection of an indicator for Ecosystem Services and also to the question of scale or, more precisely, (i) the spatial extent represented by the map; (ii) the scale of the cartographic representation (in other words, the ratio between distances on the map and the actual distances); and finally (iii) the spatial grain, that is, the smallest spatial entity that is represented (Grêt-Regamey et al. 2014) (see section 3.4.1). Extent, scale and grain are not completely independent – any increase in the mapped extent most often involves a smaller scale and coarser spatial grain (Herold et al. 2002; Gómez-Zotano et al. 2018). Consequently, while Ecosystem Services can be mapped across different areas, from the highly local to the global, and at multiple spatial resolutions, the majority of existing maps cover local extents (10^2–10^3 km^2) and regional extents (10^3–10^5 km^2) with spatial resolutions being of the order of 0.01 km^2. This is related to the greater availability of the data used for mapping services across these extents and at these resolutions (Martínez-Harms and Balvanera 2012; Malinga et al. 2015; Englund et al. 2017). However, thanks to remote-sensing and the use of digital soil mapping, there has been an ever-increasing availability of multiple spatial data (climate, topography, hydrography, land use, soil types and properties, etc.) at a global level, with fine resolution (smaller than 0.01 km^2). Consequently, the dependencies between extent, scale and grain have become somewhat looser and allowed the creation of continental or global Ecosystem Services maps at a relatively fine resolution (Malinga et al. 2015).

The choice of whether to map supply, use or demand of services becomes all the more important given that these different elements usually have specific spatial distributions (Baró et al. 2016). However, it must also be observed that the vast majority of projects today focus on mapping supply (Czucz et al. 2020).

7.6.2. *How to map?*

There are many approaches to mapping Ecosystem Services (Martínez-Harms and Balvanera 2012; Andrew et al. 2015; Malinga et al. 2015; Englund et al. 2017).

These can be classified based on several criteria (Figure 7.3) depending on (i) the existence of measurements for the Ecosystem Services, (ii) whether these measurements fall within or outside the zone to be mapped, (iii) the basic spatial entity used to represent the Ecosystem Services and finally (iv) the type of relationship between the predictive variables and the service indicator to be predicted (Martínez-Harms and Balvanera 2012).

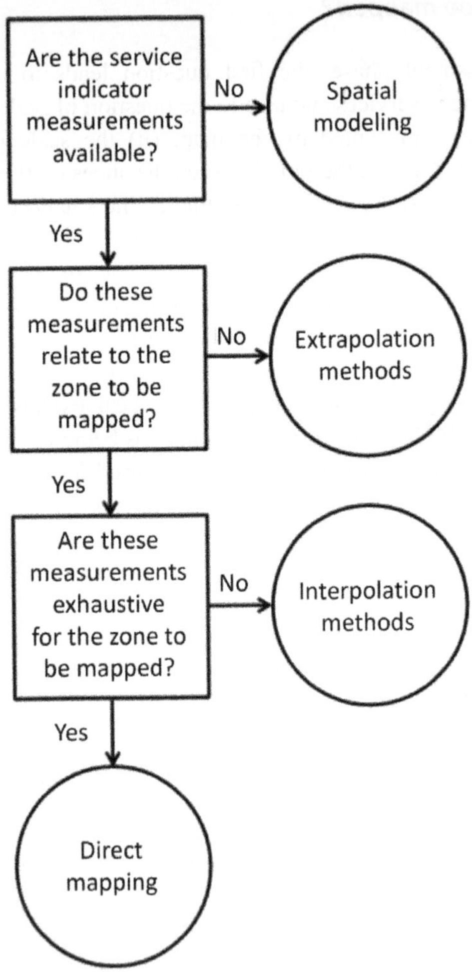

Figure 7.3. *Decision-tree to choose a mapping strategy for Ecosystem Services based on the availability of data*

7.6.2.1. *Direct mapping*

In a few, relatively rare cases, the chosen indicator for the Ecosystem Service is directly measured and is available in databases that cover the whole region to be mapped, or it can be directly calculated from a combination of various data that are available. For instance, the mapping of the biomass production service can be carried out based on agricultural and forest production statistics (Hölting et al. 2019). Mapping the climate regulating service is based on a computation of organic carbon stored in soils using data on the soil type, its bulk density, the organic carbon concentrations and the coarse elements in different horizons (Eigenbrod et al. 2010; Makovníková et al. 2018), and the mapping of some cultural services may be based on the location or frequentation of touristic places (Eigenbrod et al. 2010). An advantage, as well as a limitation, of these direct mapping approaches is that they use data that are already available. Such methods are relatively simple and quick to implement. The quality of the data, which are most often obtained through institutional environmental monitoring programs or reporting (Andrew et al. 2015), is recognized and documented. Depending on the acquisition frequency and temporal depth, using this approach may also make it possible to track the evolution of the supply of Ecosystem Services across time, including retrospectively. Nevertheless, using only available data highly restricts the choice of indicator selected to represent the service, or the spatial and temporal resolution of the mapped products. Indeed, using available data often leads to prioritizing state indicators for ecosystems (especially for regulating services) or indicators of goods or benefits that include a large contribution from the economic and social system (especially for production services and cultural services) as a substitute for the (potential or actual) supply of services (Heink et al. 2016; Czucz et al. 2020). In terms of spatial entity, data such as statistics on agricultural or forest production are often aggregated at the level of the administrative entity (Maes et al. 2011; Hölting et al. 2019; Mayer et al. 2021), which is not functionally very useful, and which may prove to be too coarse to characterize the spatial distribution of services (Andrew et al. 2015). Finally, in temporal terms, while certain data are annually updated (e.g. certain statistics on agricultural and forest production), others are updated less regularly. For instance, the Corine Land Cover database, on land use, is updated every 4 years, while others are very rarely updated (e.g. data on soils).

7.6.2.2. *Extrapolation methods*

Extrapolation methods are among the most widely used, especially for mapping the monetary value of Ecosystem Services, as they are relatively simple to implement. They are based on extrapolating the value (biophysical or monetary) of an Ecosystem Service over an entire, predefined spatial entity. In the majority of situations, the basic spatial entities are classes of land use or land cover, however they may also integrate climatic, topographic or pedological information (Jacobs et al. 2015). The values attributed to each cartographic unit are either directly taken from literature (Costanza et al. 1997b) or obtained from experts (Burkhard et al. 2009; Jacobs et al. 2015; Roche and Campagne 2019). The global map of the monetary value of Ecosystem Services proposed by Costanza et al. (1997b) was thus based on the generalization of well-documented monetization of Ecosystem Services in model ecosystems, and the generalization of these values to all similar ecosystems on a global scale. These methods, called *look-up tables* or *matrix approaches*, offer many advantages. They require little data, and most of these data, especially land cover and land use data, are widely available (Andrew et al. 2015). They are also highly integrative, allowing the simultaneous and coherent mapping of a wide range of Ecosystem Services, including cultural services, in the most diverse ecosystems, or again, the supply and demand of services (Burkhard et al. 2012). Finally, the results can be easily understood when stakeholders are included in the process of selecting the indicators and the values to be extrapolated (Burkhard et al. 2012; Campagne and Roche 2018). However, they are also subject to criticism. First of all, these methods generate generalization errors. These are related to the hypotheses made regarding the representativeness of the extrapolated value, and the spatial and temporal invariance of the value of services within each cartographic unit (Eigenbrod et al. 2010; Andrew et al. 2015; Jacobs et al. 2015; Campagne and Roche 2018). Furthermore, when the extrapolated values are obtained from expert-based estimations, they are highly dependent on the modalities involved in collecting these opinions and are only semi-quantitative. Such maps are therefore difficult to validate or even compare with other kinds of maps (Campagne and Roche 2018).

7.6.2.3. *Interpolation methods*

This family of methods is based on the existence of a set of measurements (often point measurements and limited in number) of the indicator to be mapped within the region to be mapped. It thus involves using available data on the variable to be predicted and, if required, on a set of secondary predictive variables that are better known (so-called covariates) in order to construct and train a model that can estimate the value of the indicator at any point in the zone of study. This family comprises of a wide variety of methods (e.g. see section 3.4.3.1). The construction of the model can be based on the single variable to be predicted or on the variable to be predicted and a set of covariates. The spatial structure of the variables may be taken into account by the model (geostatistical model) or may not be taken into account (statistical model). The model may also be a deterministic model or a probabilistic one. Finally, the covariates may or may not be spatially exhaustive over the zone to be mapped. These mapping methods depend on the preexisting availability of Ecosystem Service measurements and are quite rare (Sumarga and Hein 2014; Ungaro et al. 2014). On the other hand, they present the advantage of allowing a systematic mapping of the uncertainties associated with predictions (Ungaro et al. 2014).

7.6.2.4. *Spatial modeling*

Spatial models do not require preexisting measurements of Ecosystem Services, except for calibration or validation. Ecosystem Services are therefore estimated by the modeling at every point in the region to be mapped based on the predictive variables that describe the biophysical characteristics (e.g. climate, topographic, hydrographic or pedological data) or the socioeconomic characteristics (land cover, land use and land management) of the areas to be mapped (Andrew et al. 2015). These methods are most often based on two more or less consecutive steps: (i) first dividing the zone to be mapped into elementary spatial entities and (ii) the assessment of Ecosystem Services in each elementary spatial entity.

Elementary spatial entities may correspond to the division of the zone to be mapped into a regular square mesh, whose surface, that is, the grain or resolution of the final map of Ecosystem Services, is most often defined based on the resolution of the data used to assess the Ecosystem Services (Schulp et al. 2012; Guerra et al. 2016). In other cases, the elementary spatial entities of the model are obtained by dividing the zone to be mapped based on climatic, topographic, hydrological and pedological conditions, or on land cover and land use conditions (Stürck et al. 2014; Perez-Soba et al. 2016; Obiang Ndong et al. 2021). The Ecosystem Service assessment step therefore relies on using a model that can estimate the service indicator based on predictive variables. While certain models are truly spatial models (Francesconi et al. 2016; Guerra et al. 2016; Marques et al. 2021), in the sense that they model the relations between different elementary spatial units (e.g. neighborhood relations), the vast majority are aspatial models, which do not include the explicit relationships between the different elementary spatial units of the model (Obiang Ndong et al. 2020). In the latter case, the mapping is then obtained through a repetition, for each elementary spatial unit, of one of the service modeling approaches described in section 7.5.2.4. The models used to map Ecosystem Services can thus be expert-based models (Choquet et al. 2021; Rabot et al. 2022; Scammacca et al. 2023), statistical models (Choquet et al. 2021; Rabot et al. 2022; Ros et al. 2022) or mechanistic models (Lugato et al. 2014; Choquet et al. 2021; Obiang Ndong et al. 2021).

Using spatial-modeling approaches to map services is generally considered to be preferable to extrapolation methods (Martínez-Harms and Balvanera 2012; Andrew et al. 2015; Englund et al. 2017). The spatial-modeling approaches make it possible to quantitatively relate the different Ecosystem Services to the key biophysical variables that determine the spatial distributions of these services, and which can play key roles in the territorial management of Ecosystem Services. However, they require a lot of data (most of which are often unavailable), many specific technical skills, and are limited to the few rare Ecosystem Services for which models are already available. Furthermore, as these are often implemented in the absence of any direct measurement of the indicators used, these methods are almost never validated (Schulp et al. 2014; Englund et al. 2017). Finally, even when the models used have been calibrated and validated beforehand, the performance of these approaches is still dependent on the quality of the input data, especially in terms of spatial coverage and resolution.

7.6.2.5. *Conclusion*

As with the quantification step, there is a wide variety of approaches used to map Ecosystem Services, from relatively easy and quick methods such as approaches using look-up tables based on land cover and land use data to the spatial modeling approach using spatial models. It is possible to leverage this diversity to tailor the mapping strategy to a specific goal and to specific human or technical resources (Grêt-Regamey et al. 2017b). However, compared to the quantification step, mapping poses some specific problems. In the first place, there is the difficulty of the availability of service indicators or the predictive variables required to compute the indicators, and this difficulty grows with the increase in the extent to be mapped and the finesse of the desired spatial precision. Consequently, the mapping of Ecosystem Services relies, at least in the short term, only on available data, whose nature or spatial precision limits, among other things, the possibility of mapping Ecosystem Services, the type of Ecosystem Services that can be mapped and the use of the maps that are produced. While statistics may be available on a national, pan-European, or international scale to map all or part of provisioning or regulating (especially climate) services, this is not the case for most other services. Furthermore, even when the data exist, its spatial precision is not always adequate, at least for certain objectives. For example, in their study on the impact of the spatial precision of soil data used to map Ecosystem Services, Scammacca et al. (2023) estimated that traditionally available soil data allowed the mapping of the spatial distribution of Ecosystem Services, over the territory being studied with a spatial grain of the order of around 10 km^2, which is potentially sufficient for land-use planning at a regional or national level, but too coarse to be used for local land-use planning. In this context, recent developments in digital soil mapping approaches that attempt to provide exhaustive spatial data over large areas with fine spatial resolution could bring comprehensiveness and the spatial precision of the data. Although at present these new mapping products only concern a limited number of variables that describe the physical, chemical or biological state of ecosystems (Mulder et al. 2016; Poggio et al. 2021), nothing prevents us from considering that such products could eventually include indicators of ecosystem functions or services. In any case, the physical, chemical or biological state variables are variables that can already be used to quantify and map Ecosystem Services.

7.7. Applying the concept of Ecosystem Services to the soil component of ecosystems

7.7.1. *Soils have largely been forgotten about in ecosystems approaches*

Although the role of soils in supplying Ecosystem Services has been explicitly considered by certain precursors of the concept of Ecosystem Services (notably Daily (1997)), they were most often neglected or only partially integrated into most conceptual frameworks and classifications of Ecosystem Services (Dominati et al. 2010; Baveye et al. 2016). Soils are therefore primarily considered as one of the elements in an ecosystem whose formation is assured by the functioning of the ecosystem (Millennium Ecosystem Assessment 2005), or whose quality over time must be maintained by the functioning of the ecosystem (Haines-Young and Potschin 2018). Consequently, soils are most often integrated into the self-maintaining function (Millennium Ecosystem Assessment 2005) and are considered as one of the products of the functioning of the ecosystem rather than as a component that directly contributes to the provision of a service. Additionally, while some of the direct contributions from soils and soil properties can be integrated into services such as the biomass production service, water production service, or the service of regulating hydrological or erosion control, other services provided by soils, such as the control of compaction, are absent in most classifications, including the most recent, exhaustive and harmonized classifications such as the latest version of CICES (Paul et al. 2021). Furthermore, in the rare cases where soils and their properties are assumed to contribute to the supply of Ecosystem Services, their role is limited to the contribution of their biological component. The erosion control is thus considered to be an essentially biotic service in the CICES classification, and is defined as the reduction of soil loss related to the stabilizing action of living organisms (Haines-Young and Potschin 2018). While this definition emphasizes on the stabilizing role of vegetation and fauna in the soil, it neglects the role played by abiotic components in the soils in regulating erosion, such as the granulometric distribution, or the mineralogical nature of the soil constituents (Adhikari and Hartemink 2016). In this sense, the integration of “abiotic” services in the latest version of the CICES classification can be thought of as a step forward in considering soils as direct contributors to different provisioning, regulating or cultural services. At the same time, abiotic services are far from being defined as precisely as biotic services (Paul et al. 2021), thereby limiting the possibility of integrating the contribution of the abiotic component of soils. At a more fundamental level, while differentiating between biotic and abiotic services may make sense in order to distinguish the contribution of the living part of ecosystems from the abiotic components (water, wind, solar energy, etc.), this distinction is not very functional in the case of soils. Soils represent an interface between the biotic and abiotic components of ecosystems, and their properties are dependent on a variety of inextricably linked abiotic and biotic properties. Let us take the example of soil structure: this depends on the physical and chemical properties of the constituents of the soil as much as it depends on the activity of multiple living organisms in the soil and, in turn, modulates the supply of multiple provisioning or regulating services (Adhikari and Hartemink 2016). Finally, even when it seems possible to integrate at least part of the contribution that soils make to the supply of abiotic Ecosystem Services, that is, abiotic cultural services, the very broad definition of these services as "all natural, abiotic elements that carry a cultural dimension" does not make it possible to distinguish between the specific contribution of soils from that of geological or geomorphological factors, which are in fact cited much more frequently (Haines-Young and Potschin 2018).

Two additional reasons may contribute to the limited recognition of the role played by soils and, more generally, the abiotic components of ecosystems. The first, which is a conceptual problem, goes back to the very concept of Ecosystem Services, which was initially created to quantify the value brought in by biodiversity and promote engagement that would protect biodiversity. In this context, the climatic, geological and pedological components of ecosystems were simply factors that modulated the ecosystem's biotic components, which alone contributed to the direct supply of services (Baveye et al. 2016). The second reason, a practical one, is related to the low availability of data and tools that allow us to assess the impact of the spatial variability of soils and their properties on Ecosystem Services, especially when contrasted with extrapolation approaches based on land use data. The soil science community's contribution is thus focused on (i) recognizing and conceptualizing the role of soils, with their biotic and abiotic components, as direct contributors to the supply of Ecosystem Services, (ii) the definition and classification of services related to soils and finally (iii) the development and evaluation of tools that can assess the contributions of soils to multiple services.

7.7.2. *New conceptual frameworks to reposition soils within the cascade model*

While the origin and recent development of the concept of Ecosystem Services go back to the early 2000s, it took another decade for the emergence of the first conceptual frameworks adapted to soils (Robinson et al. 2009; Dominati et al. 2010). What is required therefore is a radical repositioning of the place of soil within the production chain of Ecosystem Services. We need to move from a reductionist view of soils limited to the benefits produced by different self-maintaining services (Millennium Ecosystem Assessment 2005) or regulating services (Haines-Young and Potschin 2018) toward a more holistic vision in which soils are both a product of the functioning of the ecosystem as well as a component of the "ecosystem conditions" (Figure 7.2) contributing directly to a large part of Ecosystem Services and the well-being of human societies (Robinson et al. 2009; Dominati et al. 2010). Analogous to the definition of natural capital (Costanza et al. 1997b), soil is conceptualized as stocks (of solid, liquid, gas, energy and organization) that govern matter and energy flows, and the services are the direct contributions that these make to human well-being (Robinson et al. 2009; Dominati et al. 2010). In the logical continuation of these preliminary projects, other projects sought to provide information on the different categories of the conceptual cascade model, linking properties, processes, functions, services and benefits (Adhikari et Hartemink 2016; Greiner et al. 2017; Paul et al. 2021). Similar to what may be observed with Ecosystem Services in general, there is no harmonized definition even today for these different categories and the boundaries between them are still intensely debated (Adhikari and Hartemink 2016; Jónsson and Davíðsdóttir 2016; Paul et al. 2021).

7.7.3. *What definitions can be used for Ecosystem Services related to or obtained from soils?*

In literature, we find many terms that refer to services related to the functioning of soils, including soil services, soil-provided Ecosystem Services, soil-related Ecosystem Services, soil-based Ecosystem Services or, again, soil-supported Ecosystem Services. These different terms express the coexistence of at least two conceptions of these services, in which the soils and their properties occupy a different role in the production chain of Ecosystem Services (Bartkowski et al. 2018).

The first of these, with terms such as "soil services", or "soil-provided Ecosystem Services", suggests that the soil is the true producer of Ecosystem Services (Bartkowski et al. 2018; Fossey et al. 2020) at least a part of which can be distinct from, and added to, the Ecosystem Services provided by other components of the ecosystem. In the second conceptualization, the terms "soil-related ecosystem services" or "soil-based Ecosystem Services" suggest that the soils, in combination with other components of the ecosystem, contribute to all or part of the services provided by the whole ecosystem (Bartkowski et al. 2018; Fossey et al. 2020).

Although there has been no unanimous and definitive consensus on this question, and although these terms are used interchangeably in many situations, some kind of agreement does seem to have emerged around the concept of services related to the functioning of soils, defined as the subset of Ecosystem Services provided by the ecosystem whose production is directly and quantifiably dependent on soil properties, processes and functions (Paul et al. 2021). In parallel to the recognition of this subset of soil-related services, some authors have recently suggested defining a second, larger subset of services, *Ecosystem Services affected by (agricultural) soil management*, which brings together all services affected by soil management practices (Paul et al. 2021). Although there is quite a large intersection between both of these subsets, some services belong to only one or the other. For example, pollination services or esthetic experiences are not directly linked to soil properties, but are impacted by different soil management modalities, especially in an agricultural context, such as soil tillage or the crop protection methods (Paul et al. 2021). Conversely, a service such as the supply of mineral substances is directly linked to soil properties and does not depend on soil management modalities. In their attempt to bring together work centered on soil and the CICES classification, within the 89 service classes listed in the CICES classification, Paul et al. (2020) identified 29 service classes that were directly linked to soils and 40 service classes affected by soil management modalities, emphasizing (if the point still needed to be made!) how important soils were in the supply of Ecosystem Services.

7.7.4. *Assessing and mapping soil-related Ecosystem Services: where are we at present?*

In parallel to projects that aimed to draw attention to soils as essential contributors to Ecosystem Services, many projects sought to develop methods to assess soil-related services. Initially, the focus was on identifying dependencies between soil properties and soil-related services (Dominati et al. 2010; Adhikari and Hartemink 2016). Often defined qualitatively and based on expert opinions (Rutgers et al. 2012; Adhikari and Hartemink 2016) identifying soil determinants in Ecosystem Services gave rise to the earliest approaches to assessing soil-related Ecosystem Services through the use of proxy variables (Rutgers et al. 2012; van Wijnen et al. 2012; Hewitt et al. 2015). Subsequently, other approaches were developed with the aim of assessing soil-related services through empirical models (Calzolari et al. 2016; Debeljak et al. 2019; Choquet et al. 2021) or mechanistic models (Dominati et al. 2014; Dominati et al. 2016; Demestihas et al. 2018; Choquet et al. 2021; Ellili-Bargaoui et al. 2021; Obiang Ndong et al. 2021). Finally, the earliest decision-making tools were developed based on the assessment of soil-related Ecosystem Services for urban planning (Anne et al. 2018).

Nonetheless, in their synthesis of approaches used for the assessment and mapping of soil-related Ecosystem Services, Greiner et al. (2017) showed that only about ten soil-related services were actually assessed at present. Chief among these were biomass production, water production or habitat production for biodiversity, as well as global climate regulation, hydrological control or the regulation of concentrations of different organic and inorganic compounds. Soil-related cultural services, on the other hand, were rarely quantified as, without exception, there was a lack of clearly established quantitative relationships between soil properties and these services (Motiejunaite et al. 2019). While these limitations are not specific to soil-related services and are common to the assessment of all Ecosystem Services (Martínez-Harms and Balvanera 2012; Obiang Ndong et al. 2020), projects related to soil-related Ecosystem Services very rarely look at more than three services (Greiner et al. 2017) while those studying Ecosystem Services as a whole often look at more than three services simultaneously (Obiang Ndong et al. 2020). While the quantification of relationships between different services is studied for Ecosystem Services as a whole, this is largely yet to be explored for the subset of soil-related services.

Finally, in accordance with Ecosystem Services as a whole, the assessment and mapping of soil-related Ecosystem Services benefits today from the large number of approaches developed as a response to the diversity in definitions, indicators and assessment and mapping methods. Thus, although still rare, recent projects have focused on (i) directly measuring some soil-related Ecosystem Services (Chalhoub et al. 2020) or (ii) evaluating the impact of assessment methods or the type of soil data used on the levels and spatial distribution pattern of these services (Choquet et al. 2021; Scammacca et al. 2023). The double objective here is to obtain better information on the application conditions and the performance of various approaches used for assessing and mapping soil Ecosystem Services, eventually contributing to the harmonization of these approaches.

7.8. Conclusion

Ecologists and economists initially developed the concept of Ecosystem Services as a strategic and pragmatic option to engage decision-makers in the protection of biodiversity and the ecosystems that housed this biodiversity. By proposing a radically different vision of relations between human societies and their environment in a functional, multidimensional and utilitarian sense, which was at least partially in phase with certain societal expectations, and expressed in language (economic terms) that was acceptable to the dominant political and economic institutions, this concept rapidly gained popularity to the extent that it was perceived as a “magic” concept, likely to optimize the allocation and management of all natural resources from the local to a global level. In just about two decades, this interest resulted in a multitude of projects that sought to define, classify, assess and map Ecosystem Services, and finally operationalize ecosystem service approaches. This abundance of research is today both a strength and a weakness. It is an asset because the diversity in definitions, and in assessment and mapping approaches makes it possible to bring in some flexibility, which is very necessary when adapting a generic concept to a specific context and specific objectives. Its weakness lies in the fact that the abundance of initiatives has been accompanied by extremely diverse conceptual frameworks, classifications and definitions of Ecosystem Services or indicators, and the methods used to quantify and map these services. This has caused more confusion than it has helped in adopting and operationalizing the concept. Furthermore, despite this diversity, all Ecosystem Services approaches suffer to various extents from different limitations.

In the first place, despite the advances made in the past two decades, the lack of any harmonized and stable definition of the concept of Ecosystem Services and of the different individual Ecosystem Services limits the salience of any approach used to quantify, map and, where relevant, monetize Ecosystem Services as it is only possible to quantify a clearly defined concept. In the second place, approaches to assessing Ecosystem Services present varying degrees of development between families of services, and between individual services within each family. On the whole, there is also only a partial integration of cultural services. However, the relevance of the concept of Ecosystem Services resides entirely in its ability to establish a balance between all benefits derived from the functioning of ecosystems, especially between traditional provisioning services and the newer cultural and regulating services. The desire to adapt assessment and mapping methodologies to the limited and differentiated availability of data and tools, with the aim of immediate operationalization, often creates a gap between what is supposed to be measured and what is actually measured, thus limiting the actual scope and relevance of Ecosystem Services approaches. Finally, while the spatial dimension of the concept of Ecosystem Services was integrated early on as an essential question, even today there is very little information on the temporal evolution of service bundles (e.g. due to climate, land use or land management changes) despite the importance of these data for public or private decision-making.

The various advantages and limitations described earlier are exacerbated in the case of the subset of soil-related services. Indeed, using an Ecosystem Services approach makes it possible to envisage a paradigm shift from the traditional view of soils, which was largely centered on the concept of fertility (inherited from chemical-dominated agriculture in the 19th century) and on the optimization of the single service of biomass production, toward a multidimensional vision that can bring together biomass production, environmental quality and the conservation of biodiversity. This kind of change carries the possibility of allowing soils and soil management to become one of the tools in the agroecological transition and to meet multiple challenges related to the artificialization or degradation of soil. Nonetheless, integrating soils into Ecosystem Services approaches comes up directly against the following obstacles: (i) the very nature of soils, which are interfaces between biotic and abiotic components in ecosystems, with abiotic components still very much the poor cousin in the Ecosystem Services approach, despite being dynamic, and (ii) the lack of data and assessment tools, a particularly glaring lack in the domain of soil.

Over the past two decades, while the concept of Ecosystem Services has indisputably promoted the consideration of dependencies between the functioning of ecosystems and the well-being of human societies, several challenges must still be overcome to make it an operation tool that can be used for various public policies related to ecosystem management. Challenges related to the harmonization of concepts and methods used for assessment, mapping and monetization of services seem to be particularly critical for the triple objective of (i) enabling knowledge gathering, (ii) facilitating comparisons across space and time, and finally (iii) increasing user confidence. This is the focus of many research projects carried out at a European level, such as SERENA (2021–2024), MINOTAUR (2021–2024) and BENCHMARK (2023–2028), to name just a few. A few challenges go beyond the scope of the framework of Ecosystem Services, strictly speaking. For instance, the availability of data, whether direct measurements of services or the multiple variables used in indirect assessments, which is still a major restriction on the development of assessment and mapping approaches that can be concretely implemented.

7.9. Acknowledgments

The reflections summarized here were developed during (i) the project Assessing Ecosystem Services in transitioning ecosystems (ASSETS, 2016–2019) coordinated by Benoit Gabrielle and Patricia Garnier, funded by the Agence Nationale de la Recherche (ANR-11-LABX-0034); (ii) the doctoral thesis authored by Choquet (2017–2020), funded by the ADEME; (iii) the projects Stocktaking for agricultural soil quality and Ecosystem Services indicators and their reference values (SIREN, 2021–2022) coordinated by Faber and Taylor, and Soil Ecosystem services and soil threats modelling and mapping (SERENA, 2022–2025) coordinated by Cousin, Calzolari and Ungaro, funded by EJP-Soil, as well as (iv) various sessions in the doctoral course "Integrated assessment of ecosystem Services in periurban and rural areas", which received financial support from Agreenium in 2018 and 2021, and from EJP-Soil in 2023. The authors extend their heartfelt gratitude to all of the participants in these different projects and actions for their varied, and sometimes unexpected, contributions to the development of the reflections presented here.

7.10. References

Adhikari, K. and Hartemink, A.E. (2016). Linking soils to ecosystem services – A global review. *Geoderma*, 262, 101–111.

Andrew, M.E., Wulder, M.A., Nelson, T.A., Coops, N.C. (2015). Spatial data, analysis approaches, and information needs for spatial ecosystem service assessments: A review. *GIScience & Remote Sensing*, 52(3), 344–373.

Anne, B., Geoffroy, S., Cherel, J., Warot, G., Marie, S., Noël, C.J., Louis, M.J., Christophe, S. (2018). Towards an operational methodology to optimize ecosystem services provided by urban soils. *Landscape and Urban Planning*, 176, 1–9.

Bagstad, K.J., Semmens, D.J., Waage, S., Winthrop, R. (2013). A comparative assessment of decision-support tools for ecosystem services quantification and valuation. *Ecosystem Services*, 5, E27–E39.

Bagstad, K.J., Cohen, E., Ancona, Z.H., McNulty, S.G., Sun, G. (2018). The sensitivity of ecosystem service models to choices of input data and spatial resolution. *Applied Geography*, 93, 25–36.

Baró, F., Palomo, I., Zulian, G., Vizcaino, P., Haase, D., Gómez-Baggethun, E. (2016). Mapping ecosystem service capacity, flow and demand for landscape and urban planning: A case study in the Barcelona metropolitan region. *Land Use Policy*, 57, 405–417.

Bartkowski, B., Hansjurgens, B., Mockel, S., Bartke, S. (2018). Institutional economics of agricultural soil ecosystem services. *Sustainability*, 10(7), 2447.

Baveye, P.C. (2017). Quantification of ecosystem services: Beyond all the "guesstimates", how do we get real data? *Ecosystem Services*, 24, 47–49.

Baveye, P.C., Baveye, J., Gowdy, J. (2016). Soil "ecosystem" services and natural capital: Critical appraisal of research on uncertain ground. *Frontiers in Environmental Science*, 4(41). doi: 10.3389/fenvs.2016.00041.

Baveye, P.C., Chalhoub, M., Choquet, P., Montagne, D. (2018). Is the focus on "ecosystems" a liability in the research on nature's services? *Frontiers in Ecology and Evolution*, 6(226). doi: 10.3389/fevo.2018.00226.

Bennett, E.M., Peterson, G.D., Gordon, L.J. (2009). Understanding relationships among multiple ecosystem services. *Ecology Letters*, 12(12), 1394–1404.

Boumans, R., Roman, J., Altman, I., Kaufman, L. (2015). The Multiscale Integrated Model of Ecosystem Services (MIMES): Simulating the interactions of coupled human and natural systems. *Ecosystem Services*, 12, 30–41.

Boyd, J. and Banzhaf, S. (2007). What are ecosystem services? The need for standardized environmental accounting units. *Ecological Economics*, 63(2–3), 616–626.

Braat, L.C. and de Groot, R. (2012). The ecosystem services agenda: Bridging the worlds of natural science and economics, conservation and development, and public and private policy. *Ecosystem Services*, 1(1), 4–15.

Bunemann, E.K., Bongiorno, G., Bai, Z.G., Creamer, R.E., De Deyn, G., de Goede, R., Fleskens, L., Geissen, V., Kuyper, T.W., Mader, P. et al. (2018). Soil quality – A critical review. *Soil Biology & Biochemistry*, 120, 105–125.

Burkhard, B., Kroll, F., Müller, F., Windhorst, W. (2009). Landscapes' capacities to provide ecosystem services – A concept for land-cover based assessments. *Landscape Online*, 15, 1–22.

Burkhard, B., Kroll, F., Nedkov, S., Müller, F. (2012). Mapping ecosystem service supply, demand and budgets. *Ecological Indicators*, 21, 17–29.

Burkhard, B., Kandziora, M., Hou, Y., Müller, F. (2014). Ecosystem service potentials, flows and demands-concepts for spatial localisation, indication and quantification. *Landscape Online*, 34, 1–32.

Calzolari, C., Ungaro, F., Filippi, N., Guermandi, M., Malucelli, F., Marchi, N., Staffilani, F., Tarocco, P. (2016). A methodological framework to assess the multiple contributions of soils to ecosystem services delivery at regional scale. *Geoderma*, 261, 190–203.

Campagne, C.S. and Roche, P.K. (2018). May the matrix be with you! Guidelines for the application of expert-based matrix approach for ecosystem services assessment and mapping. *One Ecosystem*, 3, e24134.

Chalhoub, M., Gabrielle, B., Tournebize, J., Chaumont, C., Maugis, P., Girardin, C., Montagne, D., Baveye, P.C., Garnier, P. (2020). Direct measurement of selected soil services in a drained agricultural field: Methodology development and case study in Saclay (France). *Ecosystem Services*, 42. doi: 10.1016/j.ecoser.2020.101088.

Chaudhary, S., McGregor, A., Houston, D., Chettri, N. (2015). The evolution of ecosystem services: A time series and discourse-centered analysis. *Environmental Science & Policy*, 54, 25–34.

Choquet, P., Gabrielle, B., Chalhoub, M., Michelin, J., Sauzet, O., Scammacca, O., Garnier, P., Baveye, P.C., Montagne, D. (2021). Comparison of empirical and process-based modelling to quantify soil-supported ecosystem services on the Saclay plateau (France). *Ecosystem Services*, 50, 101332.

Corvalan, C.F., Hales, S., McMichael, T. (2005). *Ecosystems and Human Well-being: Synthesis*. Island Press, Washington, D.C.

Costanza, R., Cumberland, J.H., Daly, H., Goodland, R., Norgaard, R.B. (1997a). *An Introduction to Ecological Economics*. CRC Press, Boca Raton.

Costanza, R., d'Arge, R., de Groot, R., Farber, S., Grasso, M., Hannon, B., Limburg, K., Naeem, S., O'Neill, R.V., Paruelo, J. et al. (1997b). The value of the world's ecosystem services and natural capital. *Nature*, 387(6630), 253–260.

Costanza, R., de Groot, R., Braat, L., Kubiszewski, I., Fioramonti, L., Sutton, P., Farber, S., Grasso, M. (2017). Twenty years of ecosystem services: How far have we come and how far do we still need to go? *Ecosystem Services*, 28, 1–16.

Crossman, N.D., Burkhard, B., Nedkov, S., Willemen, L., Petz, K., Palomo, I., Drakou, E.G., Martín-Lopez, B., McPhearson, T., Boyanova, K. et al. (2013). A blueprint for mapping and modelling ecosystem services. *Ecosystem Services*, 4, 4–14.

Czucz, B., Haines-Young, R., Kiss, M., Bereczki, K., Kertesz, M., Vari, A., Potschin-Young, M., Arany, I. (2020). Ecosystem service indicators along the cascade: How do assessment and mapping studies position their indicators? *Ecological Indicators*, 118, 106729.

Daily, G.C. (1997). *Nature's Services: Societal Dependance on Natural Ecosystems*. Island Press, Washington, D.C.

Debeljak, M., Trajanov, A., Kuzmanovski, V., Schröder, J., Sandén, T., Spiegel, H., Wall, D.P., Van de Broek, M., Rutgers, M., Bampa, F. et al. (2019). A field-scale decision support system for assessment and management of soil functions. *Frontiers in Environmental Science*, 7, 115.

Demestihas, C., Plenet, D., Genard, M., de Cortazar-Atauri, I.G., Launay, M., Ripoche, D., Beaudoin, N., Simon, S., Charreyron, M., Raynal, C. et al. (2018). Analyzing ecosystem services in apple orchards using the STICS model. *European Journal of Agronomy*, 94, 108–119.

Díaz, S., Pascual, U., Stenseke, M., Martín-López, B., Watson, R.T., Molnár, Z., Hill, R., Chan, K.M.A., Baste, I.A., Brauman, K.A. et al. (2018). Assessing nature's contributions to people. *Science*, 359(6373), 270–272.

Dominati, E., Patterson, M., Mackay, A. (2010). A framework for classifying and quantifying the natural capital and ecosystem services of soils. *Ecological Economics* 69(9), 1858–1868.

Dominati, E., Mackay, A., Green, S., Patterson, M. (2014). A soil change-based methodology for the quantification and valuation of ecosystem services from agro-ecosystems: A case study of pastoral agriculture in New Zealand. *Ecological Economics*, 100, 119–129.

Dominati, E., Mackay, A., Bouma, J., Green, S. (2016). An ecosystems approach to quantify soil performance for multiple outcomes: The future of land evaluation? *Soil Science Society of America Journal*, 80(2), 438–449.

Ehrlich, P.R. and Ehrlich, A.H. (1981). *Extinction: The Causes and Consequences of the Disapperance of Species*. Random House, New York.

Eigenbrod, F., Armsworth, P.R., Anderson, B.J., Heinemeyer, A., Gillings, S., Roy, D.B., Thomas, C.D., Gaston, K.J. (2010). The impact of proxy-based methods on mapping the distribution of ecosystem services. *Journal of Applied Ecology*, 47(2), 377–385.

Ellili-Bargaoui, Y., Walter, C., Lemercier, B., Michot, D. (2021). Assessment of six soil ecosystem services by coupling simulation modelling and field measurement of soil properties. *Ecological Indicators*, 121, 107211.

Englund, O., Berndes, G., Cederberg, C. (2017). How to analyse ecosystem services in landscapes – A systematic review. *Ecological Indicators*, 73, 492–504.

European Commission (2014). Mapping and assessment of ecosystems and their services: An analytical framework for ecosystem assessments under Action 5 of the EU Biodiversity Strategy to 2020: Discussion paper – Final. Technical Report, Publications Office of the European Union, Luxembourg.

Fish, R., Church, A., Winter, M. (2016). Conceptualising cultural ecosystem services: A novel framework for research and critical engagement. *Ecosystem Services*, 21, 208–217.

Fossey, M., Angers, D., Bustany, C., Cudennec, C., Durand, P., Gascuel-Odoux, C., Jaffrezic, A., Pérès, G., Besse, C., Walter, C. (2020). A framework to consider soil ecosystem services in territorial planning. *Frontiers in Environmental Science*, 8, 28.

Francesconi, W., Srinivasan, R., Pérez-Miñana, E., Willcock, S.P., Quintero, M. (2016). Using the Soil and Water Assessment Tool (SWAT) to model ecosystem services: A systematic review. *Journal of Hydrology*, 535, 625–636.

Geijzendorffer, I.R., Cohen-Shacham, E., Cord, A.F., Cramer, W., Guerra, C., Martín-López, B. (2017). Ecosystem services in global sustainability policies. *Environmental Science & Policy*, 74, 40–48.

Gómez-Baggethun, E., de Groot, R., Lomas, P.L., Montes, C. (2010). The history of ecosystem services in economic theory and practice: From early notions to markets and payment schemes. *Ecological Economics*, 69(6), 1209–1218.

Gómez-Baggethun, E., Barton, D.N., Berry, P., Dunford, R., Harrison, P.A. (2016). Concepts and methods in ecosystem services valuation. In *Routledge Handbook of Ecosystem Services*, Potschin, M.B., Haines-Young, R., Fish, R., Kerry Turner, R. (eds). Routledge, London.

Gómez-Zotano, J., Riesco-Chueca, P., Frolova, M., Rodríguez-Rodríguez, J. (2018). The landscape taxonomic pyramid (LTP): A multi-scale classification adapted to spatial planning. *Landscape Research*, 43(7), 984–999.

Gray, M. (2018). The confused position of the geosciences within the "natural capital" and "ecosystem services" approaches. *Ecosystem Services*, 34, 106–112.

Greiner, L., Keller, A., Grêt-Regamey, A., Papritz, A. (2017). Soil function assessment: Review of methods for quantifying the contributions of soils to ecosystem services. *Land Use Policy*, 69, 224–237.

Grêt-Regamey, A., Weibel, B., Bagstad, K.J., Ferrari, M., Geneletti, D., Klug, H., Schirpke, U., Tappeiner, U. (2014). On the effects of scale for ecosystem services mapping. *PLoS One* 9(12), e112601.

Grêt-Regamey, A., Sirén, E., Brunner, S.H., Weibel, B. (2017a). Review of decision support tools to operationalize the ecosystem services concept. *Ecosystem Services*, 26, 306–315.

Grêt-Regamey, A., Weibel, B., Rabe, S.E., Burkhard, B. (2017b). A tiered approach for ecosystem services mapping. In *Mapping Ecosystem Services*, Burkhard, B. and Maes, J. (eds). Pensoft Publishers, Sofia.

Guerra, C.A., Maes, J., Geijzendorffer, I., Metzger, M.J. (2016). An assessment of soil erosion prevention by vegetation in Mediterranean Europe: Current trends of ecosystem service provision. *Ecological Indicators*, 60, 213–222.

Haines-Young, R. and Potschin, M.B. (2018). Common International Classification of Ecosystem Services (CICES) V5.1 and guidance on the application of the revised structure. Paper, Fabis Consulting Ltd, Nottingham.

Hein, L., Bagstad, K.J., Edens, B., Obst, C.G., de Jong, R., Lesschen, J.P. (2016). Defining ecosystem assets for natural capital accounting. *PLoS One*, 11, e0164460.

Heink, U., Hauck, J., Jax, K., Sukopp, U. (2016). Requirements for the selection of ecosystem service indicators – The case of MAES indicators. *Ecological Indicators*, 61, 18–26.

Herold, M., Clarke, A., Menz, G., Allee, M. (2002). A multi-scale framework for mapping and analysis of the spatial and temporal pattern of urban growth. In *Geoinformation for European-wide Integration*, Benes, T. (ed.). Millpress, Rotterdam.

Hewitt, A., Dominati, E., Webb, T., Cuthill, T. (2015). Soil natural capital quantification by the stock adequacy method. *Geoderma*, 241, 107–114.

Hölting, L., Jacobs, S., Felipe-Lucia, M.R., Maes, J., Norström, A.V., Plieninger, T., Cord, A.F. (2019). Measuring ecosystem multifunctionality across scales. *Environmental Research Letters*, 14(12), 124083.

Jackson, B., Pagella, T., Sinclair, F., Orellana, B., Henshaw, A., Reynolds, B., McIntyre, N., Wheater, H., Eycott, A. (2013). Polyscape: A GIS mapping framework providing efficient and spatially explicit landscape-scale valuation of multiple ecosystem services. *Landscape and Urban Planning*, 112, 74–88.

Jacobs, S., Burkhard, B., Van Daele, T., Staes, J., Schneiders, A. (2015). 'The Matrix Reloaded': A review of expert knowledge use for mapping ecosystem services. *Ecological Modelling*, 295, 21–30.

Jónsson, J.Ö.G. and Davíðsdóttir, B. (2016). Classification and valuation of soil ecosystem services. *Agricultural Systems*, 145, 24–38.

Lavorel, S., Bayer, A., Bondeau, A., Lautenbach, S., Ruiz-Frau, A., Schulp, N., Seppelt, R., Verburg, P., Teeffelen, A.V., Vannier, C. et al. (2017). Pathways to bridge the biophysical realism gap in ecosystem services mapping approaches. *Ecological Indicators*, 74, 241–260.

Le Clec'h, S., Oszwald, J., Decaens, T., Desjardins, T., Dufour, S., Grimaldi, M., Jegou, N., Lavelle, P. (2016). Mapping multiple ecosystem services indicators: Toward an objective-oriented approach. *Ecological Indicators*, 69, 508–521.

Lee, H. and Lautenbach, S. (2016). A quantitative review of relationships between ecosystem services. *Ecological Indicators*, 66, 340–351.

Loomis, J.B., Kroeger, T., Richardson, L.A., Casey, F. (2008). A benefit transfer toolkit for fish, wildlife, wetlands, and open space. *Western Economics Forum*, 7(2) [Online]. Available at: https://waeaonline.org/western_econ_forum/a-benefit-transfer-toolkit-for-fish-wildlife-wetlands-and-open-space/.

Lugato, E., Bampa, F., Panagos, P., Montanarella, L., Jones, A. (2014). Potential carbon sequestration of European arable soils estimated by modelling a comprehensive set of management practices. *Glob. Change Biol.*, 20(11), 3557–3567.

Maes, J., Paracchini, M., Zulian, G. (2011). A European assessment of the provision of ecosystem services: Towards an atlas of ecosystem services. Technical Report, Publications Office of the European Union, Luxembourg.

Makovníková, J., Pálka, B., Širáň, M., Houskova, B., Kanianska, R., Kizekova, M. (2018). Ecosystem services in differently used agroecosystems along a climatic gradient in Slovakia. *Open Journal of Ecology*, 8, 623–645.

Malinga, R., Gordon, L.J., Jewitt, G., Lindborg, R. (2015). Mapping ecosystem services across scales and continents – A review. *Ecosystem Services*, 13, 57–63.

Marques, S.M., Campos, F.S., David, J., Cabral, P. (2021). Modelling sediment retention services and soil erosion changes in Portugal: A spatio-temporal approach. *ISPRS International Journal of Geo-Information*, 10(4), 262.

Martínez-Harms, M.J. and Balvanera, P. (2012). Methods for mapping ecosystem service supply: A review. *International Journal of Biodiversity Science, Ecosystem Services & Management*, 8(1–2), 17–25.

Mayer, A., Kaufmann, L., Kalt, G., Matej, S., Theurl, M.C., Morais, T.G., Leip, A., Erb, K.-H. (2021). Applying the human appropriation of net primary production framework to map provisioning ecosystem services and their relation to ecosystem functioning across the European Union. *Ecosystem Services*, 51, 101344.

Monnoyer-Smith, L. (2016). EFESE : l'essentiel du cadre conceptuel. Assessment program, Ministère de l'Écologie, Paris [Online]. Available at: https://www.notre-environnement.gouv.fr/IMG/pdf/thc3a9ma20-2d245.pdf.

Motiejunaite, J., Borja, I., Ostonen, I., Bakker, M.R., Bjarnadottir, B., Brunner, I., Irsenaite, R., Mrak, T., Oddsdottir, E.S., Lehto, T. (2019). Cultural ecosystem services provided by the biodiversity of forest soils: A European review. *Geoderma*, 343, 19–30.

Mouchet, M.A., Paracchini, M.L., Schulp, C.J.E., Stürck, J., Verkerk, P.J., Verburg, P.H., Lavorel, S. (2017). Bundles of ecosystem (dis)services and multifunctionality across European landscapes. *Ecological Indicators*, 73, 23–28.

Mulder, V.L., Lacoste, M., Richer-de-Forges, A.C., Arrouays, D. (2016). GlobalSoilMap France: High-resolution spatial modelling the soils of France up to two meter depth. *Science of the Total Environment*, 573, 1352–1369.

Nahlik, A.M., Kentula, M.E., Fennessy, M.S., Landers, D.H. (2012). Where is the consensus? A proposed foundation for moving ecosystem service concepts into practice. *Ecological Economics*, 77, 27–35.

Natural Capital Project (2022). InVEST 3.13 user's guide. Guide, Standford University, University of Minnesota, Chinese Academy of Sciences, The Nature Conservancy, World Wildlife Fund, and Stockholm Resilience Center [Online]. Available at: https://naturalcapitalproject.stanford.edu/software/invest.

Obiang Ndong, G., Therond, O., Cousin, I. (2020). Analysis of relationships between ecosystem services: A generic classification and review of the literature. *Ecosystem Services*, 43, 101120.

Obiang Ndong, G., Villerd, J., Cousin, I., Therond, O. (2021). Using a multivariate regression tree to analyze trade-offs between ecosystem services: Application to the main cropping area in France. *Science of the Total Environment*, 764, 142815.

Obriot, F., Stauffer, M., Goubard, Y., Cheviron, N., Peres, G., Eden, M., Revallier, A., Vieublé-Gonod, L., Houot, S. (2016). Multi-criteria indices to evaluate the effects of repeated organic amendment applications on soil and crop quality. *Agriculture, Ecosystems & Environment*, 232, 165–178.

van Oudenhoven, A.P.E., Schröter, M., Drakou, E.G., Geijzendorffer, I.R., Jacobs, S., van Bodegom, P.M., Chazee, L., Czúcz, B., Grunewald, K., Lillebø, A.I. et al. (2018). Key criteria for developing ecosystem service indicators to inform decision making. *Ecological Indicators*, 95, 417–426.

Pascual, U., Balvanera, P., Díaz, S., Pataki, G., Roth, E., Stenseke, M., Watson, R.T., Başak Dessane, E., Islar, M. et al. (2017). Valuing nature's contributions to people: The IPBES approach. *Current Opinion in Environmental Sustainability*, 26–27, 7–16.

Paul, C., Kuhn, K., Steinhoff-Knopp, B., Weißhuhn, P., Helming, K. (2021). Towards a standardization of soil-related ecosystem service assessments. *European Journal of Soil Science*, 72(4), 1543–1558.

Perez-Soba, M., Elbersen, B., Kempen, M., Braat, L., Staristky, I., Van Der Wijngaart, R., Kaphengst, T., Andersen, E., Germer, L., Smith, L. et al. (2016). *Agricultural Biomass as Provisioning Ecosystem Service: Quantification of Energy Flows*. Technical Report, Publications Office of the European Union, Luxembourg.

Pickard, B.R., Daniel, J., Mehaffey, M., Jackson, L.E., Neale, A. (2015). EnviroAtlas: A new geospatial tool to foster ecosystem services science and resource management. *Ecosystem Services*, 14, 45–55.

Plieninger, T., Bieling, C., Fagerholm, N., Byg, A., Hartel, T., Hurley, P., Lopez-Santiago, C.A., Nagabhatla, N., Oteros-Rozas, E., Raymond, C.M. et al. (2015). The role of cultural ecosystem services in landscape management and planning. *Current Opinion in Environmental Sustainability* 14, 28–33.

Poggio, L., de Sousa, L.M., Batjes, N.H., Heuvelink, G.B.M., Kempen, B., Ribeiro, E., Rossiter, D. (2021). SoilGrids 2.0: Producing soil information for the globe with quantified spatial uncertainty. *Soil*, 7(1), 217–240.

Potschin, M.B. and Haines-Young, R.H. (2011). Ecosystem services: Exploring a geographical perspective. *Progress in Physical Geography: Earth and Environment*, 35(5), 575–594.

Potschin-Young, M., Haines-Young, R., Görg, C., Heink, U., Jax, K., Schleyer, C. (2018). Understanding the role of conceptual frameworks: Reading the ecosystem service cascade. *Ecosystem Services*, 29, 428–440.

Rabin, S.S., Alexander, P., Henry, R., Anthoni, P., Pugh, T.A.M., Rounsevell, M., Arneth, A. (2020). Impacts of future agricultural change on ecosystem service indicators. *Earth Syst. Dynam.*, 11(2), 357–376.

Rabot, E., Guiresse, M., Pittatore, Y., Angelini, M., Keller, C., Lagacherie, P. (2022). Development and spatialization of a soil potential multifunctionality index for agriculture (Agri-SPMI) at the regional scale. Case study in the Occitanie region (France). *Soil Security*, 6, 100034.

Raudsepp-Hearne, C. and Peterson, G.D. (2016). Scale and ecosystem services: How do observation, management, and analysis shift with scale – Lessons from Québec. *Ecology and Society*, 21(3), 16.

Raudsepp-Hearne, C., Peterson, G.D., Bennett, E.M. (2010). Ecosystem service bundles for analyzing tradeoffs in diverse landscapes. *Proceedings of the National Academy of Sciences*, 107(11), 5242–5247.

Remme, R.P., Schröter, M., Hein, L. (2014). Developing spatial biophysical accounting for multiple ecosystem services. *Ecosystem Services*, 10, 6–18.

Rendon, P., Steinhoff-Knopp, B., Burkhard, B. (2022). Linking ecosystem condition and ecosystem services: A methodological approach applied to European agroecosystems. *Ecosystem Services*, 53, 101387.

Robinson, D.A., Lebron, I., Vereecken, H. (2009). On the definition of the natural capital of soils: A framework for description, evaluation, and monitoring. *Soil Science Society of America Journal*, 73(6), 1904–1911.

Roche, P.K. and Campagne, C.S. (2019). Are expert-based ecosystem services scores related to biophysical quantitative estimates? *Ecological Indicators*, 106, 105421.

Ros, G.H., Verweij, S.E., Janssen, S.J.C., De Haan, J., Fujita, Y. (2022). An open soil health assessment framework facilitating sustainable soil management. *Environmental Science & Technology*, 56(23), 17375–17384.

Ruhl, J.B. and Salzman, J.E. (2007). The law and policy beginnings of ecosystem services. *Journal of Land Use and Environmental Law*, 22, 157.

Rutgers, M., van Wijnen, H.J., Schouten, A.J., Mulder, C., Kuiten, A.M.P., Brussaard, L., Breure, A.M. (2012). A method to assess ecosystem services developed from soil attributes with stakeholders and data of four arable farms. *Science of the Total Environment*, 415, 39–48.

Scammacca, O., Sauzet, O., Michelin, J., Choquet, P., Garnier, P., Gabrielle, B., Baveye, P.C., Montagne, D. (2023). Effect of spatial scale of soil data on estimates of soil ecosystem services: Case study in 100 km^2 area in France. *European Journal of Soil Science*, 74(2), e13359.

Schröter, M., Barton, D.N., Remme, R.P., Hein, L. (2014). Accounting for capacity and flow of ecosystem services: A conceptual model and a case study for Telemark, Norway. *Ecological Indicators*, 36, 539–551.

Schröter, M., Remme, R.P., Sumarga, E., Barton, D.N., Hein, L. (2015). Lessons learned for spatial modelling of ecosystem services in support of ecosystem accounting. *Ecosystem Services*, 13, 64–69.

Schulp, C.J.E., Alkemade, R., Klein Goldewijk, K., Petz, K. (2012). Mapping ecosystem functions and services in Eastern Europe using global-scale data sets. *International Journal of Biodiversity Science, Ecosystem Services & Management*, 8(1–2), 156–168.

Schulp, C.J.E., Burkhard, B., Maes, J., Van Vliet, J., Verburg, P.H. (2014). Uncertainties in ecosystem service maps: A comparison on the European scale. *PLoS One*, 9(10), e109643.

Sherrouse, B.C., Clement, J.M., Semmens, D.J. (2011). A GIS application for assessing, mapping, and quantifying the social values of ecosystem services. *Applied Geography*, 31(2), 748–760.

Storie, R. (1978). Storie index soil rating. Special publication 3203, University of California, Oakland [Online]. Available at: https://anrcatalog.ucanr.edu/Details.aspx?itemNo=3203#DigitalDetailSection.

Stürck, J., Poortinga, A., Verburg, P.H. (2014). Mapping ecosystem services: The supply and demand of flood regulation services in Europe. *Ecological Indicators*, 38, 198–211.

Sumarga, E. and Hein, L. (2014). Mapping ecosystem services for land use planning, the case of Central Kalimantan. *Environmental Management*, 54(1), 84–97.

Therond, O., Tichit, M., Tibi, A., Accatino, F., Biju.-Duval, L., Bockstaller, C., Bohan, D., Bonaudo, T., Boval, M., Cahuzac, E. et al. (2017). Volet "écosystèmes agricoles" de l'Évaluation Française des Écosystèmes et des Services Écosystémiques. Report, INRA, Paris.

Turner, K.G., Anderson, S., Gonzales-Chang, M., Costanza, R., Courville, S., Dalgaard, T., Dominati, E., Kubiszewski, I., Ogilvy, S., Porfirio, L. et al. (2016). A review of methods, data, and models to assess changes in the value of ecosystem services from land degradation and restoration. *Ecological Modelling*, 319, 190–207.

Ungaro, F., Zasada, I., Piorr, A. (2014). Mapping landscape services, spatial synergies and trade-offs. A case study using variogram models and geostatistical simulations in an agrarian landscape in North-East Germany. *Ecological Indicators*, 46, 367–378.

Villa, F., Bagstad, K.J., Voigt, B., Johnson, G.W., Portela, R., Honzak, M., Batker, D. (2014). A methodology for adaptable and robust ecosystem services assessment. *PLoS One*, 9(3), e91001.

Villamagna, A.M., Angermeier, P.L., Bennett, E.M. (2013). Capacity, pressure, demand, and flow: A conceptual framework for analyzing ecosystem service provision and delivery. *Ecological Complexity*, 15, 114–121.

Vorstius, A.C. and Spray, C.J. (2015). A comparison of ecosystem services mapping tools for their potential to support planning and decision-making on a local scale. *Ecosystem Services*, 15, 75–83.

Westman, W.E. (1977). How much are nature's services worth? *Science*, 197(4307), 960–964.

van Wijnen, H.J., Rutgers, M., Schouten, A.J., Mulder, C., de Zwart, D., Breure, A.M. (2012). How to calculate the spatial distribution of ecosystem services – Natural attenuation as example from The Netherlands. *Science of the Total Environment*, 415, 49–55.

Wischmeier, W.H. and Smith, D.D. (1978). *Predicting Rainfall Erosion Losses: A Guide to Conservation Planning*. US Department of Agriculture, Washington, D.C.

Xu, S., Liu, Y., Wang, X., Zhang, G. (2017). Scale effect on spatial patterns of ecosystem services and associations among them in semi-arid area: A case study in Ningxia Hui Autonomous Region, China. *Science of the Total Environment*, 598, 297–306.

8

Mediterranean Agricultural Soils

Emmanuelle VAUDOUR
UMR ECOSYS, AgroParisTech, INRAE, Université Paris Saclay, Palaiseau, France

8.1. Introduction

8.1.1. *Physical and human frameworks and the pedogenesis process in Mediterranean regions*

The Mediterranean region is usually circumscribed to the biogeographical area where olive trees (*Olea europea* L.) are grown (Terral 1999). This corresponds to the region lying between the latitudes 25° and 45° N and, strictly speaking, encompasses the regions bordering the Mediterranean, which are characterized by a long, dry summer (≥3 months), a mild winter (monthly average temperatures higher than 8–10°C) as well as heavy precipitation. Subtropical regions on the Western front of any continent have a comparable, Mediterranean climate (California, central Chile, the Cape province, southern Australia, central Mexico).

In the Mediterranean region, sensu stricto, the agricultural soils are located in a remarkable physical and human geographical context. This context evolved over a long time, both in terms of the morphogenesis of the landscape over 250 million years (Mohrange and Gassend 2007) as well as in terms of human presence: the Mediterranean basin is one of the main centers of the Neolithic transition and the emergence of agriculture, which was born within what is today the eastern Mediterranean regions and the Middle East (Guilaine 2008). The growth of agriculture shaped the development of the classic civilizations that left their mark on

Agricultural Soil Science,
coordinated by Yves COQUET and Joël MICHELIN. © ISTE Ltd 2025.

the history of the Mediterranean world and its exchanges with other regions, in the unique space it occupied between three continents. The ancient agricultural revolution gave rise to rain-fed grain farming, with associated pastoralism and livestock rearing (agro–sylvo–pastoralism). The Greeks and Romans were especially renowned for their agronomic knowledge and techniques, the Etruscans were renowned for their hydraulic planning and expertise in drainage, the Egyptians were renowned for their irrigation systems (Mazoyer and Roudart 2002; Robert 2004). Mediterranean agricultural soils were used in rural landscapes that often had a shared heritage, characterized by old land use patterns, family-run sites, shared culinary traditions (the "Mediterranean diet"), and the long-standing practice of silvopasture (the Spanish *dehesas*), agroforesty (Latin *hortus*, or Provençal *oullières*) and irrigated horticulture (the Spanish *huertas*, or *oasis* in the Maghreb).

The general context of Mediterranean pedogenesis is marked by the fragmentation and complexity of the relief and geomorphological structure as well as the diversity in the types of geomorphological structure: steep plains and mountainous blocks that are interwoven in such a way that the landscape rarely remains homogeneous for more than 50 km (Birot and Dresch 1953). The orogenesis of many mountainous arcs, such as those seen in the eastern Mediterranean, the undersea "accretionary prism" (Huguen and Mascle 2005), corresponds to the ancient continental borders of the Thetys or Paleo-Mediterranean Sea, formed through the movement of the African and Eurasian tectonic plates. We can identify ancient massifs that are not very deformed, such as the Anti-Atlas or the Iberian Meseta, and folded massifs (which may or may not be flattened), which are largely from the Tertiary and Quaternary ages. Thus, Mediterranean soils often have Plio-Quaternary heritage, such as the Villafranchian alluvial terraces (Vaudour 1983) and the calcium landscapes formed by the karst massifs, such as the Languedoc and Ardèche plateaus, the Vaucluse plateau and the Sainte-Baume, Montrieux, Calanques and Bonifacio massifs in France (Nicod 1995). Carbonate rocks, which do not occur frequently on a global scale, form parent materials that are widely available in the Mediterranean region and constitute one of the unique features of this region: carbonate balances are of great importance in Mediterranean soils (Vaudour 2003a).

The alternation of the dry summer seasons and wet seasons marked by violent precipitation leads to the following unique features, referred to as climatically related landform assemblages, which compare them to arid regions: (i) the frequency of steep slopes, which have led to remarkable walls formed of dry rocks, or terrace farming; (ii) the clear-cut structural forms, with areas of hard rock ending in steep rocky areas over depressions formed of soft rock that is perfectly flat or have small slopes (erosion glacis). Due to the stability of the structural shapes, such as the

Plio-Villafranchian alluvial plateaus, pedogenesis within these morphological units is the result of consecutive paleo-climates, leading up to the present-day climate (Vaudour 1980, 1983; Fédoroff and Courty 2013). One specific pedogenetic process that has majorly influenced the characteristics of Mediterranean soils, and which occurred throughout a climatic sequence, is geochemical weathering. This is the chemical transformation of the parental material, whose primary minerals undergo a transformation in their crystalline structure, are subject to hydrolysis, or are dissolved. This reaction may produce leachates, which are then drained away from the profile (lixiviation process) and may recombine, leading to the formation of assemblages of secondary clay minerals. Geochemical weathering depends not only on the mineralogical composition of the parent material, which determines its alterability (Figure 8.1) but also on temperature, redox potential and the humidity of the environment (Paquet 1970; Pédro 1984; Legros 2007) as well as the lateral, upstream to downstream transfers of products of alteration (Paquet 1983).

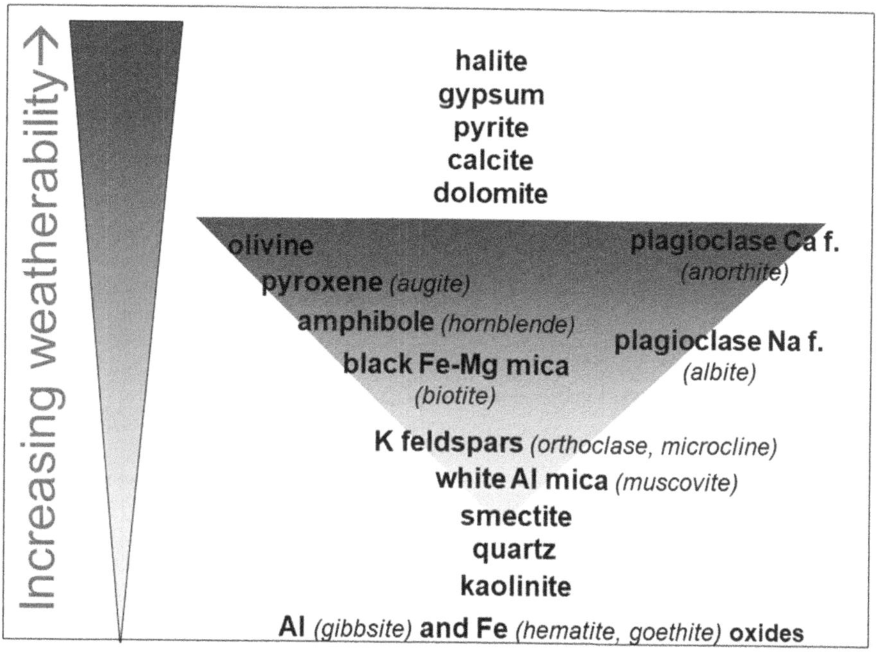

Figure 8.1. *The order of weatherability of the main minerals (Goldich 1938; Legros 2007)*

A key driving factor in geochemical weathering is the vertical percolation of water through soil, assessed through climatic drainage (Bourgeon and Pédro 1992).

This is defined in over a month as the difference D_i between precipitation P_i and potential evapotranspiration ETP_i (in mm) (equation [8.1]) and the annual sum is D minus 100 mm, that is, a quantitative estimate of the water reserves in the soil as established at the start of the rainy season for a depth of 1 m (equation [8.2]).

$$D_i = P_i - ETP_i \tag{8.1}$$

$$D = \sum_{i=1}^{i=12} D_i - 100 \tag{8.2}$$

Median climatic drainage D_m (mm)	Primary mineral	Chemical equation	Secondary mineral
$0 < D_m < 200$	Potassium feldspar	$2KAlSi_3O_{8(s)} + 2H^+_{(aq)} + 4H_2O_{(l)}$ Orthoclase-microcline → $Al_2O_{10}Si_4(OH)_{2(s)} + 2K^+_{(aq)} + 2H_4SiO_{4(aq)}$ Pyrophyllite Octahedral substitution of Al by Mg: montmorillonite Tetrahedral substitution – Si by Al, octahedral substitution – Al by Fe: beidellite $Al_2O_{10}(Si_{3,7}Al_{0.3})(OH)_{2(s)}$	Clay 2/1
$200 < D_m < 400$	Plagioclase feldspar	$2CaAlSi_2O_{8(s)} + 10H^+_{(aq)} + H_2O_{(l)}$ Anorthite → $Al_2O_5Si_2(OH)_{4(s)} + 2Ca^{2+}_{(aq)} + 2H_4SiO_{4(aq)}$ Kaolinite	Clay 1/1
	Potassium feldspar	$2KAlSi_3O_{8(s)} + 2H^+_{(aq)} + 9H_2O_{(l)}$ Orthoclase-microcline → $Al_2O_5Si_2(OH)_{4(s)} + 2K^+_{(aq)} + 4H_4SiO_{4(aq)}$ Kaolinite	Clay 1/1
	Plagioclase feldspar	$2CaAlSi_2O_{8(s)} + 10H^+_{(aq)} + 6H_2O_{(l)}$ Anorthite → $2Al(OH)_{3(s)} + 2Ca^{2+}_{(aq)} + 4H_4SiO_{4(aq)}$ Gibbsite	Al oxide
$400 < D_m$	Potassium feldspar	$2KAlSi_3O_{8(s)} + 2H^+_{(aq)} + 14H_2O_{(l)}$ Orthoclase-microcline → $2Al(OH)_{3(s)} + 2K^+_{(aq)} + 6H_4SiO_{4(aq)}$ Gibbsite	Al oxide

Table 8.1. *Equation for the hydrolysis of primary feldspar minerals (Pédro 1984; Legros 2007)*

The most easily alterable minerals are the salts (halite, gypsum and calcite), which are found in greater quantities in soils where drainage is zero or even negative. In climatic regions where D is episodically positive, and its median value over a climatic series, D_m does not exceed 400 mm, for example, the Indian peninsula (Bourgeon and Pédro 1992), 2/1 clay structures (such as smectites) and 1/1 structures (Kaolinite) are formed. Kaolinite is found in greater quantities when D is high and frequent. Table 8.1 illustrates the acid hydrolysis equations that occur with primary minerals present in granites, such as feldspars, depending on the median drainage, D_m. When D_m is low, then depending on the alterability of feldspar, we either obtain a bisiallitization leading to 2/1 smectite clay formation (such as montmorillonite) or a monosiallization leading to the formation of 1/1 clays such as kaolinite.

In current Mediterranean conditions, this results in clay assemblages corresponding to a partial desilification: one of the products of the reaction is silicic acid. This reaction occurs in a medium that ranges from being slightly to very acidic. It takes place faster when the medium is already acidic and when the primary mineral is more alterable.

8.1.2. *Zonality of Mediterranean soils*

During a conference on Mediterranean soils, which took place between September 12 and 17, 1966 in Madrid as part of the activities of the International Soil Sciences Society, a decision was adopted by those present (although not unanimous) that the term "Mediterranean" would be dropped from soil nomenclature (Segalen 1967; Vaudour 1969; Yaalon et al. 1997; Mancini 2002). The reason for this decision was that the Mediterranean region was "not a climatic entity", as there were many different climates observed here, with large variations in annual precipitation (P) and temperature, accompanied by large differences between soils. The various Mediterranean climates are as follows (Le Houérou 1982, 2004): Saharan ($P < 100$ mm), arid ($100 < P < 400$ mm), semi-arid ($400 < P < 600$ mm), sub-humid ($600 < P < 800$ mm), humid ($800 < P < 1200$ mm), perhumid and Alpine ($P > 1200$ mm). These climates have variants based on the minimum average temperature in the coldest months. The Mediterranean zone is also distinguished by how its paleo-climates influenced the pedogenesis of its soils, which was an aspect that had not yet been well studied at the time of the decision taken in 1966 (Mancini 2002).

Despite dropping the term "Mediterranean", it must be noted that this adjective proves to be indispensable in bibliography search engines such as Google Scholar, the Web of Science, and Persée, especially as the terms designating the soils differ depending on the taxonomic references used. The corpus of scientific publications focusing specifically on pedogenesis, the geographic distribution and erosion of Mediterranean soils mainly dates to the period from 1960 to 1995, with remarkable contributions from French researchers across an intersection of pedology, geomorphology, sedimentology, stratigraphy and the geochemistry of surface formations (e.g. Boulaine 1957, 1978; Durand 1963; Lamouroux 1967; Millot et al. 1969, 1979; Vaudour 1969, 1975, 1979, 1980, 1983; Barrière and Michaux 1970; Paquet 1970; Ruellan 1970; Arnal 1971; Barrière 1971; Bottner 1972; Paquet and Vaudour 1974; Dachary 1975; Barbaroux and Bousquet 1976; Lamouroux et al. 1977; Boulaine 1978; Penven et al. 1981; Paquet 1983; Vaudour et al. 1989; Fédoroff 1997). In the same period, other research was carried out in Israel (e.g. Dan et al. 1969, 1982; Yaalon et al. 1974; Singer et al. 1998), Spain (e.g. Torrent et al. 1980, 1983; Bech Borras et al. 1982; Ibañez et al. 1994; Bech et al. 1997) and in Italy (e.g. Boero and Schwertmann 1987, 1989; Boero et al. 1992). At the time, the volume of French publications on the subject was not recognized, especially in English publications: in a synthesis looking into how Mediterranean soils differed specifically, Yaalon (1997) mentions only three authors from the "French school", while briefly stating that the concept of "fersiallitic soil" had emerged from the French school. Contrary to the decision taken in 1966, this question of the specific or zonal characteristic of Mediterranean soils remained open. It is still an underdeveloped field of study, as the publications that followed, till date, have focused on soil management and degradation, as can be seen in the topics covered by the last Conference on Soils with Mediterranean type of Climate, held in Beirut in 2010 (http://www.emwis.org/thematicdirs/events/2009/06/10th-international-meeting-soils-mediterranean) as well as in more recent syntheses (Costantini and Lorenzetti 2013; Costantini et al. 2016; Ferreira et al. 2022).

Soil sequences are successions of soils with different morphologies, depending on the duration of the pedogenesis (chronosequence), their position on a slope (toposequence) or a change in the parent material (lithosequence) (Baize 2016). In the Mediterranean basins, soil sequences have been described in Algeria (Boulaine 1957, 1978; Durand 1963; Penven 1980), Lebanon (Lamouroux 1967; Darwish and Zuraik 1997), Morocco (Millot et al. 1969; Bronger and Bruhn-Lobin 1997), Spain, in the Iberian peninsula (Vaudour 1977, 1979, 1980; Bech et al. 1982, 1997; Vandour et al. 1989; Ibañez et al. 1994; Alonso-Zarza et al. 1998; Gonzalez-Martin et al. 2007; Badía et al. 2009; Wagner et al. 2012), Balearic islands (Muhs et al. 2010; Wagner et al. 2012), in the Mediterranean South of France (Barrière and Michaux 1970; Arnal 1971; Barrière 1971; Bottner 1972; Bornand and Chamley 1974; Bornand 1978; Vaudour 2003b), Portugal (Dachary 1975), Italy (Boero and Schwertmann 1987), Sicily (Egli et al. 2020), Greece (Barbaroux and Bousquet 1976; Yassoglou et al. 1997), Turkey (Atalay 1997) and in Israel (Dan et al. 1969,

1982; Yaalon et al. 1974; Singer et al. 1998). Soils in the Mediterranean region bear the mark of their environment, so much so that "the same landscapes can be seen from Palestine to Italy, from Syria to Castille, from Morocco to Greece" (Boulaine 1975) and there are, of course, soil sequences that have a zonal significance (Millot et al. 1969; Barrière 1971; Yaalon et al. 1974; Boulaine 1975; Vaudour 1980; Ibañez et al. 1994): (i) soil sequences from brown to red, so-called fersiallitic soils or Fersialsols according to the French soil taxonomy (AFES 2008), resulting from geochemical weathering, in which the pedological horizons differ in the degree of weathering and the assemblages of clay minerals, the thickness and degree of evolution toward the base of the soil profile (*solum*), depending on drainage and both the duration of the evolution and the lithological facies of the parent substrate(s); (ii) soil sequences marked by the presence of calcareous accumulations (Calcarisols; AFES 2008) or gypsum accumulations (Gypsosols; AFES 2008) whose thickness, depth, morphological characteristics, mineralogical composition and degree of evolution toward the top of the soil profiles vary based on aridity, the endorheism, the duration of evolution and the characteristics of slopes and the upstream fersiallitic sequences; (iii) soils marked by sodization or salinization in an endorheic context or depending on the inflow of irrigation water.

8.2. Fersiallitic red soils

8.2.1. *Definition and formation processes*

Based on work carried out by the "French school" on Mediterranean soils, the French classification of soils that was developed in 1967 by the *Commission de pédologie et de cartographie des sols* (CPCS) (Commission for Soil Pedology and Cartography) individualized fersiallitic soils and has, since, grown into the vast reference work Fersialsols of the French Taxonomy or *Référentiel Pédologique* (Bresson 2008). Fersiallitic soils are the result of a mild weathering of primary minerals where 2/1 clays are predominant, partially heritage and partially neoformed, with a high individualization of iron oxi-hydroxides. This is the fersiallitization process. There is no exact equivalent for Fersialsols in other soil classification systems. However, their references can be translated in the global classification, the *World Reference Base* or WRB (FAO 2014) (Table 8.2). There are approximate terms such as "terra rossa" soils on hard limestone or "Rhodoxeralfs" in the US Soil Taxonomy (Soil Survey Staff 1999) or "Duplex soils" in Australia. Fersialsols are defined by the presence of a diagnostic horizon resulting from the fersiallitization process: the fersiallitic horizon, denoted by FS. Assessing that a horizon is an FS is based on the following five criteria:

i) the dominant presence of 2/1 bisiallitic "smectite" clay;

ii) a red color with a hue of at least 5YR on the Munsell color scale (or even redder, 2.5YR) along with a high chroma, greater than 3.5;

iii) a free-iron/Fe_t ratio that is greater than 50%, where Fe_t is the total iron content (Baize and Jabiol 2011);

iv) a clearly defined angular polyhedral structure, with millimetric microstructure and gleaming faces;

v) base saturation.

Référentiel Pédologique (AFES 2008)	World Reference Base (FAO 2014)
Carbonated Fersialsols	Haplic Cambisols (Calcaric, Chromic)
Calcic Fersialsols	Haplic Cambisols (Eutric, Chromic)
Unsaturated Fersialsols	Haplic Cambisols (Eutric or Dystric, Chromic)
Eluvial Fersialsols	Haplic Luvisols (Chromic) or Albic Alisols (Chromic)

Table 8.2. *Equivalence between Fersialsols and other denominations (Bresson 2008, modified)*

"Free iron" denotes forms of iron that are not integrated within the crystalline networks of the soil minerals. Soil minerals are either primary minerals present in the parent materials, such as pyroxenes, amphiboles or magnetite, or they are secondary minerals formed in the soil, such as clay where iron most often occupies the octahedral sites (montmorillonite and ferriferous illite, vermiculite, chlorite and interstratified minerals) (Segalen 1971). Free iron includes amorphous or crystallized mineral forms, and the complex organic forms of iron. Free iron in amorphous form, with or without low crystallization, or in organic form (Fe_o) may be dosed in different ways, from the slowest (oxalate) to the quickest (easily extractable iron from the Segalen method) (Table 8.3). Iron in the form of crystallized iron oxides (hematite *Hm* α-Fe_2O_3, goethite *Gt* FeOOH), denoted by Fe_d is dosed with the reduction reagent citrate–bicarbonate–dithionite as per the Mehra-Jackson method (1960), while the dosage of the total iron (Fe_t) requires strong acids.

Fersiallitization directly leads to the extraction of the iron bound to the clays, or the iron from the primary minerals, or may even be accompanied by earlier steps, called ferrugination (Segalen 1971), in which the iron is separated from the minerals that make up the rocks and is individualized into new forms. The crystallized forms are responsible for the red (due to hematite) and ochre (due to goethite) colors we see.

The process of turning red is the "rubefaction" process, which results in the formation of hematite through the dehydration of iron oxihydroxides formed from the released iron. When only goethite is present, the color is yellow (hues from 10YR to 2.5Y on the Munsell color scale), while the presence of hematite, even at levels as low as 1% free iron, is enough to impart a redder 5YR color (Torrent et al. 1983).

The degree of rubefaction may be evaluated using the RR redness rating defined by Torrent et al. (1983) (equation [8.3]).

$$RR = \frac{(10-H)\times C}{V} \quad [8.3]$$

where H is the Munsell hue, C (chroma) is the saturation and V (value) is the clarity.

Dosage	pH	Fractions of iron	Common designation	Notation	Author(s)
Oxalate	3	Amorphous, that is, poorly crystallized and/or organic	Amorphous iron	Fe_o	Tamm (1922) reviewed by Schwertmann (1964)
Ethylenediaminetetraacetic acid (EDTA)	7.5–10.5	Amorphous, that is, poorly crystallized and easily extractable	Extractable iron	Fe_{EDTA}	Borgaard et al. (1979)
HCl 6 N and NaOH 0.5 N		Amorphous, that is, poorly crystallized and easily extractable	Easily extractable iron	FFE	Segalen (1968)
Citrate–bicarbonate–dithionite (CBD) reducing solution	7	Total free iron, especially crystallized iron in the form Fe^{III}	Free iron	Fe_d	Mehra and Jackson (1960)
Strong acids HCl-HNO_3 at 100°C or HF		All the iron in the sample	Total iron	Fe_t	Various authors

Table 8.3. *Main methods of iron extraction, their dosage and interpretations (from Segalen 1971; Stucki et al. 1988; Baize and Jabiol 2011)*

Region	Statistical criterion	Altitude (m)	*P* (mm)	*T* (°C)	No. of dry months/year	*RR*	Average size (nm)		Al substitution (mole %)		Fe_d	Fe_t	Fe_d/Fe_t	Hm/(Hm + Gt)
							Gt	Hm	Gt	Hm	(g.kg^{-1})	(g.kg^{-1})		
Sardinia (N_{FS} = 7)	Median	75	655	16.2	5.8	8	26	26	9	4	36.5	58.5	0.71	0.49
	Minimum	2	511	15.9	4.7	6	19	20	4	0	28.5	43.7	0.52	0.24
	Maximum	763	791	20.3	6.6	11.2	49	53	16	14	87.6	108.6	0.81	0.65
North-East Italy (N_{FS} = 5)	Median	150	1,415	13.2	11	8	19	12	13	1	51.8	70.6	0.74	0.71
	Minimum	5	1,078	12.8	0	6	17	11	10	0	33.9	62.6	0.54	0.60
	Maximum	852	3,313	14.2	2.7	11.2	21	12	17	2	53.3	70.8	0.77	076

Table 8.4. *Characteristics of color and the minerology of iron oxides for "terra rossa" Fersialsols in Sardinia and North-East Italy (Veneto and Frioul) (from Boero et al. 1992). NFS, numbers of samples that meet the Fersialsol criteria among the soils described by the authors; Hm, Hematite; Gt, goethite*

Location	Soil type	Depth (cm)	Fraction <2 µm	Hue (YR)	*RR*	Fe_o (g.kg^{-1})	Fe_d (g.kg^{-1})	Fe_o/Fe_d	Fe_t (g.kg^{-1})	Fe_o/Fe_t	Hm/(Hm + Gt)
Mount Carmel	RE1	0–10	No	10.0	0.0	0.7	4.6	0.16	38.8	0.12	
	RE1	10–25	No	11.0	–0.3	0.6	4.5	0.14	40.4	0.11	
	RE1	10–25	Yes	7.7	1.7	0.2	13.4	0.11	70.0	0.19	0.3
	RE2	0–10	No	10.0	0.0	1.1	13.4	0.08	48.7	0.28	
	RE2	10–25	No	9.3	0.5	1.1	13.7	0.08	50.0	0.27	
	RE2	10–25	Yes	4.9	5.5	2.2	26.1	0.08	81.3	0.32	0.4
	FS1	0–10	Yes	4.4	6.6	2.4	60.3	0.04	91.7	0.66	0.6
	FS1	10–25	Yes	4.2	7.2	2.2	60.9	0.04	96.2	0.65	0.6
	FS2	0–10	No	4.1	7.7	2.0	39.0	0.05	64.4	0.61	
	FS2	10–25	No	4.0	7.9	1.8	37.5	0.05	68.6	0.55	0.6
Galilee	RE3	0–5	Yes	7.7	1.2	2.1	17.3	0.12	37.8	0.46	0.5
	RE4	0–5	Yes	5.9	3.9	3.8	32.1	0.11	55.5	0.58	0.6
	FS3	0–5	Yes	5.1	5.5	3.1	40.5	0.08	53.3	0.76	0.6
	FS4	0–5	Yes	4.8	6.1	3.5	42.1	0.08	56.1	0.75	0.6

Table 8.5. *Colors and forms of iron between Rendosols and Fersialsols from the same landscape units in Israel (Singer et al. 1998, adapted). Hm, Hematite; Gt, goethite*

The redness rating, *RR*, is strongly correlated (R^2 = 0.8) with the hematite content: the values for a sample of 100 European soils whose hematite content is between 0 and 5% lie between 0 and 11, while a saturation effect plateaus at 12 for Brazilian soil with a hematite content that is higher than 5% (Torrent et al. 1983). For example, the range of RR values is 6–11 for Italian Fersialsols from hard limestone, with a median Fe_d/Fe_t ratio of 0.7 (Table 8.4) (Boero et al. 1992). For soils in Greece, the RR does not really exceed 8.5 and is negatively correlated with the mean annual pluviometry (Yassoglou et al. 1997). It is the same for Fersialsols from hard limestone, compared with Rendosols for the sites in Israel (Table 8.5) (Singer et al. 1998): (i) the *RR* values lie between –0.3 and 7.9 for all soils together, with a median of 6.6 for Fersialsols; (ii) the median ratio of free iron to total iron (Fe_d/Fe_t) is 0.65, while it is only 0.28 for Rendosols; the median Fe_d content in Rendosols is almost one-fourth that of the Fe_d content of Fersialsols. The relevant threshold for the Fe_d/Fe_t ratio depends on the nature of the parent material and has proven to be much smaller (around 0.3) for FS horizons from basalts, amphibole gneiss, ferrous chlorite schists, etc. (Bresson 2008). This is why a ratio of the easily extractable iron (Segalen method) to the total iron that is generally greater than 0.3 in FS horizons is also recommended, even if this approach can be improved. Although the Fe_d dosage still risks partly solubilizing the iron contained within certain minerals, the FFE dosage attacks poorly crystallized forms of iron, especially the very fine, small crystals (crystallites) that are generally seen with hematite.

Rubefaction, or the formation of hematite, is one of the key factors in the pedogenesis of Fersialsols. The prerequisite for rubefaction to occur is ferrugination, which is promoted by the wet season, in which iron is released, leading to the formation of ferrihydrites (Segalen 1971). Ferrugination is facilitated when (i) the parent material is rich in iron-containing minerals; (ii) the wet season is adequate; (iii) there is rapid turnover of organic matter; (iv) the soil has a pH value of between 6 and 8, making the iron insoluble. Rubefaction itself is a phenomenon that is still very poorly described in real conditions. In the laboratory, thermal transformation between 200°C and 400°C produces hematite from goethite, akaganite, ferrihydrite ($Fe_5(OH)O_7$, $4H_2O$) and maghemite (γ-Fe_2O_3) (Schwertmann, in Stucki et al. 1988). The occurrence of hematite is not very dependent on climatic condition as hematite can be the iron oxide that predominates in situations characterized by a mean annual temperature varying between 13°C and 20°C, or characterized by mean annual pluviometry of 500–3,000 mm, as observed for various Fersialsols formed from limestone (type "terra rossa") in two contrasting regions in Italy (Table 8.4) (Boero et al. 1992). The pedoclimate is more salient in explaining the occurrence of hematite: Singer et al. (1998) observed that Fersialsols dry out and heat up quicker than Rendosols. However, the climate affects the size of the hematite crystals as well as the rate of aluminum substitution. Hematite and goethite are crystallized much better in a hot and dry (xeric) environment. Using Mössbauer spectrometry, the size of hematite crystals is seen to be much smaller than goethite crystals (Figure 8.2) and the size increases as the climate grows hotter and drier (xeric), as in Sardinia, for example, when contrasted with Friuli Venezia Giulia in Northern Italy (Table 8.4) (Boero et al. 1992). The ratio of Al substitution in goethite to Al substitution in hematite tends to increase in humid environments due to the very low rate of aluminum substitution in hematite (Table 8.5). The rate of aluminum substitution in goethite is variable, between 4 and 17%, while hematite is rarely substituted, which makes it more susceptible to dissolution by a reducing agent (Peterschmidt et al. 1996). A possible explanation for this low rate of substitution in hematite in Northern Italy could be that the vermiculization and rubefaction processes are competing for Al (Boero et al. 1992). According to Singer et al. (1998), the Hm/(Gt+Hm) ratio, which is higher for Fersialsols than Rendosols, may be used as an indicator for the pedoclimate (Table 8.4).

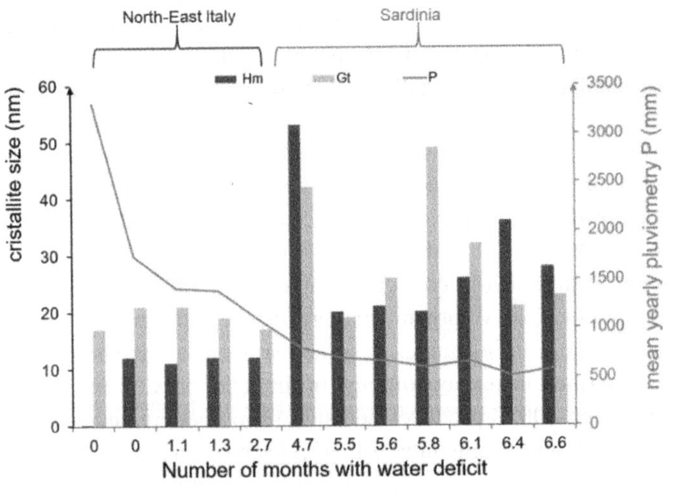

Figure 8.2. *Size of iron oxides for "terra rossa"-type Fersialsols in Sardinia and North-East Italy, depending on the number of dry months annually (Boero et al. 1992, adapted). Hm, Hematite; Gt, goethite. For a color version of this figure, see www.iste.co.uk/coquet/agricultural.zip*

8.2.2. *Geographic distribution and notable sequences*

Fersialsols may be formed from a variety of parent materials: fixed, hard rocks which may or may not be limestone, surficial formations of various origins, on the condition that (i) a slightly acidic ambience and adequate climatic drainage allow geochemical alteration, at least locally, following the dissolution of limestone; and (ii) the alteration medium is aerated and draining in order to allow rubefaction. For these reasons, Fersialsols are generally not formed on marl, given the lack of adequate drainage (Legros 2007). The fersiallitic weathering is limited, as it results in a partial desilification, and the mineralogical assemblage (which is also a result of the heritage of the parent material) is typically characterized by a ratio $SiO_2/Al_2O_3 \geq 2$ (Lamouroux 1971).

8.2.2.1. *"Terra rossa" Fersialsols formed from hard limestone*

The dissolution of limestone through positive climatic drainage on sedimentary limestone rocks leads to the formation of a loose alteration cortex, releasing primary minerals that are altered and whose non-carbonate residues, between 1 and 8% of the rock (Lamouroux 1967; Yaalon 1997) contain iron, silica and aluminum. These get further concentrated as the cortex is decarbonated to neoform clays (montmorillonite) and iron oxides.

The predominant mechanism here is pellicular weathering, defined as follows by Lamouroux (1967): "in an aerated, humid environment, clayey soils that are not calcareous or only slightly calcareous and rich in montmorillonite shrink when they dry, tearing away a thin film from the surface of the carbonate rocks that are in contact with them" (Figure 8.3).

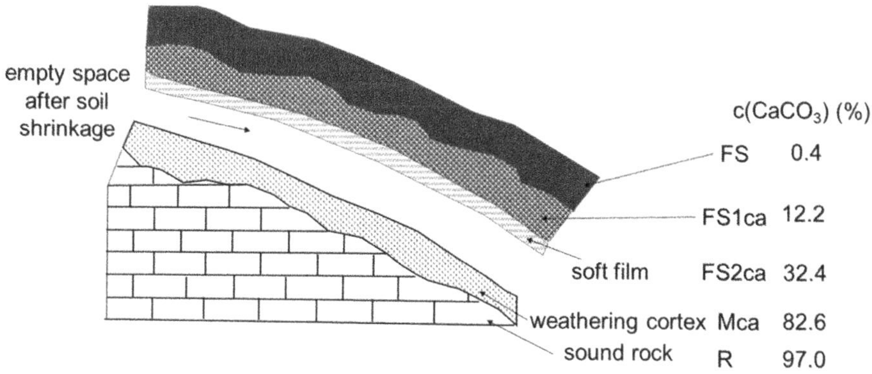

Figure 8.3. *Pellicular weathering (adapted from Lamouroux 1967). The notations are taken from the Référentiel Pédologique (AFES 2008)*

The decarbonation phase requires a minimum rainfall: below 300 mm in the southern Maghreb, the soils are no longer decarbonated and the clayification is more discrete. Consequently, the soils remain brown (Legros 2007). The thickness of the film (pellicule) that is torn away ranges from a few micrometers to tens of micrometers. Hypothesizing that the limestone is hard and pure, with 1–2% of insoluble residues, and a tearing velocity of 10-40 μm/year, 10–40 cm would have had to be dissolved during the Holocene (10,000 years) in order to form 1–8 mm of pedological material. By extrapolating this process over a long period of 2 Ma, the formation of 50 cm of fersiallitic soil would require the dissolution of a considerable thickness – 50 m of limestone rock. Thus, it has been proposed that the present-day thickness of fersiallitic soils is also from contributions from allochthonous, colluvial, alluvial and wind-borne materials (Yaalon 1997). Based on the ratios of trace elements, Zr/Hf, La/Yb, Cr/Sc and Th/Ta, analyzed in the soils of the Balearic Islands, Muhs et al. (2010) interpret the origin of the silicate fractions of the "terra rossa" Fersialsols of Mallorca as mainly being derived from wind dust from the Sahara. However, on stable structural surfaces, such as alluvial plateaus, the soils are indeed the result of pedogenesis over a long duration (>1.5 Ma).

8.2.2.2. *Fersialsols from fluvio-glacial pebbles*

Older Fersialsols are part of soil chronosequences in alluvial terraces that have been studied in the Orb and the Rhône valleys in France, in the Lower Moulouya in Morocco and in Jarama and Henares in Spain (Vaudour 1983). The age of the alluvial terraces was related to the stages of fersiallitic evolution (Arnal 1971; Barrière 1971), which can go up to 2.5 Ma (Villafranchian or Plio-Pleistocene) for the Rhône (Bornand 1978; Vaudour 1983) (Figure 8.4). The initial stage involves the decarbonation of fluvio-glacial pebbles, containing rocks of various nature (Dx): the pebbles are silicated (Dsi) because the decarbonation concentrates the silicious elements, especially in quartzite stones, which are more resistant. Fersiallitization that results from a slightly acidic environment induces a weathering front that gradually sinks toward the base of the solum, accompanied by increasing clayification and rubefaction (stages 1–4).

There is no corrosion of quartz in the alterite at the base of the soil (Dsi), very little in the median part of the soil (FS horizon), and considerable corrosion, with small cavities, as we approach the surface (Legros 2007). The "clay + hematite" combination maintains good structure, aeration and oxygen penetration (FS horizon). Smectites, which are dominant in the early stages, decrease in proportion toward the top of the FS horizon, giving way to kaolinite, which becomes dominant in the final stages.

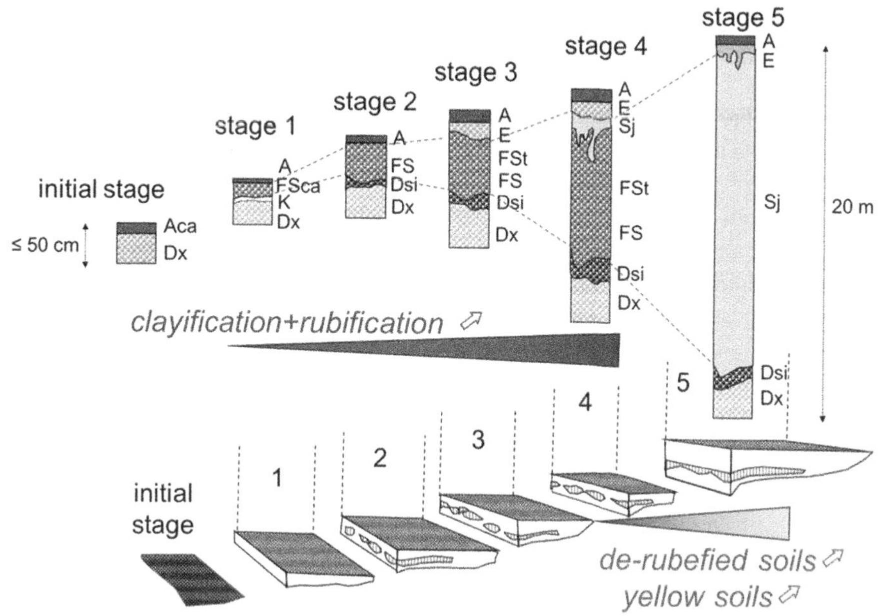

Figure 8.4. *Chronosequence of fersiallitic soils in the Rhone Valley (Bornand 1978, adapted). The notations are taken from the Référentiel Pédologique (AFES 2008). For a color version of this figure, see www.iste.co.uk/coquet/agricultural.zip*

Although the primary minerals remain, as well as the clays that are formed from weak transformations (interstratified illites), the clay assemblages undergo progressive desilicification accompanied by leaching of the bases (lixiviation). This results in gradual acidification, which promotes argilluviation processes (E horizon atop an FSt horizon): small clays, such as kaolinite, migrate toward the bottom and the soil structure is prone to compaction. The Fersialsols that are thus formed are related to eluvial Fersialsols (Table 8.1). The accumulation of clay and the compaction (FSt) contribute to water saturation in winter, which itself facilitates the reduction of iron. Hydromorphic conditions induce de-rubefaction, a differential removal of hematite in favor of goethite, and the development of a "xanthomorphic" character (denoted by "j"): the matrix acquires a yellow or orange hue (7.5 YR or yellower) and the redistribution of iron and manganese form fine, glossy black coatings on the faces of aggregates and/or very small, black nodules (Bresson 2008), with the easily extractable iron content remaining high. If the phenomenon continues, goethite itself is eliminated and the color changes to gray or white (E horizon). The final stages of yellowing and of the soil degradation (with glosses) are called "ultisolic" and when compared with the stages reached after a shorter period in tropical conditions, they are related to tropical, ferruginous soils. However, the kaolinite in Mediterranean regions is never crystallized to this extent (Paquet 1970).

As with the pedogenesis of Fersialsols from limestones, fersiallitic weathering corresponds to a massive loss of material, where "60–80% of the initial material has been exported" (Bornand 1978). The considerable thickness of the stage 5 soil also raises questions about the nature of the parent material (were allochtonous materials brought in?) and on the cumulative phenomena over such a long time period (did the past Villafranchian climates lead to this evolution?), as well as the eventual hydromorphy of the initial stage for the pebbles initially located in a poorly drained downstream area, which have been brought into inverted relief following the incision of the valley (Vaudour 1983). If alluvial terraces are dated using the degree of fersiallitic evolution, inconsistencies emerge between the northern and southern Mediterranean shores (Vaudour 1983): this would require refining the dating of the detrital pebbles.

8.2.2.3. *Fersialsols from non-carbonated rocks*

Red soils, in general, appear in regions that grow more arid as the carbonate content of the substrates decreases (Legros 2007). Therefore, Fersialsols formed from shale correspond to dry regions, since the alterite is rich in iron and in clays inherited from the transformation of micas (illites) and can be directly rubefied. However, it is essential that the small quantity of bases that are present does not get leached in order to avoid acidification and a drop in pH, which would promote iron reduction. Volcanic rocks are rich in bases, which thus facilitates rubefaction if the pH is maintained at a suitable level (Legros 2007).

The Fersialsols described in the Alentejo palaeozoic peneplain in southern Portugal (Dachary 1975) correspond to advanced stages with argilluviation: the presence of an illuviation horizon (FSt), leached and with a low base saturation rate (between 30 and 50%). The climatic conditions are those of a semi-arid climate (P = 550 mm, T = 16°C; four dry months). Despite a low pH of 5 and a high clay content (38–54%), iron reduction does not take place in the well-rubefied FSt horizon, probably due to the good porosity, related to the maintenance of the structure (prismatic macrostructure with an angular polyhedral substructure).

For Fersialsols developed in the Mediterranean mountains, with sub-humid climate, intense hydrolysis transforms the phyllites (illite and chlorite) into vermiculite, "through ions being drawn off from the interfolial spaces and octahedral cavities" (Paquet 1970). This vermiculitization is observed in the Beni Snassen in Morocco (Paquet 1970) and in the soils in north-east Italy, and is interpreted as being concomitant with rubefaction (Boero et al. 1992).

An extreme case is that of the sandstone on the northern edges of the Sahara, which tends to appear whenever the average annual pluviometry exceeds 100 or 150 mm (Legros 2007).

8.2.3. *Agronomic potential*

FS horizons typically have properties that are conducive to cultivation and fertility: (i) a high cation exchange capacity (CEC), of the order of 25–30 cmol.kg^{-1} (e.g. Vaudour et al. 2017), which can form a reservoir of nutrient cations, such as K^+ and Mg^{2+}; (ii) well-distributed porosity that allows the aeration of the root system; (iii) easily extractable iron content (iron EDTA $\geq$ 17 mg.kg^{-1}; Vaudour et al. 2017) that guarantees the formation of chlorophyll pigments and phenolic compounds. This also creates a high phosphorus exchange capacity (Bresson 2008). For perennial crops, such as grapevines, some of the most remarkable viticulture lands are characterized by Fersialsols. For example, the high alluvial soil of the Châteauneuf-du-Pape (Vaudour 2003b), whose depth of over 10 m (Figure 8.5) guarantees deep roots systems, which promotes plant resilience against water stress during the summer.

Figure 8.5. *Wine-growing Fersialsol formed from fluvio-glacial pebbles from the Plio-Pleistocene High Alluvial Level in the southern Rhône region (Vaudour 2003b)*

8.3. Vertisols

Although non-specific, Vertisols are frequently present in Mediterranean and simili-Mediterranean regions, where they generally develop in downstream areas, typically located in a poorly drained lowland.

8.3.1. *Definition and formation processes*

Vertisols (AFES 2008; FAO 2014) are soils that are very rich in smectites and have the following characteristics:

i) a high 2/1 bisiallitic, "smectite" clay content (40–60%);

ii) dominant fine clays (<0.2 µm);

iii) highly stable, organic matter content of the order of 1–2%;

iv) "black" color;

v) alternating shrinking/swelling (linear extension coefficient or the "coefficient of linear extensibility" index COLE >0.09);

vi) shiny and oblique *slickenside* faces, and a microtopography of the soil surface with alternating micro-monticules and micro-depressions, called *gilgai*.

Solum	Depth (cm)	Munsell color	Clay (%)	$CaCO_3$ (%)	MO (%)	pH	Clay minerals	CEC (cmol.kg^{-1})	Detritus
Detritus	0–40	10YR2/2	48.5	27	3.6	7.6	montmorillonite >>kaolinite>illite>>attapulgite		LAv
Detritus	40–80	10YR2/2	54.5	24	4.1	7.7	montmorillonite >>kaolinite>illite>>attapulgite		SV
Detritus	80–140	10YR2/2	41.7	35	3.5	7.7	montmorillonite >>kaolinite>illite>>attapulgite		V
Clay 5	0–30	10YR3/3	64.5	7	2.7	7.6	montmorillonite>>kaolinite>illite		LAv
Clay 5	30–70	2.5Y4/4	70.7	7	1.6	7.6	montmorillonite >>kaolinite and illite		SV
Clay 5	70–100	2.5Y4/4	60	5	1.0	7.6	montmorillonite >>kaolinite and illite		V
JV	0–90	10YR3/3	42	6	1.3	7.7	illite>>montmorillonite>>kaolinite		Av/SV
JV	90–140	10YR3/3	36	7	1.8	7.5	illite>>montmorillonite>>kaolinite		Sv
JV	140–180	10YR7/3	34	40	0.5	7.8	montmorillonite>>illite and attapulgite>>kaolinite		K
Córdoba	0–30	–	69.4	8	1.7	7.7	–	47	LAv
Córdoba	30–60	–	70.5	9	1.2	7.6	–	37	SV
Córdoba	60–90	–	78.7	7	0.9	7.8	–	30	V
Detritus	0–40	10YR2/2	48.5	27	3.6	7.6	montmorillonite >>kaolinite>illite>>attapulgite		LAv
Detritus	40–80	10YR2/2	54.5	24	4.1	7.7	montmorillonite >>kaolinite>illite>>attapulgite		SV

Table 8.6. *Examples of Vertisols (FAO 2015) in Sicily ("Detritus" and "Clay 5", Egli et al. 2020) and in Spain ("JV", Paquet and Vaudour 1974; "Córdoba", Lopez-Bellido et al. 2016)*

The thickness varies from 60 to 100 cm. These are also called *Black Cotton soils* in the United States, *regur* in India and *black turf soils* in South Africa.

In thc topographic situations where Vertisols are described, the clays are either inherited or neoformed due to the concentrations of silicic acid solutions from upstream (Paquet 1970). The parent material is not a coarse material but soft and fine (Yaalon 1997). These may be Pliocene evaporitic clays in Sicily (Egli et al. 2020), Mio-Pliocene gypsum formations around Madrid (Paquet and Vaudour 1974; Vaudour 1977), or thick silt in Andalusia (Lopez-Bellido et al. 2016) (Table 8.6). In Vertisols, where the confined and ion-rich environment combines all the conditions required for durability and preservation, the stable clay mineral in equilibrium is montmorillonite (Paquet 1970), while the illites inherited from the parent material do not undergo any transformation in the solum.

According to the Référentiel Pédologique, classification as Vertisols requires the presence of two horizons with the following characteristics: an SV horizon (structural vertic) that can be unsaturated, and the vertic horizon (V), located below, which starts at a depth of less than 10 cm (AFES 2008). Consequently, the characteristics of the JV solum (Table 8.6) make it a Vertisol according to the FAO nomenclature but lie outside the definition of the Référentiel Pédologique.

8.3.2. *Agronomic potential*

Vertisols in the Mediterranean are most often used to grow cereals and possibly for vines, and often without irrigation. In India, they are used to cultivate cotton. Vertic horizons have a high CEC, and the shrinkage cracks facilitate water recharging, which promotes fertility. However, the shrinkage/swelling properties are not conducive to the movement of machinery. The low permeability of Vertisols slows down water recharging, while the cracks increase the area for exchange between water and air within the solum, which intensifies water loss through evaporation (Lopez-Bellido et al. 2016). More generally, the cracks influence the dynamics of the hydric, gaseous and thermal exchanges, thereby directly impacting the crop yield, in interaction with the system of agriculture, especially soil tilling. Restricting soil tillage would have beneficial effects on crop yield.

8.4. Soils with limestone accumulations or Calcarisols

Soils with limestone accumulations are typical of Mediterranean and simili-Mediterranean regions and these generally develop in downstream regions that may or may not be poorly drained.

8.4.1. *Definition and formation processes*

Soils with limestone accumulations, limestone slabs or limestone crusts are defined by the predominant concentrations of secondary carbonates. We also speak of calcretes, caliche, limestone crusts, carapaces and cuirasses. Formation processes involve carbonate chemical equilibria (equations [8.4]–[8.11]), summarized by equations [8.8] and [8.10].

$$CO_{2(g)} \leftrightarrows CO_{2(l)} \quad [8.4]$$

$$CO_{2(l)} + H_2O_{(l)} \leftrightarrows H_2CO_{3(l)} \quad [8.5]$$

$$H_2CO_3 + H_2O \leftrightarrows H_3O^+ + HCO_3^- \quad pK\,(25°C) = 6.37 \quad [8.6]$$

$$CaCO_3 + H_2CO_{3(l)} \leftrightarrows Ca^{2+} + 2HCO_3^- \quad [8.7]$$

$$CaCO_3 + CO_{2(l)} + H_2O_{(l)} \leftrightarrows Ca^{2+} + 2HCO_3^- \quad [8.8]$$

$$HCO_3^- + H_2O \leftrightarrows H_3O^+ + CO_3^{2-} \quad pK\,(25°C) = 10.25 \quad [8.9]$$

$$Ca^{2+} + 2\,HCO_3^- \leftrightarrows CaCO_3\downarrow + CO_{2(g)} + H_2O_{(g)} \quad [8.10]$$

$$Ca^{2+} + CO_3^{2-} \leftrightarrows CaCO_3\downarrow \quad pK\,(25°C) = 9.0 \quad [8.11]$$

Initially, the CO_2 in contact with water forms carbonic acid, which decomposes into H_3O^+ and hydrogen carbonate or bicarbonate HCO_3^- and then into CO_3^{2-}. If the partial pressure of CO_2 (pCO_2) decreases and/or if the HCO_3^- ions are leached, the acidification process becomes irreversible. Carbonic acid contributes to the dissolution of limestone in coarse elements, or the limestone held within the matrix. Second, processes that promote the saturation of the bicarbonate or carbonate solution (which may have migrated from the top of the solum (called the *per descensum* process) or through capillarity (called a *per ascensum* process), or which may have migrated laterally into the landscape) lead to the formation of clusters, then nodules and then the influx of $CaCO_3$. The formation of these different types of accumulation is chiefly controlled by the following processes (Freytet and Verrechia 1989; Legros 2007): (i) evaporation within the solum, which grows more intense and occurs closer to the surface as the aridity of the environment increases; (ii) water absorption through roots (leading to "rhizoconcretions"); (iii) the facilitating presence of coarse particles with the water sheet saturating the lower surface of these particles more easily; and (iv) "biomineralization" through the presence of microorganisms such as algae and cyanophyceae (leading to the wavy or stromatolite-like biogenic film or laminar crust). The epigenesis of indurated $CaCO_3$ or calcrete is another mechanism – this is the replacement of a mineral by calcite.

Soils with $CaCO_3$ accumulations chiefly come under Calcarisols in the Référentiel Pédologique (AFES 2008), according to which a continuous, non-indurated calcic horizon (Kc) or a continuous, indurated, petrocalcic horizon (Km) must start at least 35 cm from the surface, and must be at least 10 cm thick for the Kc and at least 5 cm thick for the Km. Soils with $CaCO_3$ accumulations come under Calcisols in the WRB (FAO 2014), in which these accumulations must occur within the first 100 cm. There may be a horizon of discontinuous accumulations, called a *calcic horizon*, or a continuous horizon, called a *petrocalcic horizon* (calcrete). The calcic horizon is defined by the secondary accumulation of calcium carbonate $CaCO_3$, either in the diffuse form of millimetric calcite particles dispersed in the network, or in discontinuous concentrations in the form of pseudomycelium, coatings (cutanes), soft or indurated nodules, or veins. We must not confuse the WRB term "Calcisol" and the adjective "calcic" with the terms in the Référentiel Pédologique, where a calcic horizon (denoted by "ci") is non-carbonated, but saturated with calcium, and a Calcisol is a solum characterized by Aci/Sci superposition.

8.4.2. *Geographic distribution and notable sequences*

The geographic distribution and the profiles of the soils with $CaCO_3$ accumulations are depicted in Figures 8.6–8.8, for northeastern Morocco (Millot et al. 1969; Ruellan 1970; Paquet 1983). In the semi-arid Triffa plain, and the arid Zebra plain, located in the north-east and north-west of the Beni-Snassen mountain ranges, respectively, and separated from the lower Moulouya valley, the soils are formed from alluvial–colluvial gravels, notably Quaternary and Plio-Quaternary. As with the oldest fan terraces, the mountain massifs are the site of fersiallitization (Figure 8.6) and, regardless of the age of the deposit, carbonate accumulations form on the gravels on the slopes and the plains. The K horizons begin at 30–50 cm in the Zebra plains and slightly deeper (40–70 cm) in the Triffa plains (Ruellan 1970).

In contrast to the geochemical weathering front, the carbonation front intensifies vertically, from the bottom to the top of the solum, with the accumulation starting as diffuse shapes, then clusters and nodules, and then becoming increasingly continuous and indurated toward the top, until, on the oldest surface (Villafranchian slope) a specific compact concrete pavement is formed (Figure 8.7). Millot et al. (1969) refer to the combination of "incrustation, crust and pavement" as a "carapace", also referred to as a "calcrete" by Alonso-Zarza et al. (1998).

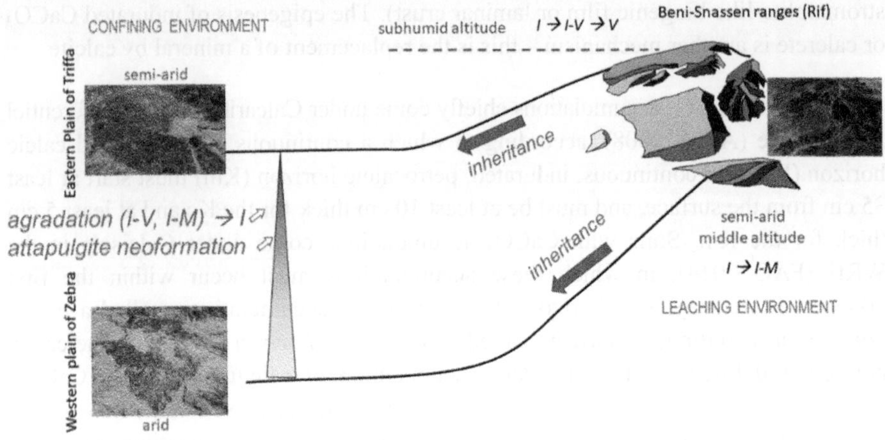

Figure 8.6. *Altitudinalclimosequence for soils with calcite accumulations in northeastern Morocco (Paquet et al. 1969, adapted). I, illite; V, vermiculite, M, montmorillonite. Taken from true-color Google Earth® satellite images. For a color version of this figure, see www.iste.co.uk/coquet/agricultural.zip*

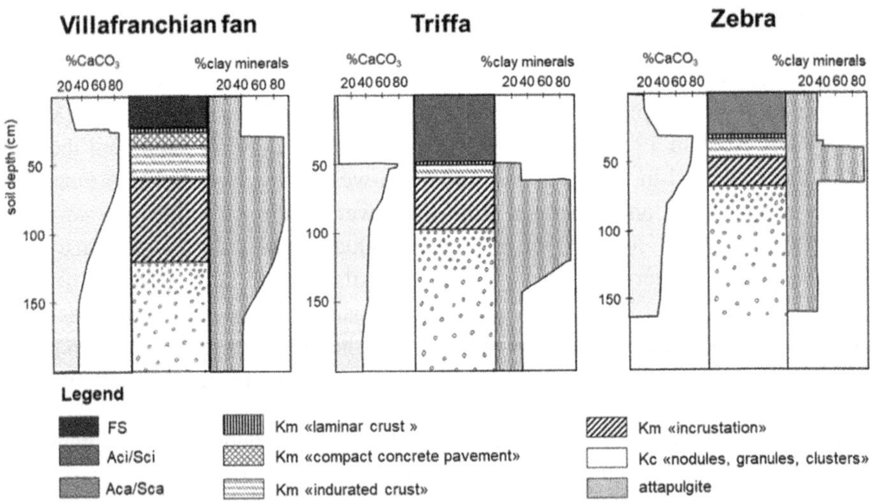

Figure 8.7. *Soils with $CaCO_3$ accumulations in northeastern Morocco (Millot et al. 1969, adapted). For a color version of this figure, see www.iste.co.uk/coquet/agricultural.zip*

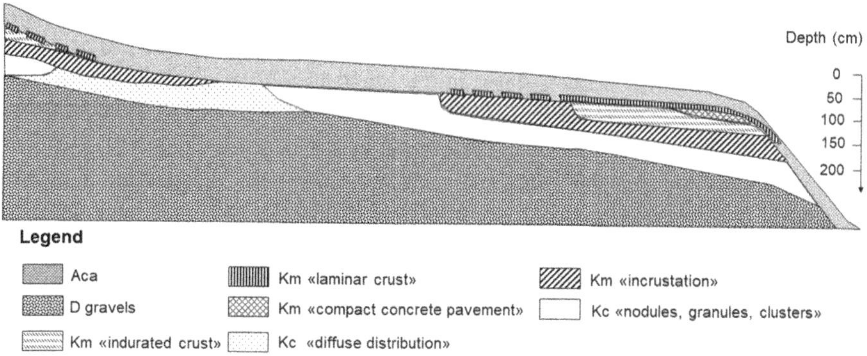

Figure 8.8. *Lateral shifts between the various types of limestone accumulation horizons in northeastern Morocco (Ruellan 1970, adapted). For a color version of this figure, see www.iste.co.uk/coquet/agricultural.zip*

The intensification of $CaCO_3$ accumulation leading to the formation of the carapace is also described laterally, along the slope, upstream to downstream, and as a function of the slope (Figure 8.8). In addition, carbonates dissolve better in cold water, which has a higher CO_2 content than warm water. Karst, caused by the dissolution of limestone, are well-developed phenomena in mountains. The lateral shifts are also related to the heritage of products of geochemical weathering, which are leached from upstream to downstream and are then recombined into attapulgite (or palygorskyte) $Mg(Al_{0.5-1}Fe_{0-0.5})Si_4O_{10}(OH)\cdot 4H_2O$, a fibrous, clay mineral that is specifically observed in carbonate accumulations in arid and semi-arid environments. An evaporative medium that is rich in silica and magnesium, with a high pH, is conducive to the formation of this mineral (Paquet 1983; Pozo and Calvo 2018). Attapulgite constitutes 40% of the clay fraction in the nodular zone and reaches it maximum value (90–95%) in the incrustations. It then decreases to 40% in either the "indurated crust" or the "concrete pavement", where it is epigenized by calcite (Figure 8.7). Attapulgite tends to be unstable and disappears in surface horizons or in wetter conditions, where it may transform into smectite (Paquet 1983).

The magnitude of carbonate accumulations cannot be explained by the *per descensum* process alone, especially for accumulations referred to as "crusts" and "pavement". In relief units with geomorphological variations, the vertical succession describes polyphase soils that are the result of reciprocal relationships between geomorphogenesis and pedogenesis over several climatic sequences, or climosequences (Millot et al. 1969; Vaudour 1975, 1980, 2003a; Alonso-Zarza et al. 1998).

Beyond the lateral shifts, the calcretes are polygenic and polyphase formations. They are partly fossils, which can be linked to humid, swampy environments, as described in the semi-endorheic basin in the Montes de Toledo (Vaudour 1975), or linked to a shallow water table, as in the Sorbas basin in Andalusia (Candy and Black 2009). These "result from interference in the deposition phases, and from the transformation of the sediment during premature diagenesis, including biomineralization processes" (Vaudour 2003a). In the Beni Snassen, therefore, the ribboned crusts of the terraces in the Sefrou wadi, which underwent U/Th dating, indicate repeated chronostratigraphic inversions, suggesting successive "formation and dissociation" (Mezrhab 1994). The alluvial slopes of the Campo de Cartagena (Spain) also indicate a complex history. The soil sequences here include buried paleoFersialsols (Alonso-Zarza et al. 1998). In addition, while carbonate accumulation tends to increase over time, the rate at which the carbonate covering on coarse elements increases varies as a function of climatic fluctuations (Badía et al. 2009), along with "micritization" (the dissolution–reprecipitation process). The more arid and colder periglacial phases are associated with a slower growth of "calcretes" (Candy and Black 2009).

8.4.3. *Agronomic potential*

Certain soils with carapace accumulations may prove to come under both Calcisols and Fersialsols. These therefore benefit from characteristics that promote the fertility of the fersiallitic horizons. However, in the most arid conditions, the tillable horizon has quite a small thickness and, consequently, the storage capacity is also small. Lateral roots are liable to dissolve the indurated part of the carapace, releasing protons in the vicinity of the rhizosphere. The carapace is therefore not impassable. On the contrary, around pulverulent nodules, the lateral roots can no longer play the role of a buffer due to the predominance of what is called "active" calcium carbonate, that is, calcium carbonate that reacts with ammonium oxalate. This is chiefly composed of fine fractions of $CaCO_3$, smaller than 20 μm (clay + fine silt). The presence of active calcium carbonate interferes with the ferric iron supply which causes iron chlorosis, a physiological problem in plants. The risk of iron chlorosis is amplified when deep ripping breaks through a carapace when trying to increase the explorable depth, allowing roots to reach the pulverulent clusters rich in "active" calcium carbonate. The carapace is, first and foremost, a chemical constraint against root implantation, especially for perennial plants with deep root systems, such as grapevines (Vaudour et al. 2005).

Additionally, carbonate crusts may sometimes have both calcite and dolomite, or even gypsum. The presence of dolomite is likely to cause an excess of magnesium in the soil solution. Due to the antagonism between Mg^{2+} and K^+ ions in the nutrition of plants like grapevines, this causes an imbalance in the potassium supply and may even cause potassium deficiency. In the context of soils with high pH, the sulfate that may be released by gypsum ($CaSO_4$, $2(H_2O)$) is not identified as problematic unless it is associated with sodium chloride, NaCl, or with sodium Na^+. These salts are more soluble than gypsum and, therefore, more likely to enter solutions. They are characteristic of salty or sodic soils.

8.5. Salisols and Sodisols

This section briefly describes the characteristics of Salisols and Sodisols, which are of interest in the Mediterranean regions. These soils develop in this region when the climatic drainage (equations [8.1] and [8.2]) is repeatedly and predominantly negative: (i) in endorheic or semi-endorheic basins; (ii) due to the salts in the irrigation water; (iii) around the saltwater Mediterranean lagoons; and (iv) due to saliferous parent materials such as sedimentary rocks of marine origin, or detrital marine and laguno-continental formations from the Oligocene and Neogene periods.

In these diverse conditions, which may sometimes occur together, soluble salts get concentrated and salinization occurs.

8.5.1. *Definition and formation processes*

Salsodic solums, that is, containing salts or alkali, are characterized by (AFES 2008): (i) the presence of a certain quantity of salts, either soluble in the soil solution or as precipitate in the horizon itself (salic horizon); (ii) the presence, on the exchange complex of the horizon, of a relatively large quantity of sodium as compared to other adsorbed cations (sodic horizon). In all cases, Na^+ is the most frequent cation. The overall quantity of the salts is assessed by measuring the electrical conductivity (EC) of a paste made of a saturated soil sample, obtained by kneading at least 300 g of fine, dried earth with distilled water, and then taking this paste up to the Atterberg liquid limit, in order to obtain 50 cm^3 of the extraction solution. The earth/water ratio in the paste made in this way varies depending on the texture. For example, the ratio is 1/2 for a clayey texture and 1/5 for a sandy sample (AFES 2008).

Saline soils generally include soils whose EC is greater than 4 dS/m and the total cation quantity is greater than the CEC (USDA 1954). According to the Référentiel Pédologique, Salisols are characterized by the presence of a salic horizon at a depth of less than 60 cm. Salic horizons (denoted by SaA, SaS, SaC, SaY and possibly

SaBT, SaH and SaK) contain salts that are more soluble than gypsum, such as NaCl (halite), Na_2SO_4 (mirabilite), KCl (sylvite), $MgCl_2$, $CaCl_2$, $MgSO_4$ (epsomite), $NaHCO_3$ (bicarbonate of soda), $NaNO_3$ (nitronatrite) as well as complex salts. They are distinguished by an EC value that is greater than or equal to 8 dS/m, at 25°C, at least once a year. They may be carbonated or bicarbonated (pH > 8.5), or even chloro-sulfated or nitrated (pH < 8.5 and CE ≥ 15 dS/m at 25°C).

Sodic soils, sometimes called alkaline soils, denote soils in which there is a high proportion of exchangeable sodium within the CEC (*exchangeable sodium percentage* or ESP), greater than the arbitrary threshold of 15% (USDA 1954). Sodisols are characterized by the presence of a sodic horizon at a depth of less than 60 cm and by the absence, or near absence, of soluble salts, leading to poor maintenance of structure. Sodic horizons (denoted by NaA, NaS, NaBT, NaC) are distinguished by an ESP ratio >15% and by a dispersion of the sodic clay. This causes degradation of the structure, becoming massive, with low intra-aggregate porosity. The term "solonetzic" corresponds to the presence of an eluvial horizon. A solonetzic horizon lies atop an illuvial horizon (NaBT) enriched in sodium and clay, with a typically prismatic structure and low intra-aggregate porosity. Solodized horizons correspond to a degraded solonetzic stage: a desaturation in the upper horizons corresponds to a drop in pH to values between 4 and 5. Under the combined effect of acidity and oxidation–reduction, the clay minerals are degraded and the E horizon is bleached.

Référentiel Pédologique (AFES 2008)	World Reference Base (FAO 2014)
SALISOLS CHLORO-SULFATÉS (Chloro-sulfated salisols)	Solonchaks (Chloridic or Sulfatic)
SALISOLS CARBONATÉS (Carbonated salisols)	Solonchaks (Carbonatic)
SODISOLS INDIFFÉRENCIÉS (Undifferentiated Sodisols)	Solonetz
SODISOLS SOLONETZIQUES (Solonetzic sodisols)	Haplic Solonetz
SODISOLS SOLODISÉS (Solodized sodisols)	Solonetz (Albic)
SODISALISOLS	Endosalic Solonetz
SALISODISOLS	Episalic Solonetz

Table 8.7. *The equivalence between the names for Salsodic solums (AFES 2008)*

Salisols correspond to Solonchaks from the World Soil Classification (FAO 2014) (Table 8.7) and to Salorthids in the Soil Taxonomy (Soil Survey Staff 1999), while solonetzic Sodisols and solodized Sodisols correspond, respectively, to Solonetz and Solods.

Salsodic soils have solums with both salic and sodic horizons and these are called either Sodisalisols or Salisodisols, based on whether the shift from the surface to the depths is a shift toward salic horizons (Sodisalisols) or sodic horizons (Salisodisols). According to the World Soil Classification, these soils are Solonetz, accompanied by the qualifiers "endosalic" or "episalic" to indicate whether the salic horizons are located at depth or on the surface (Table 8.7).

Solum	Depth (cm)	EC (dS/m)	ESP (%)	Clay (%)	Silt (%)	Sand (%)	$CaCO_3$ (%)	Gypsum (%)	MO (%)	pH	Na+ (cmol/kg)	CEC (cmol.kg^{-1})	Horizon
Sérignan T1	0–40	7.0	9.2	21.3	59.4	19.3	5.7		3.87	8	1.18	12.9	LA
Sérignan T1	40–60	7.4	11.7	18.3	60.5	21.2	6.5		3.04	8.2	1.45	12.4	SJ_p
Sérignan T1	60–90	8.0	12.8	17.8	69.5	12.7	5.9		3.11	8.3	1.44	11.3	$SaSJ_{p1}$
Sérignan T1	90–100	9.1	19.4	16.2	70.9	12.9	5.7		2.94	8.5	2.39	12.4	$SaSJ_{p2}$
Sérignan T1	>100	8.9	35.4	12.7	64.3	23	6.1		2.87	8.6	3.14	8.9	$SaSJ_{p3}$
Hazeva Profile 1	0–8	5.5	20.4	3.4	23.5	73.1	30.8	0.08		8.2	1.20	5.9	NaJs
Hazeva Profile 1	8–30	16.4	34.0	1.6	11	87.5	24.5	0.21		7.6	1.55	4.9	SaJ
Hazeva Profile 1	30–60	11.0	31.3		1.4	98.6	20.3	0.34		8.1	0.85	2.7	SaDx2
Hazeva Profile 1	60–82	16.2	32.2	2.1	3.8	94.2	19.4	0.12		7.9	1.05	3.3	SaDx3
Hazeva Profile 1	82–100	14.7	31.6	0.8	1.7	97.5	8.4	0.13		8.4	0.55	1.7	SaDx4
K5	10–50	1.7	11.7	51	46	1	40		0.9	8.2	1.49	2.1	A
K5	50–70	3.4	20.1	59	36	1	40		0.8	8.4	3.61	4.3	NaS

Table 8.8. *Examples of salsodic solums in Languedoc, in the Orb delta (Sérignan T1; Bless et al. 2022), in northern Tunisia in the Mejerda Delta (K5; Hachicha 1998, adapted) and in Israel in the Arava Valley (Hazeva Profile 1; Dan et al. 1982)*

Table 8.8 illustrates examples of salsodic soils used for growing grapevines (Sérignan) or irrigated for cereals and fodder crops (K5), or not yet used for growing crops (Hazeva Profile 1). It can be seen that the textures are very diverse, predominantly silty (Sérignan), sandy (Hazeva) or even clayey, dominated by montmorillonite (K5).

8.5.2. *Geographic distribution and notable sequences*

Soil salinization and sodization are complex processes that are closely linked to surface and underground flows. Apart from geochemical weathering upstream, geothermal springs, transport (rain, streams, underground or coastal water tables, wind) and accumulation in the soils through the evaporative effect of the winds or through the decomposition of organisms, are at origin of soluble salts and their varied concentrations in soils (Hamdi-Aissa et al. 2004; Khatteli et al. 2016).

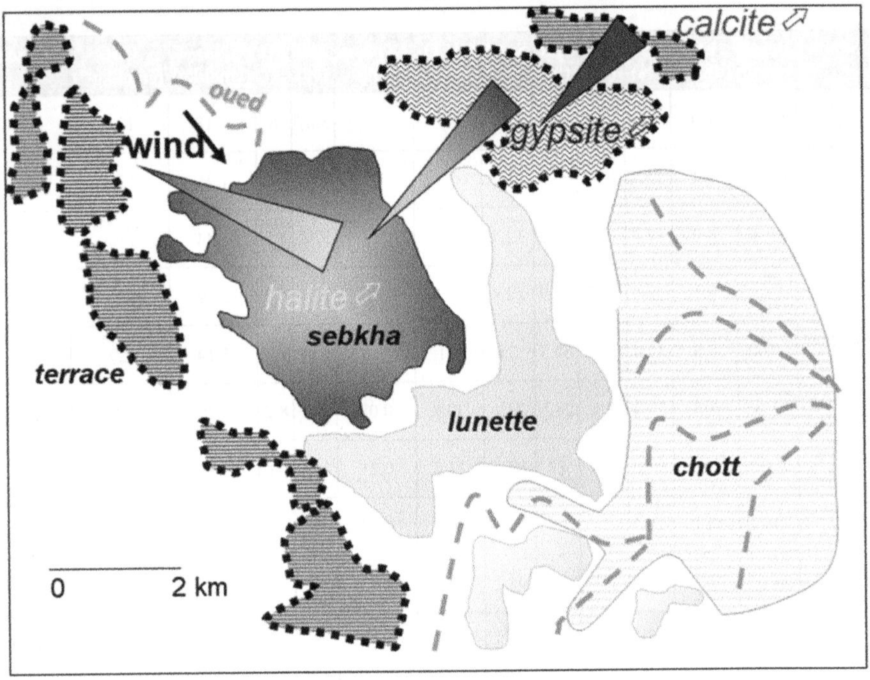

Figure 8.9. *Evaporative sequence around an endorheic depression (based on the El Kourzia sabkha and its lunette, northern Tunisia (Coque 1979), modified). For a color version of this figure, see www.iste.co.uk/coquet/agricultural.zip*

In closed, endorheic depressions, salsodic soils develop in the center of the depression, where saline efflorescences appear on the soil surface within the saltwater table fluctuation zone (Coque 2002). Wind deflation denudes the banks against direction of the active winds, while the lunette is indicative of the correlated accumulation of pseudo-sands made of salt crystals (Figure 8.9). Many lunettes have been described in Tunisia, notably by Coque (1979). Moving centripetally, from the glacis slopes and terraces around the sebkha, the chemical composition of the soil solution changes from being dominated by calcium, to sulfate, to the predominance of sodium chloride salts as we move toward the center of the depression (Figure 8.9). In the hyper-arid Saharan climate, as in the Ouargla basin in southern Algeria, soils with gypsum accumulations (Gypsosols) develop along with chloro-sulfated Salisols (Hamdi-Aissa et al. 2004). Data on thermodynamic equilibrium indicate that the saline solutions are governed by the following sequence: calcite > gypsum > mirabilite > halite. Using X-ray diffraction and micromorphology, Hamdi-Aissa et al. (2004) observed (i) forms of dehydrated gypsum related to a decrease in water activity (<1) and an increase in the ionic force of the concentrated solutions (>3 cmol.kg^{-1} of H_2O); (ii) the presence of diverse sulfates (notably, glauberite $Na_2Ca(SO_4)_2$ and bloedite $Na_2Mg(SO_4)_2{\cdot}4H_2O$), which are more stable in this environment than simple sulfates; and (iii) celestite crystals ($SrSO_4$) in the form of inclusions and on the periphery of the gypsum crystals. The range of daily and seasonal thermal values in the Saharan climate lends itself to the dissolution/precipitation process for gypsum, generating dehydrated gypsum whose presence is a pedological criterion for differentiating between arid and hyperarid environments (Hamdi-Aissa et al. 2004).

8.5.3. *Agronomic potential*

Salsodic soils in Mediterranean environments and the arid borders of these regions result in a combination of chemical constraints (saline toxicity), physical constraints (degraded structure), as well as hydric stress, and, often, hydromorphy (root asphyxia). If irrigation systems are available, one of the possible solutions is to drain the salts, which makes it possible to alleviate the hydric stress while also reducing the quantity of salts in the solum. However, bringing in irrigation water, such as geothermal saltwater from the Saharan aquifer, results in increased salinization, which was clearly seen after four years in the irrigated areas around Tunisia (Hachicha et al. 2012).

Setting up an irrigation system requires vigilance across multiple aspects (Figure 8.10), with respect to (i) the equipment required, the type of infrastructure and the associated installation and maintenance costs; (ii) compatibility with the objectives of the crop production in terms of yield and pest control; (iii) adapting agricultural practices that may remove the need for the system, or which might be required in conjunction with implementation of the irrigation system; (iv) the technical monitoring of the inflow volume, the frequency of the supply and fertilizers; (v) the quality of the water (salt content, presence of pollutants) and the soils that are being irrigated; and finally (vi) conflicts over water use that are inherent to the

Mediterranean Basin, given the proximity of urbanized areas and the development of tourism.

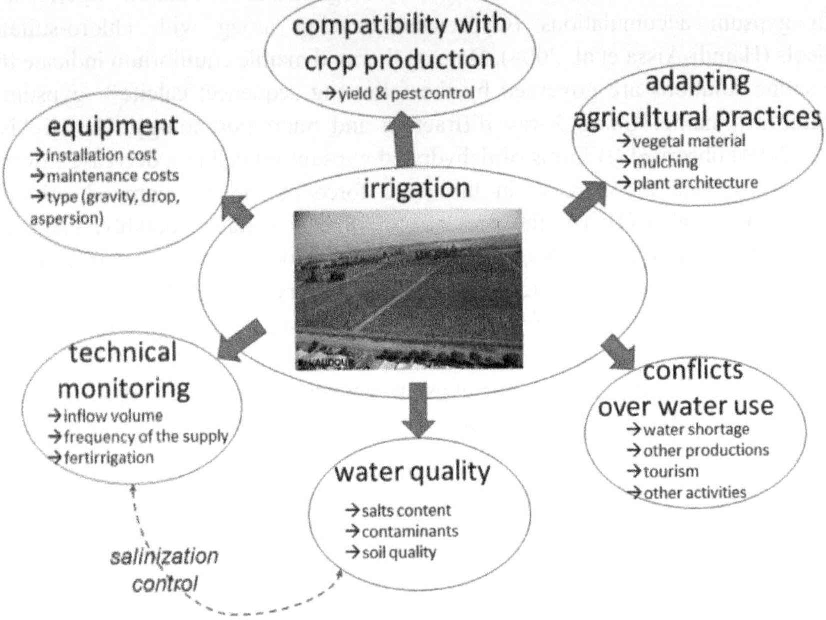

Figures 8.10. *Things to be vigilant about with respect to irrigation in the Mediterranean regions and their arid borders. For a color version of this figure, see www.iste.co.uk/coquet/agricultural.zip*

The use of irrigation requires adopting crops that are more salt tolerant. Consequently, in the Saharan context, in the geothermal areas of the Tozeur region of Tunisia, in the first year, when the soil is virgin, the chosen crop is melons, then for the next three-to-five years it is tomatoes, and finally cucumbers for one or two years (Hachicha et al. 2012). In addition, soils are improved with sand from the dunes, there is the summer lessivage of salts, and rows of crops are moved into greenhouses.

8.6. Conclusion

Mediterranean agricultural soils have some advantages and many constraints, requiring patient know-how and a cautious control of their weaknesses. Subject to demographic pressure, industrial development, deforestation, overgrazing, fires and agricultural changes, which lead to urban sprawl and artificialization, they have been prey to multiple threats, especially from the 1960s onward (de Réparaz 2001). All of these are contributing factors to the degradation of the soil. In semi-arid and arid contexts, these degradation processes are called "desertification" (Costantini and Lorenzetti 2013; Stolte et al. 2015; Costantini et al. 2016; Khatteli et al. 2016; Ferreira et al. 2022). Among the desertification processes, erosion is the best documented and it is facilitated by the presence of slopes, low organic matter content (generally <2% in Mediterranean regions) and aggressive rains (Ferreira et al. 2022). Mediterranean agricultural soils are particularly susceptible to erosion when there is thinly scattered plant cover (Cerdan et al. 2010), which is often the case on soils used in viticulture, leading to nutrient loss (Vaudour et al. 2015). The length of the dry summer season varies around the Mediterranean and governs the permanence of salts in the soil profile, leading to saline and/or sodic soils, or carbonated soils. Given the extent of the irrigated regions and faulty drainage systems, salinization tends to increase, especially in Spain and Italy. However, this phenomenon has not yet been adequately studied (Ferreira et al. 2022).

8.7. References

AFES (2008). *Référentiel pédologique*. Éditions Quæ, Versailles [Online]. Available at: https://www.afes.fr/wp-content/uploads/2017/11/Referentiel_Pedologique_2008.pdf.

Alonso-Zarza, A.M., Sanz, M.E., Calvo, J.P., Estévez, P. (1998). Calcified root cells in Miocene pedogenic carbonates of the Madrid Basin: Evidence for the origin of *Microcodium* b. *Sedimentary Geology*, 116, 81–97. doi: 10.1016/S0037-0738(97)00077-8.

Arnal, H. (1971). Les sols polygonaux étirés et sols striés d'âge Würmien de Landun (Gard). *Bulletin de l'Association française pour l'étude du quaternaire*, 8(3), 151–160. doi: 10.3406/quate.1971.1178.

Atalay, I. (1997). Red Mediterranean soils in some karstic regions of Taurus mountains, Turkey. *Catena*, 28, 247–260.

Badía, D., Martí, C., Palacio, E., Sancho, C., Poch, R.M. (2009). Soil evolution over the Quaternary period in a semiarid climate (Segre river terraces, northeast Spain). *Catena*, 77, 165–174.

Baize, D. (2016). *Petit lexique de pédologie*. Éditions Quæ, Versailles.

Baize, D. and Jabiol, B. (2011). *Guide pour la description des sols*. Éditions Quæ, Versailles.

Barbaroux, L. and Bousquet, B. (1976). Contribution à l'étude des sols méditerranéens. Les sols rouges du domaine hellénique en Grèce occidentale. Signification climatique et stratigraphique. *Géologie Méditerranéenne*, 3(4), 237–250. doi: 10.3406/geolm.1976.980.

Barrière, J. (1971). Utilisation des paléosols comme éléments de datation des formations quaternaires. *C. R. Acad. Sc., Paris*, 273, 310–313.

Barrière, J. and Michaux, J. (1970). La formation détritique villafranchienne de la Mosson (Montpellier) et son paléosol. *Bulletin de l'Association française pour l'étude du quaternaire*, 7(2–3), 93–103. doi: 10.3406/quate.1970.1151.

Bech, J., Rustullet, J., Garrigo, J., Tobías, F.J., Martínez, R. (1997). The iron content of some red Mediterranean soils from of northeast Spain and its pedogenic significance. *Catena*, 28, 211–229.

Bech Borras, J., Fédoroff, F., Quantin, P., Segalen, P. (1982). Etude des sols fersiallitiques lessivés formés sur des arènes granitiques de la Selva (Catalogne, Espagne). *Cah. ORSTOM, Ser. Pédol.*, 19(3), 233–256.

Birot, P. and Dresch, J. (1953). *La Méditerranée et le Moyen-Orient. Tome I. La Méditerranée occidentale*. Presses Universitaires de France, Paris.

Bless, A.E.S., Colin, F., Crabit, A., Follain, S. (2022). Soil aggregate stability in salt-affected vineyards: Depth-wise variability analysis. *Land*, 11, 541. doi: 10.3390/land11040541.

Boero, V. and Schwertmann, U. (1987). Occurrence and transformations of iron and manganese in a colluvial terra rossa toposequence of northern Italy. *Catena*, 14, 519–531. doi: 10.1016/0341-8162(87)90003-8.

Boero, V. and Schwertmann, U. (1989). Iron oxide mineralogy of terra rossa and its genetic implications. *Geoderma*, 44, 319–327. doi: 10.1016/0016-7061(89)90039-6.

Boero, V., Premoli, A., Melis, P., Barberis, E., Arduino, E. (1992). Influence of climate on the iron oxide mineralogy of Terra Rossa. *Clays and Clay Minerals*, 40, 8–13. doi: 10.1346/CCMN.1992.0400102.

Bornand, M. (1978). Altération des matériaux fluvio-glaciaires, genèse et évolution des sols sur terrasses quaternaires dans la moyenne vallée du Rhône. Thesis, Université Montpellier II – Sciences et Techniques du Languedoc [Online]. Available at: https://tel.archives-ouvertes.fr/tel-00992061.

Bornand, M. and Chamley, H. (1974). Sur les minéraux argileux des terrasses Pléistocènes du confluent Rhône-Isère. *Bulletin de l'Association française pour l'étude du quaternaire*, 11(2), 61–64. doi: 10.3406/quate.1974.1240.

Bottner, P. (1972). Évolution des sols en milieu carbonaté. La pédogenèse sur roches calcaires dans une séquence bioclimatique méditerranéo-alpine du Sud de la France. Thesis, Institut de Géologie – Université Louis-Pasteur, Strasbourg [Online]. Available at: https://www.persee.fr/doc/sgeol_0302-2684_1972_mon_37_1.

Boulaine, J. (1957). *Etude des sols des plaines du Chélif*. Thesis, University of Algiers.

Boulaine, J. (1975). *Géographie des sols*. Presses Universitaires de France, Paris.

Boulaine, J. (1978). Les sols calfersiques. *Cah. ORSTOM, Sér. Pédol.*, 16(3), 265–291.

Bourgeon, P. and Pédro, G. (1992). Rôle majeur du drainage climatique dans la différenciation altéritique et pédologique des sols des régions chaudes. Exemple du passage sols fersiallitiques-sols ferrallitiques au Sud du Karnataka (Inde). *Comptes Rendus de l'Académie des Sciences. Série 2*, 314, 717–725 [Online]. Available at: https://gallica.bnf.fr/ark:/12148/bpt6k6293252h/f733.item#.

Bresson, L.M. (2008). Fersialsols. In *AFES, 2008. Référentiel Pédologique*. Éditions Quæ, Versailles.

Bronger, A. and Bruhn-Lobin, N. (1997). Paleopedology of Terrae Rossae – Rhodoxeralfs from calcarenites from NW Morocco. *Catena*, 28, 279–295.

Candy, I. and Black, S. (2009). The timing of quaternary calcrete development in semi-arid southeast Spain: Investigating the role of climate on calcrete genesis. *Sedimentary Geology*, 218, 6–15. doi: 10.1016/j.sedgeo.2009.03.005.

Cerdan, O., Govers, G., Le Bissonnais, Y., Van Oost, K., Poesen, J., Saby, N., Gobin, A., Vacca, A., Quinton, J. Auerswald, K. et al. (2010). Rates and spatial variations of soil erosion in Europe: A study based on erosion plot data. *Geomorphology*, 122, 167–177. doi: 10.1016/j.geomorph.2010.06.011.

Coque, R. (1979). Sur la place du vent dans l'érosion des milieux arides. L'exemple des lunettes (bourrelets éoliens) de la Tunisie. *Méditerranée*, 35, 15–21. doi: 10.3406/medit.1979.1892.

Coque, R. (2002). *Géomorphologie*. Armand Colin, Paris.

Costantini, E.A.C. and Lorenzetti, R. (2013). Soil degradation processes in the Italian agricultural and forest ecosystems. *Italian Journal of Agronomy*, 8, 28. doi: 10.4081/ija.2013.e28.

Costantini, E.A.C., Branquinho, C., Nunes, A., Schwilch, G., Stavi, I., Valdecantos, A., Zucca, C. (2016). Soil indicators to assess the effectiveness of restoration strategies in dryland ecosystems. *Solid Earth*, 7, 397–414. doi: 10.5194/se-7-397-2016.

Dachary, M.C. (1975). Genèse actuelle des sols rouges sur schistes de la région de Béja (Alentejo, Portugal). *Science du Sol, Bulletin de l'AFES*, 4, 231–248.

Dan, J., Yaalon, D.H., Koyumdjisky, H. (1969). Catenary soil relationships in Israel, 1. The netanya catena on coastal dunes of the sharon. *Geoderma*, 2, 95–120. doi: 10.1016/0016-7061(69)90018-4.

Dan, J., Yaalon, D.H., Moshe, R., Nissim, S. (1982). Evolution of reg soils in Southern Israel and Sinai. *Geoderma*, 28, 173–202. doi: 10.1016/0016-7061(82)90002-7.

Darwish, T.M. and Zurayk, R.A. (1997). Distribution and nature of red Mediterranean soils along an altitudinal sequence. *Catena*, 28, 191–202.

Durand, J.H. (1963). Les croûtes calcaires et gypseuses en Algérie : formation et âge. *Bull. Soc. Géol. Fr.*, 7(5), 959–968.

Egli, M., Plötze, M., Tikhomirov, D., Kraut, T., Wiesenberg, G.L.B., Lauria, G., Raimondi, S. (2020). Soil development on sediments and evaporites of the Messinian crisis. *Catena*, 187, 104368. doi: 10.1016/j.catena.2019.104368.

FAO (2014). *World Reference Base for Soil Resources: International Soil Classification System for Naming Soils and Creating Legends for Soil Maps*. FAO, Rome.

Fédoroff, N. (1997). Clay illuviation in red Mediterranean soils. *Catena*, 28, 171–189.

Fédoroff, N. and Courty, M.-A. (2013). Revisiting the genesis of red Mediterranean soils. *Turkish Journal of Earth Sciences*, 22(3), 359–375. doi: 10.3906/yer-1205-10.

Ferreira, C.S.S., Seifollahi-Aghmiuni, S., Destouni, G., Ghajarnia, N., Kalantari, Z. (2022). Soil degradation in the European Mediterranean region: Processes, status and consequences. *Science of the Total Environment*, 805, 150106. doi: 10.1016/j.scitotenv.2021.150106.

Freytet, P. and Verrecchia, E. (1989). Les carbonates continentaux du pourtour méditerranéen : microfaciès et milieux de formation. *Méditerranée*, 68, 5–28. doi: 10.3406/medit.1989.2613.

Goldich, S. (1938). A study in rock-weathering. *The Journal of Geology*, 46(1). doi: 10.1086/624619.

González Martín, J.A., Rubio Fernández, V., García Giménez, R., Jiménez Ballesta, R. (2007). Red palaeosols sequence in a semiarid Mediterranean environment region. *Environ. Geol.*, 51, 1093–1102. doi: 10.1007/s00254-006-0400-4.

Guilaine, J. (2008). *Les racines de la Méditerranée et de l'Europe*. Fayard, Paris.

Hachicha, M. (1998). Mise en valeur de sols salés du nord de la Tunisie- organisation, fonctionnement et évolution des sols sous conditions naturelles et sous drainage et irrigation aux eaux saumâtres. Doctoral Thesis, Ecole Nationale Supérieure Agronomique de Rennes, Rennes. doi: 10.13140/RG.2.1.5019.4323.

Hachicha, M., Khaldi, R., Mougou, A. (2012). Irrigation avec l'eau géothermale salée dans le Sud tunisien. *Etude et Gestion des Sols*, 19, 91–103.

Hamdi-Aissa, B., Valles, V., Aventurier, A., Ribolzi, O. (2004). Soils and brine geochemistry and mineralogy of hyperarid desert playa, Ouargla Basin, Algerian Sahara. *Arid Land Research and Management* 18, 103–126. doi: 10.1080/15324980490279656.

Huguen, C. and Mascle, J. (2005). La ride méditerranéenne (Méditerranée orientale) : apports de la cartographie multifaisceaux à l'analyse morphologique d'un prisme en accrétion-collision. Géomorphologie : relief, processus, environnement. *Groupe français de géomorphologie (GFG)*, 2, 91–104.

Ibañez, J.J., Perez González, A., Jiménez Ballesta, R., Saldaña, A., Gallardo-Díaz, J. (1994). Evolution of fluvial dissection landscapes in Mediterranean environments – quantitative estimates and geomorphological, pedological and phytocenotic repercussions. *Zeitschrift für Geomorphologie*, 38(1), 105–119.

Khatteli, H., Ramadan Ali, R., Bergametti, G., Bouet, C., Hachicha, M., Hamdi-Aissa, B., Labiadh, M., Montoroi, J.-P., Podwojewski, P., Rajot, J.-L. et al. (2016). Soils and desertification in the Mediterranean region. In *The Mediterranean Region under Climate Change*, Moatti, J.-P. and Thiébault, S. (eds). IRD Éditions, Bondy. doi: 10.4000/books.irdeditions.23994.

Lamouroux, M. (1967). Contribution à l'étude de la pédogenèse en sols rouges méditerranéens. *Science du Sol*, 2, 55–85.

Lamouroux, M. (1971). Etude de sols formés sur roches carbonatées (Pédogénèse fersiallitique au Liban). *Cah. ORSTOM, Ser. Pédol.*, 9(3), 365–370.

Lamouroux, M., Loyer, J.Y., Bouleau, A., Janot, C. (1977). Formes du fer des sols rouges et bruns fersiallitiques. Applications de la spectrométrie Mössbauer. *Cah. ORSTOM, Ser. Pédol.*, 15(2), 199–210.

Le Houérou, H.N. (1982). The arid bioclimates in the Mediterranean isoclimatic zone. *Ecologia mediterranea*, 8(1–2), 103–114. doi: 10.3406/ecmed.1982.1937.

Le Houérou, H.N. (2004). *Atlas of Climatic Diagrams for the Isoclimatic Mediterranean Zones*. H.N. Le Houérou, Montpellier.

Legros, J.P. (2007). *Les grands sols du monde*. Presses Polytechniques Romandes, Lausanne.

Lopez-Bellido, R.J., Muñoz-Romero, V., Lopez-Bellido, F.J., Guzman, C., Lopez-Bellido, L. (2016). Crack formation in a Mediterranean rainfed Vertisol: Effects of tillage and crop rotation. *Geoderma*, 281, 127–132. doi: 10.1016/j.geoderma.2016.07.001.

Mancini, F. (2002). The study of Mediterranean soils: A difficult task. In *7th International Meeting on Soils with Mediterranean Type of Climate (Selected Papers)*. CIHEAM, Bari.

Mazoyer, M. and Roudart, L. (2002). *Histoire des agricultures du monde : du Néolithique à la crise contemporaine*. Le Seuil, Paris.

Mezrhab, A. (1994). Croûtes calcaires, travertins et paléoenvironnements quaternaires dans les Bni Iznassen (Maroc oriental). Thesis, Université d'Aix-Marseille, Marseille.

Millot, G., Paquet, H., Ruellan, A. (1969). Néoformation de l'attapulgite dans les sols à carapace calcaire de la Basse Moulouya (Maroc oriental). *C. R. Acad. Sc., Paris,* 268, 2771–2774.

Millot, G., Bocquier, G., Boulet, R., Chauvel, A., Leprun, J.-C., Nahon, D., Paquet, H., Pédro, G., Rognon, P., Ruellan, A. et al. (1979). Géochimie de la surface, pédogenèse, aplanissements et formes du relief dans les pays méditerranéens et tropicaux. In *Phénomènes de transport de matière dans l'écorce terrestre. ATP - CNRS. Action thématique Programmée du Centre National de la Recherche Scientifique. Résultats scientifiques.* Institut de Géologie – Université Louis-Pasteur, Strasbourg [Online]. Available at: https://www.persee.fr/doc/sgeol_0302-2684_1979_thm_53_1_1964.

Mohrange, C. and Gassend, J.M. (2007). *Méditerranée, 250 millions d'années d'évolution*. Les Editions de La Nerthe, Toulon.

Muhs, D.R., Budahn, J, Avila, A., Skipp, G., Freeman, J., Patterson, D. (2010). The role of African dust in the formation of Quaternary soils on Mallorca, Spain and implications for the genesis of Red Mediterranean soils. *Quaternary Science Reviews*, 29, 2518–2543.

Nicod, J. (1995). Carte géomorphologique des karsts de France. Notice d'une maquette au 1/3 000 000 environ. *Karstologia*, 25, 21–34. doi: 10.3406/karst.1995.2351.

Paquet, H. (1970). Évolution géochimique des minéraux argileux dans les altérations et les sols des climats méditerranéens et tropicaux à saisons contrastées. Service de la carte géologique d'Alsace et de Lorraine, Strasbourg [Online]. Available at: https://www.persee.fr/doc/sgeol_0080-9020_1970_mon_30_1.

Paquet, H. (1983). Stability, instability and significance of attapulgite in the calcretes of Mediterranean and tropical areas with marked dry season. In *Pétrologie des altérations et des sols. Vol. II: Pétrologie des séquences naturelles.* Institut de Géologie – Université Louis-Pasteur, Strasbourg [Online]. Available at: https://www.persee.fr/doc/sgeol_0302-2684_1983_act_72_1_2018.

Paquet, H. and Vaudour, J. (1974). Sols et paléosols argileux foncés des environs de Madrid. *Revue géographique des Pyrénées et du Sud-Ouest*, 45(3), 217–242. doi: 10.3406/rgpso.1974.3407.

Paquet, H., Ruellan, A., Tardy, Y., Millot, G. (1969). Géochimie d'un bassin versant au Maroc oriental : évolution des argiles dans les sols des montagnes et des plaines de la basse Moulouya. *C. R. Acad. Sc., Paris*, 269, 1839–1842.

Pédro, G. (1984). La genèse des argiles pédologiques. Ses implications minéralogiques, physico-chimiques et hydriques. *Sciences Géologiques. Bulletin*, 37(4), 333–347. doi: 10.3406/sgeol.1984.1677.

Penven, M.J. (1980). Observations préliminaires sur les altérites granitiques de Petite Kabylie (Algérie). *Méditerranée*, 38, 47–58. doi: 10.3406/medit.1980.1917.

Peterschmidt, E., Fritsch, E., Rajot, J.L., Herbillon, A.J. (1996). Yellowing, bleaching and ferritisation processes in soil mantle of the Western Ghâts, South India. *Geoderma*, 74(3–4), 235–253.

Pozo, M. and Calvo, J. (2018). An overview of authigenic magnesian clays. *Minerals*, 8, 520. doi: 10.3390/min8110520.

de Réparaz, A. (2001). Mutations et permanences dans la géographie des campagnes méditerranéennes (l960–2000). *Méditerranée*, 97, 5–10. doi: 10.3406/medit.2001.3225.

Robert, J.N. (2004). *Les Etrusques*. Les Belles Lettres, Paris.

Ruellan, A. (1970). *Les sols à profil calcaire différencié des plaines de la Basse Moulouya.* Mém. O.R.S.T.O.M.

Segalen, P. (1967). La conférence sur les sols méditerranéens (Madrid, septembre 1966). *Bulletin bibliographique de pédologie*, 16(2), 5–6.

Segalen, P. (1971). La détermination du fer libre dans les sols à sesquioxydes. *Cah. ORSTOM, Ser. Pédol.*, 9(1), 3–27.

Singer, A., Schwertmann, U., Friedl, J. (1998). Iron oxide mineralogy of Terre Rosse and Rendzinas in relation to their moisture and temperature regimes. *European Journal of Soil Science*, 49, 385–395.

Soil Survey Staff (1999). *Soil Taxonomy: A Basic System of Soil Classification for Making and Interpreting Soil Surveys*, 2nd edition. Natural Resources Conservation Service, U.S. Department of Agriculture Handbook, Washington, D.C. [Online]. Available at: https://www.nrcs.usda.gov/wps/portal/nrcs/main/soils/survey/class/taxonomy/.

Stolte, J., Tesfai, M., Øygarden, L., Kværnø, S., Keizer, J., Verheijen, F., Panagos, P., Ballabio, C., Hessel, R., European Commission et al. (2015). *Soil Threats in Europe*. Publications Office, Luxembourg. doi: 10.2788/828742.

Stucki, J.W., Stucki, J.W., Goodman, B.A., Schwertmann, U. (1988). *Iron in Soils and Clay Minerals*. Kluwer Academic Publishers, Dortrecht. doi: 10.1007/978-94-009-4007-9.

Terral, J.F. (1999). La morphométrie à la recherche des origines de la culture et de la domestication des arbres fruitiers : le modèle de l'olivier en Méditerranée nord-occidentale. *Revue d'Archéométrie*, 23, 101–126. doi: 10.3406/arsci.1999.979.

Torrent, J., Schwertmann, U., Schulze, D.G. (1980). Iron oxide mineralogy of some soils of two river terrace sequences in Spain. *Geoderma*, 23, 191–208. doi: 10.1016/0016-7061(80)90002-6.

Torrent, J., Schwertmann, U., Fechter, H., Alferez, F. (1983). Quantitative relationships between soil colour and hematite content. *Soil Science*, 136(6), 354–358.

USDA (1954). *Diagnosis and Improvement of Saline and Alkali Soils*. Agricultural Research Service, Agriculture Handbook No. 60, U.S. Government Printing Office, Washington D.C.

Vaudour, J. (1969). Note sur les sols méditerranéens. *Annales de Géographie*, 8, 483–493.

Vaudour, J. (1975). Encroûtements, croûtes et carapaces calcaires dans la région de Madrid. *Acta Geologica Hispanica*, 10(2), 67–70.

Vaudour, J. (1977). La région de Madrid : altérations, sols et paléosols. Thesis, Aix-en-Provence, Ophrys, Gap.

Vaudour, J. (1979). Les environs Nord et Est de Madrid. Présentation d'une carte morpho-pédologique au 1/100 000 en couleurs. *Méditerranée*, 36, 45–57. doi: 10.3406/medit.1979.2181.

Vaudour, J. (1980). Pédologie, sols, paléosols, et climats (Pedology, soils, paleosoils and climate). *Bulletin de l'Association de géographes français*, 466–467, 39–43. doi: 10.3406/bagf.1980.5148.

Vaudour, J. (1983). Vieux sols et paléosols des niveaux villafranchiens méditerranéens. *Bulletin de l'Association française pour l'étude du quaternaire*, 20(2–3), 95–102. doi: 10.3406/quate.1983.1455.

Vaudour, J. (2003a). Approche systémique et analyse séquentielle des carbonates continentaux dans les milieux méditerranéens. *Collection EDYTEM. Cahiers de géographie*, 1, 135–144. doi: 10.3406/edyte.2003.869.

Vaudour, E. (2003b). *Les terroirs viticoles*. Dunod, Paris.

Vaudour, J., Verrecchia, E., Freytet, P. (1989). Étude de quelques croûtes calcaires complexes des environs de Madrid (Páramo de la Alcarria et "Campiña" du Henares, Espagne) : complémentarité des données géomorphologiques et pétrographiques. *Méditerranée*, 68, 61–73. doi: 10.3406/medit.1989.2617.

Vaudour, E., Morlat, R., Van Leeuwen, C., Dolédec, A.F. (2005). Terroirs viticoles et sols. In *Sols et Environnement*, Girard, M.C., Walter, C., Rémy, J.C., Berthelin, J. (eds). Dunod, Paris.

Vaudour, E., Costantini, E., Jones, G., Mocali, S. (2015). An overview of the recent approaches for terroir functional modelling, footprinting and zoning. *Soil*, 1, 287–312. doi: 10.5194/soil-1-287-2015.

Vaudour, E., Leclercq, L., Gilliot, J.M., Chaignon, B. (2017). Retrospective 70 y-spatial analysis of repeated vine mortality patterns using ancient aerial time series, Pléiades images and multi-source spatial and field data. *International Journal of Applied Earth Observation and Geoinformation*, 58, 234–248. doi: 10.1016/j.jag.2017.02.015.

Wagner, S., Günster, N., Skowronek, A. (2012). Genesis and climatic interpretation of paleosols and calcretes in a Plio-Pleistocene alluvial fan of the Costa Blanca (SE Spain). *Quaternary International*, 265, 170–178.

Yaalon, D.H. (1997). Soils in the Mediterranean region: What makes them different? *Catena*, 28, 157–169. doi: 10.1016/S0341-8162(96)00035-5.

Yaalon, D.H., Brenner, I., Koyumdjisky, H. (1974). Weathering and mobility sequence of minor elements on a basaltic pedomorphic surface, Galilee, Israel. *Geoderma*, 12, 233–244. doi: 10.1016/0016-7061(74)90071-8.

Yassoglou, N., Kosmas, C., Moustakas, N. (1997). The red soils, their origin, properties, use and management in Greece. *Catena*, 28, 261–278.

9

Tropical Soils and Sustainable Management

Cécile QUANTIN[1] and Thierry BECQUER[2]
[1]*UMR GEOPS, CNRS, Université Paris-Saclay, Orsay, France*
[2]*UMR Eco&Sols, Institut Agro Montpellier, CIRAD, INRAE, IRD, Université de Montpellier, France*

9.1. Introduction

This chapter focuses on soils that have undergone strong geochemical weathering, that is, soils that have been subject to *ferrallitization*, leading to the formation of deep and ancient soils that cover around 20–30% of land globally (Tardy 1993; Nahon 2003). The soils that are most representative of the ferralitization processes are grouped together under the term *Ferralsols* (IUSS Working Group WRB 2015), *Oxisols* (Soil Survey Staff 1999), *Ferrallitisols* and *Oxydisols* (AFES 2008) or *Latossolos* (Santos et al. 2018). These soils cover 7.5 million km^2 (IUSS Working Group WRB 2015) or around 5% of land area. Apart from these highly evolved soils, there are other, less-developed tropical soils: Cambisols, Nitisols, Vertisols, Histosols and even Podzols. However, these will not be studied in detail in this chapter as these soils are not characteristic of tropical environments. Nonetheless, some of them (Cambisols, Nitisols, Podzols, etc.) share certain similarities with the more representative soils in terms of the formation processes and mineralogical assemblages, which causes them to behave similarly in part. In France, the study and understanding of how these soils function owes a lot to the work carried out by ORSTOM-IRD since the late 1940s, with research focusing on establishing an inventory of the different types of soils, their classification, soil mapping and the formation processes for these soils (Aubert and Ségalen 1966; Ségalen 1994–1995).

The ferralitization process is characteristic of the tropical context, where temperatures are high throughout the year with very little fluctuation, and there is high rainfall, generally greater than evapotranspiration, resulting in high water activity and significant drainage. Ferralsols are therefore mainly found in the humid tropics of the shields of South America (Brazil) and Africa (Democratic Republic of Congo, southern Central African Republic, Angola, Guinea and eastern Madagascar), as well as in Southeast Asia, Cuba, and New Caledonia. These soils are characterized by the presence of a ferralic B horizon (Bfl) (IUSS Working Group WRB 2015), oxic B horizon (Soil Survey Staff 1999), ferralitic B horizon (F) or oxydic B horizon (OX) (AFES 2008), with or without plinthite (redoxic trait). The mineralogy of the soil horizons reflects the intensity of the *hydrolytic weathering processes*. Thus, in the first part of this chapter, we will see that ferralitization corresponds to significant *desilicification* (loss of silica) through the weathering of alterable silicate minerals, including quartz, and also corresponds to significant *dealkalization* of the profiles, leading to a relative accumulation of less soluble elements such as iron (Fe), aluminum (Al) and manganese (Mn). This produces very specific mineralogical organization (absence of alterable primary minerals, the dominance of 1/1 clays and Al and Fe oxides) and soil profile properties (acidity, exchange properties, fertility, etc.), and generally results in a very thick solum. This thickness, that is, up to several meters thick, is the combined result of climatic factors that promote the weathering of rocks and the low erosion of the profiles due to their development on ancient cratons, due to plant cover, and the absence of recent glaciations. Additionally, pedogenesis in a tropical context is accompanied by the rapid biodegradation and mineralization of organic matter.

9.2. Tropical pedogenesis: dominated by geochemical weathering

Chemical weathering is the transformation of weatherable minerals in supergene conditions (conditions prevailing on the Earth's surface) into new minerals, that is, secondary minerals, as well as solutes released into solution. For the most part, rocks are formed at depth, in thermodynamic conditions that are different from surface conditions, and their constituent minerals are, therefore, unstable in supergene conditions.

Chemical weathering occurs under the influence of bioclimatic factors (temperature, precipitation and drainage) and reagents or weathering agents: oxygen, dissolved CO_2, water and more or less complexing organic or inorganic acids. In natural environments, the chief weathering agent is the dissolution of carbon dioxide (atmospheric CO_2 or from the respiration of heterotrophic organisms) in water. This dissolution leads to H^+ being released through the following equation:

$$CO_{2(g,aq)} + H_2O \leftrightarrows HCO_3^-{}_{(aq)} + H^+{}_{(aq)}$$

Other sources of protons include the hydrolysis of iron or aluminum, the dissociation of functional oxygenated groups in organic molecules, the oxidation of pyrite, etc.

Minerals may undergo total weathering with the complete dissolution of the primary mineral. This is called "congruent dissolution", for example, calcite in the presence of water and CO_2 as per the following equation:

$$CaCO_{3(s)} + H_2O_{(aq)} + CO_{2(g)} \rightarrow Ca^{2+}{}_{(aq)} + 2HCO_3^-{}_{(aq)}$$

However, in most cases, the weathering contributes to the formation of a secondary mineral or the transformation of an initial mineral. This is called "incongruent dissolution". Most silicate minerals undergo incongruent dissolution due to the presence of Fe and Al in the crystalline networks, these elements being less soluble than the other constituent elements. Consequently, the weathering of albite in the presence of water and CO_2 can lead to the neoformation of kaolinite according to the following equation:

$$2NaAlSi_3O_{8(s)} + 2CO_{2(g,aq)} + 11H_2O \rightarrow Al_2Si_2O_5OH_{4(s)} + 4H_4SiO_{4(aq)} + 2Na^+{}_{(aq)} + 2HCO_3^-{}_{(aq)}$$

Also, the weathering of biotite can lead to the neoformation of kaolinite and goethite according to the following equation:

$$2KFeMg_2AlSi_3O_{10}(OH)_{2(s)} + 10H^+ + 1/2\ O_2 + 4\ H_2O \rightarrow Al_2Si_2O_5(OH)_{4(s)} + 2FeOOH_{(s)} + 4Mg^{2+}{}_{(aq)} + 2K^+{}_{(aq)} + 4H_4SiO_{4(aq)}$$

These partial hydrolysis reactions contribute to the release of silica (partial desilification), with the silicon (Si) that is not removed by the alteration solution being retained in the secondary phyllosilicates. They also contribute to the partial or total loss of alkalis, depending on the climatic and drainage conditions.

Primary minerals dissolve relatively slowly and the rate of dissolution is impacted by the removal of the soluble alteration products and therefore by the drainage within the profile. Thus, the entrainment of the released solutes during weathering makes it possible to keep the solution under-saturated with respect to the chief minerals allowing the dissolution to continue.

Whether they develop on felsic rocks or (ultra)mafic rocks, Ferralitic soils are characterized by the dissolution of primary minerals that only persist along the weathering front (or saprolite); by the near-total retention of Fe, Al, Ti and Mn; by the partial retention of Si; and by the very limited retention of alkalis and alkaline-earth metals.

9.3. Typical profile and differentiation of horizons and pedological units

In the initial stages of pedogenesis, meteoric water infiltrates into the fragmented, rocky material and contributes to the hydrolysis of the most weatherable minerals (plagioclases, biotite, olivine or pyroxene, depending on the nature of the rock). Alkaline elements and dissolved silica are released in solution, iron is oxidized and precipitates in the form of iron oxide *s.l.* like goethite, which imparts considerable coloring to the profile being formed, while aluminum can either be precipitated in the form of gibbsite or, if the silica concentration permits, as kaolinite (Figure 9.1). When the most weatherable minerals have been altered, secondary silicates are then weathered and lose their silicon. As this progresses, there is a growing relative accumulation of Fe and Al oxides, with a near-total elimination of silicon, as in certain Geric Ferralsols in New Caledonia (Becquer et al. 2001).

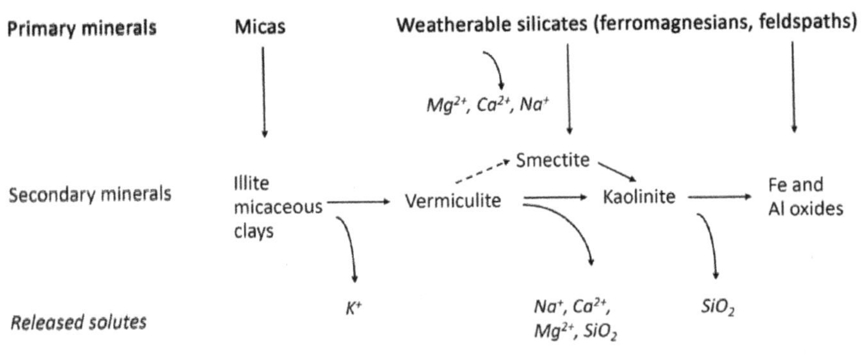

Figure 9.1. *General, simplified sequence of the weathering of minerals in soils*

The parent material of the soil (C horizon or saprolite) is thus formed after the fragmentation and alteration of the mother-rock (R) and the surficial part of the soil is considerably colonized by living organisms and is enriched by organic matter (formation of the A horizon). In tropical conditions, since there is a large quantity of water (>2,000 mm), which is generally greater than the evapotranspiration, the percolation solution is quickly renewed in the profile being formed and is under-saturated, which then maintains the dissolution of the primary minerals. The weathering front continues at depth and the percolation solution is enriched by the elements released by the weathering of alterable minerals, allowing a more significant formation of kaolinite in the soil's mineral horizons, as well as the precipitation of iron oxides. This kaolinite enrichment may contribute to decreased permeability in the horizons and, therefore, the leaching of the dissolved elements. This results in water saturation in the deep horizons, especially in the wet season, which leads to the reduction and redistribution of iron (and reduction of manganese). Characteristic hydromorphic features emerge, indicating iron segregation and a mottled zone may be formed. At the level of the aquifer watertable fluctuation zone, that is, the top of the mottled zone, there may be some level of iron accumulation (Figure 9.2).

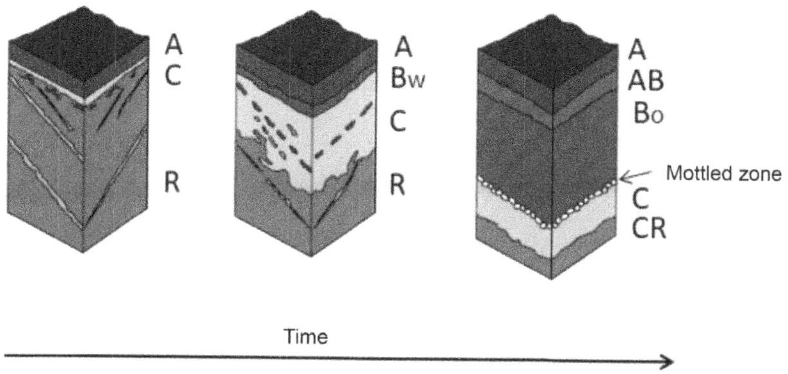

Figure 9.2. *Formation of a Ferralsol (adapted from Lepsch 2011). For a color version of this figure, see www.iste.co.uk/coquet/agricultural.zip*

In this context, the weathering front is largely saturated with water and the leaching of ions is greatly slowed, which could lead to the precipitation of smectite-type phyllosilicates. As the weathering progresses, the weathering front sinks down and the mottled zone is developed in relation with the variations in watertable levels. A "bleached" horizon may appear at the base of the mottled zone if the conditions remain anoxic due to iron loss. The B horizon develops above the mottled zone and at the level of the iron accumulation (if any). It is rich in iron and aluminum oxides (i.e. sesquioxides) and may grow to a thickness of several meters, depending on drainage and time. The induration of the iron accumulation level may be caused by the dehydration of the iron oxyhydroxides, which may be accompanied by secondary silification. In such cases, we talk about cuirasses, *ironstone*, *duricrust* or *petroferric material*. While some lateritic soil profiles may present this level of ferric induration in the subsurface (cuirass), it may be greater or absent in many other profiles.

Ferralitic soils that develop on (ultra)mafic rocks generally contain less kaolinite and the mottled zone is often less pronounced, or even absent, given that there is a less pronounced decrease in permeability. Gibbsite is absent and iron oxides are largely substituted by Al (Becquer et al. 2006).

It can be seen that when kaolinite dissolution prevails over crystallization, the soil becomes depleted of clays and may be leached at greater depths, where there may be significant gibbsite precipitate. However, since dissolved silica is actively recycled by plants, the Si on the surface, brought in through the litter and particle leaching through rain, helps in keeping the soil surface horizon oversaturated with respect to kaolinite (Lucas et al. 1993).

Finally, the activity of invertebrates, especially termites, contributes to homogenizing the upper part of soil profiles, which creates pseudo-sands, commonly found on the surface (Fritsch et al. 2002; Martinez and Souza 2020).

9.4. Principal characteristics, diversity of tropical soils

The ferralitization processes described earlier lead to the formation of a sub-surficial mineral ferralitic horizon or oxidic mineral horizon, characterized by their relative proportions of clay minerals and oxyhydroxides (AFES 2008), ferralic minerals (IUSS Working Group WRB 2015, from the Latin *ferrum*, iron, and *alumen*, aluminum), or oxidic minerals (Soil Survey Staff 1999). These relatively thick layers are the result of intense weathering, in which the granulometric clay fraction is dominated by neoformed, low activity clays (i.e. kaolinite) and by iron oxides, aluminum oxides, manganese oxides and titanium oxides, which may aggregate to form pseudo-sands and concretions. Due to the extensive weathering, the ferralic horizon contains a very small proportion of weatherable minerals and is relatively acidic (pH 4.5–6.5).

Consequently, the cation exchange capacity (CEC) is low, and the diagnostic criteria chosen are CEC < 16 $cmol_c\ kg^{-1}$ of clay (pH 7) and ECEC < 12 $cmol_c\ kg^{-1}$ of clay (IUSS Working Group WRB 2015). The color of this horizon, and more largely, of the soil, is related to the presence of iron oxyhydroxides and therefore, depending on their nature, the color varies from yellow to a reddish-shade, or even a purplish-red (10 R 3/2 to 10 R 3/4 for the Bo horizon in profile B, Figure 9.3). This richness in iron oxides, *s.l.*, which are amphoteric minerals, gives ferralitic soils a large anion exchange capacity (AEC) (Becquer et al. 2001; Qafoku et al. 2004; Feder et al. 2020). Another consequence of the specific mineralogical nature of ferralitic horizons is the absence of particle leaching in Ferralsols.

Ferralitic soils do not show large differentiation between horizons, and they generally have a poor, or non-existent, macrostructure. However, they are characterized by a very marked microstructure in the form of micro-aggregates or spherical "pseudo-sands" whose diameter is between 100 and 250 μm or even larger, partly resulting from the activity of "ecosystem engineers", ants, termites and earthworms (Lavelle et al. 1997; Schaefer 2001; Balbino et al. 2002; Martinez and Souza 2020) and which can represent 40–60% of the surface soil mass (Lavelle et al. 2020). Further, this micro-structuring gives ferralitic soils high permeability as well as a large water-retention capacity, associated with a low bulk density (Reatto et al. 2007).

Figure 9.3. *Profiles of Rhodic Ferralsol, rich in kaolinite, located in Paracatu, Brazil, developed on metapelite rock (A), Geric Ferralsol, rich in iron oxides, located in Ouénarou, New Caledonia, developed on ultramafic rock (B) and Geric Ferralsol, rich in aluminum oxides, located in Maré, New Caledonia, developed on volcanic ash and pumice (completely weathered) and resting on Corallian limestone (white) (C). For more details on A, see Burak et al. (2013), and for more details on B and C, see Becquer et al. (2001). For a color version of this figure, see www.iste.co.uk/coquet/agricultural.zip*

While the entire scientific community concurs that pedogenesis is a slow process, very few studies make it possible to date soils and specify the rate of formation of soils. Consequently, it is difficult to assess how human activity affects pedogenesis (Cornu et al. 2009). Ferralitic soils, however, contain secondary minerals, that is, minerals that form in the solum, or more broadly, in the alteration profile, and these record information about the chemical reactions and geochemical (and paleoclimatic) conditions that prevailed during their formation. Several methods can be used to date these minerals. For example, manganese oxides that contain potassium (e.g. cryptomelane) can be dated using the ^{40}Ar-^{39}Ar method (Vasconcelos 1999; Beauvais et al. 2008), iron oxides can be dated using U-Th(-Sm)/He (Shuster et al. 2005; Allard et al. 2018) and kaolinite can be dated using electronic paramagnetic resonance (EPR) (Balan et al. 2005; Allard et al. 2018).

9.5. Properties and agronomic consequences

9.5.1. *Mineralogical characteristics and exchange properties*

The fertility of the soil is significantly influenced by its exchange properties. Positive or negative electrochemical charges that can appear on the surface of constituent particles allow the retention of nutrients present in ionic form in the soil solution, thereby limiting nutrient loss through leaching. However, in the case of certain nutrients, for example, phosphorus (P), there is significant sorption onto the iron oxide surface charges, and this can excessively decrease their availability.

Surface charges on the constituents depend on two factors: the specific area of the particles and the surface properties.

There is large variation in the specific area of soil constituents. In the case of clay minerals, this area chiefly varies as a function of the interlayer spaces in the clay. Clay minerals are made up of a stack of layers (hence the name "phyllosilicates", or "layer silicates"), made up of octahedral sheets of $Al_4(OH)_6$ or $Mg_3(OH)_6$ and tetrahedral SiO_4 sheets, linked through the sharing of oxygen atoms or hydroxyl groups, and packaged into layers. The interlayer spacing, that is, the distance between two successive layers, is thus dependent on the number of sheets and the state of the interlayer zone. For clays from the kaolinite family, which dominates tropical soils, each layer is made up of two sheets: one sheet of Si atoms, located at the center of oxygen tetrahedra (tetrahedral sheet), and a sheet of Al atoms, located at the center of hydroxyl octahedra (octahedral sheet). Each layer (a tetrahedral sheet and an octahedral sheet) is separated from other layers by an interlayer space, the resulting interlayer spacing values of *00l* planes is of 7 Å for kaolinite. This interlayer space is too narrow to allow ions access. Thus, the specific area of kaolinite is only the outer surface of the stack of layers. This is of the order of 15 $m^2\ g^{-1}$ of clay (Table 9.1). In clays from the smectite family (e.g. montmorillonite), which are almost absent in Ferralsols, each layer is made up of three sheets: two tetrahedral sheets enclosing an octahedral sheet and the resulting interlayer spacing measured around 14 Å is accessible to ions. The outer surface of the stack of layers is of the order of 50 $m^2\ g^{-1}$ of clay, while the internal surface, corresponding to the interlayer space, is of the order of 750 $m^2\ g^{-1}$ of clay.

Minerals	Specific area ($m^2 g^{-1}$)	CEC ($cmol_c kg^{-1}$)	AEC ($cmol_c kg^{-1}$)
1/1 clays[1]			
Kaolinite	7–30	2–15	1[2]
Halloysite	10–45	10–40	–
2/1 clays[1]			
Talc	65–80	<1	–
Mica	60–100	10–40	–
Dioctahedral vermiculite	50–800	10–150	–
Trioctahedral vermiculite	600–800	100–200	–
Smectite	600–800	80–150	–
2/1/1 clays[1]			
Chlorite	25–150	10–40	
Aluminum oxides[2]			
Gibbsite	45	0	6[2]
Iron oxides[3]			
Goethite	8–200	–	12[2]
Hematite	2–90		
Ferrihydrite	100–400	5	80[2]

[1]Sparks (2003); Meunier (2005); [2]Parfitt (1980); [3]Cornell and Schwertmann (1996).

Table 9.1. *Specific area and charges of the chief minerals present in the soils*

The specific area of oxides varies with their degree of crystallization. Well-crystallized iron or aluminum oxides have specific areas of the order of 50–80 $m^2 g^{-1}$, while amorphous compounds have specific areas that are 10 times larger.

There are two kinds of electrochemical charges on the surface: permanent, structural negative charges and variable charges that depend on pH. Structural charges are linked to “defects” in the tetrahedral or octahedral sheets in the clay: the substitutions of Si atoms in the tetrahedra or Al in the octahedra by ions with a smaller charge (Si^{4+} by Al^{3+}, Al^{3+} by Fe^{2+}, Mg^{2+}, Ni^{2+}, etc.) produces a lack of positive charge, that is, results in the presence of negative charges in the sheet. There are quite a few such “defects” in smectite clays, which gives these clays a high CEC. On the other hand, these defects are rare in kaolinite, which therefore has a low CEC (Table 9.1). Consequently, soils that essentially comprise kaolinite clays, such as Ferralsols, will have low structural CEC.

Variables charges are linked to the ability of amphoteric molecules to become protonated or deprotonated, depending on the pH of the soil solution, through the following reactions:

– At low pH: $MOH + H^+ \leftrightarrows MOH_2^+$

– At high pH: $MOH + OH^- \leftrightarrows MO^- + H_2O$

The amphoteric components of soil are chiefly the organic compounds, iron and aluminum oxides, and, to a lesser extent, clays (on the edge of the sheets). These components have different points of zero charge (PZC). In the case of organic matter, the PZC is located at a pH of around 2–3. Protonation occurs below the PZC and positive charges are formed; deprotonation occurs above the PZC and negative charges are formed. Therefore, at the pH of soil (4–8.5), organic matter is negatively charged and consequently has a CEC. In the case of iron oxides, the PZC is located at a pH of around 8, and at the pH of Ferralsols (4–6), oxides are positively charged and therefore have an AEC. An increase in the pH of a soil results in an increased CEC and smaller AEC (and vice versa in the case of a drop in pH) (Becquer et al. 2001). Soils with high organic matter content and low-acid pH will have a high CEC. Soils with high oxide content and low pH will have a high AEC.

9.5.2. *Advanced weathering: nutrition deficiencies and aluminum toxicity*

Primary minerals are often quite rich in alkali and alkaline earth metal cations (Ca, Mg, K and Na). For example, K-feldspar ($KAlSi_3O_8$) contains 14.3% potassium; a biotite ($K(Mg,Fe)_3(OH,F)_2(Si_3AlO_{10})$) contains around 8% potassium and 7% magnesium. On the other hand, the chief secondary minerals that are found in tropical soils following the weathering of rocks (i.e. kaolinite, Fe and Al oxides) do not contain alkali or alkaline earth metal cations, as the majority of these elements are removed through drainage and leaching during the weathering process. In these soils, these elements are chiefly stored through the CEC (which can retain these cations on the electronegative charges on mineral and organic constituents) and soil organic matter (SOM), which holds the nutrients released during the mineralization or biodegradation process.

The exchangeable levels of Ca^{2+}, Mg^{2+} and K^+ in a collection of soils from Brazilian savannahs were 1.77 ± 1.91, 1.13 ± 0.68 and 0.24 ± 0.24 $cmol_c\ kg^{-1}$ soil, respectively (Vendrame et al. 2013). The contents in plant residues (which supply the SOM) from rice straw in Madagascar were as follows: 2.9 ± 0.9, 1.0 ± 0.4, and 12.8 ± 3.8 g kg^{-1} MS for Ca, Mg and K, respectively (Fanjaniaina et al. 2021). Calcium contents in soils are therefore mainly dependent on the soil CEC, while K contents in soil are mainly linked to the SOM. These low contents of alkalis and alkaline-earth metals may result in nutrient deficiencies in crop plants. In the Madagascar Highlands, calcium is the second main nutrient that limits the growth of rice, after phosphorus and above nitrogen (Raminoarison et al. 2020).

Weathering processes that eliminate alkali and alkaline earth metal cations are accompanied by soil acidification, which is the loss of exchangeable alkaline and alkaline-earth metal cations and their replacement by acid ions (H^+ and Al^{3+}) (van Breemen et al. 1983). Acidification is a natural phenomenon which may be highly accelerated by several agricultural practices. In intensive production systems, the chief cause of acidification is the use of large quantities of nitrogen fertilizers and the leaching of nitrate, along with the leaching of alkaline and alkalis-earth metal cations (Miao et al. 2011). However, a balanced application of N combined with the use of manure may prevent a drop in the soil pH and base saturation (Zeng et al. 2017). In extensive production systems, such as in sub-Saharan Africa, alkali and alkaline earth metal cations are removed with the harvest and are generally not replaced by the use of organic manure. This is a major cause of acidification (Fanjaniaina et al. 2021).

Soil acidification results in the protonation of the surface of the constituents and aluminum becoming the dominant exchangeable cation. Protons produced through various processes are largely neutralized by the alteration of soil minerals, which results in the hydrolysis of these minerals (see section 9.1). Initially, the primary minerals (such as feldspar) dissolve, which leads to the precipitation of the secondary minerals (kaolinite, gibbsite):

$$2\ KAlSi_3O_8 + 2H^+ + 9H_2O \leftrightarrows Al_2Si_2O_5(OH)_4 + 4H_4SiO_4 + 2K^+$$

Next, these secondary minerals may be altered, in turn, producing soluble aluminum:

$$Al_2Si_2O_5(OH)_4 + 6H^+ \leftrightarrows 2Al^{3+} + 2H_4SiO_4 + H_2O$$

$$Al(OH)_3 + 3H^+ \leftrightarrows Al^{3+} + 3H_2O$$

The acid ions that are produced (H^+ and Al^{3+}) compete with the alkalis and alkaline-earth metal ions for CEC adsorption. The protonation and aluminization of exchangeable cations, along with the loss of alkali and alkaline earth metal cations, result in an increase in H^+ and Al^{3+} levels on the exchange surfaces. The soil saturation rate (*V*) (the ratio of the sum of exchangeable base cations to the total CEC, expressed in %) is a good indicator of how soil acidification evolves.

The level of exchangeable Al^{3+} increases rapidly as the soil pH decreases, especially when the pH falls below 5 (Vendrame et al. 2013) (Figure 9.4).

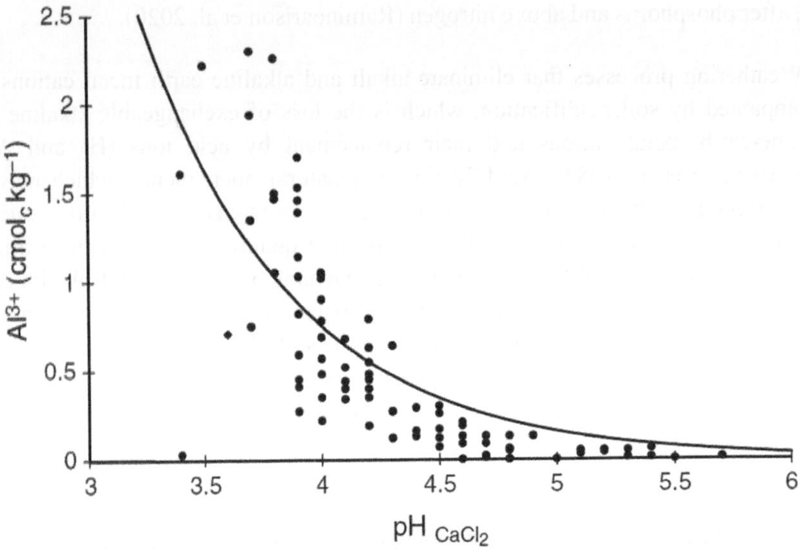

Figure 9.4. *The relation between soil pH (measured in 0.01 M $CaCl_2$) and the exchangeable aluminum level in the 0–20 cm horizon in Ferralsols in pastures in the Brazilian Cerrado region (after Vendrame et al. 2013)*

Levels of exchangeable H^+ and Al^{3+} cations are in equilibrium with the H^+ and Al^{3+} levels in the soil solution. The presence of H^+, and especially of Al, in the soil solution can be toxic to many plants, especially crop plants. Aluminum occurs in the soil solution in various forms, some of which are toxic (Al^{3+}, $AlOH^{2+}$, $Al(OH)_2^+$), while certain complex species formed with mineral ions (Al-F, Al-PO_4, Al-Si, etc.) or organic ions are mildly toxic or not toxic at all (Boudot et al. 1994). The geochemical environment within the soil solution also influences Al toxicity. It has been shown that certain cations attenuate Al toxicity by competing with the toxic Al species and reducing their ionic activity. Divalent cations are more efficient than monovalent cations at reducing aluminum toxicity ($Ca^{2+} = Mg^{2+} >>> K^+ = Na^+$). Various toxicity indicators have been proposed to account for the presence of more or less toxic forms of Al and the presence of cations that could attenuate Al toxicity. The simplest of these indices is the Ca/Al ratio in the soil solution. Other, more complex indices have also been proposed, which take into account the ionic activities of various toxic Al species and the activities of other cations in the solution (see Boudot et al. 1994).

9.5.3. *Abundance of Fe and Al oxides: implications for the availability of phosphorus*

Ferralsols are characterized by advanced weathering, resulting in the abundance of secondary minerals, especially iron oxides and aluminum oxides, which is conducive to high phosphate ion retention (Gérard 2016; Hinsinger et al. 2017). Due to various physicochemical processes, phosphate ions are adsorbed onto soil constituents carrying positive charges. This is thus particularly true of iron and aluminum oxides, whether crystallized or amorphous (see section 9.5.1). As the surfaces of these oxides get increasingly protonated as the pH decreases, their adsorption capacity increases in acidic soils (Penn and Camberato 2019). A second set of physicochemical reactions comes into play in controlling the concentration of phosphate ions in the soil solution: the precipitation and dissolution of phosphate minerals in the form of iron phosphates or aluminum phosphates in soils with very acidic pH. Iron and aluminum phosphates are precipitated when the pH decreases (Hinsinger 2001). The abundance of iron oxides and aluminum oxides, combined with the acidic pH of Ferralsols, go a large way toward explaining the limited availability of P in these soils (Hinsinger 2001; Penn and Camberato 2019).

Consequently, P has very low phytoavailability in Ferralsols. The forms of P in the soil that are absorbed by roots are orthophosphate ions dissolved in the soil solution ($H_2PO_4^-$ and HPO_4^{2-}). In tropical soils, the concentration of PO_4 ions in solution may be extremely low (0.001 mg P L^{-1}) (Randriamanantsoa et al. 2013) because of the high sorption/retention capacity of the soils (or the high AEC of Ferralsols), while the concentration of dissolved PO_4 may exceed 1 mg P L^{-1} in soils in temperate regions that are over-fertilized. When a crop is planted, the soil solution contains less than 1% of the annual P uptake. This means that more than 99% of the P taken up by the plant comes from the solid components of the soil, hence the importance of the physicochemical determinants that govern the soil-solution transfer of PO_4 ions (Morel 2002).

9.6. The sustainable management of soils with high geochemical weathering: the role of organic matter in Ferralsols

The sustainable management of soils and agrosystems depends on the various practices used for soil management (Scopel et al. 2013), plant management (Beillouin et al. 2019; Sauvadet et al. 2021) and the management of soil microorganisms (Larimer et al. 2010; Razakatiana et al. 2020). In this section, we will only examine a few aspects of organic matter management by reviewing the three points discussed in the previous section.

9.6.1. *Organic matter and exchange properties*

Organic matter in soil has a large exchange capacity and contributes to the retention of alkali and alkaline earth metal ions (see section 9.5.1). de Moraes Sá et al. (2009) showed that there was a good correlation between the soil organic carbon content (SOC) and the CEC for Brazilian Oxisols (CEC = 0.375 SOC + 2.95; R^2 = 0.70). In these soils, the CEC that depended on the mineral phase was only 2.95 $cmol_c$ kg^{-1} soil. For average SOC levels, of the order of 28 g kg^{-1} soil in the 0–20 cm horizon, the CEC is 13.5 $cmol_c$ kg^{-1} soil: 78% of the CEC is therefore due to the organic matter, while only 22% is due to the mineral components, especially kaolinite. Consequently, there is an increase in the exchangeable cations from organic matter in the soil as the soil's organic matter content increases.

Additionally, organic compounds in the soil, made up of amphoteric molecules, may get deprotonated when the soil pH increases. This results in a gradual increase in the variable negative charges as pH increases, and these charges may increase by about 25% per unit pH, according to data from Becquer et al. (2001). Soil CEC can be increased through the combined action of agricultural practices that maintain high organic matter levels and low-acid pH level. This can consequently help retain alkali and alkaline-earth metal cations in the soil's exchange complex. Practices that promote the storage of organic matter in the soil will thus play a major role in the sustainable management of Ferralsols. Also, conservation agriculture, which recycles crop residues and limits soil tillage, promotes carbon storage (Scopel et al. 2013). This is also the case for crop rotation with intercropping, which produces a lot of biomass (Blanco-Canqui et al. 2015), or again systems that integrate agriculture and livestock with the beneficial effects of grazing (Lemaire et al. 2014).

9.6.2. *Organic matter and nutrient cycling*

Returning organic residues to the soil and spreading manure have demonstratable benefits for soil quality and crop productivity (Haynes and Mokolobate 2001; Scopel et al. 2013). Organic matter is rich in various nutrients.

Crop residues are the main source of biomass in production systems where agriculture dominates. The composition of these residues is highly variable depending on the species, the parts of the plants and the phenological stage

(Palm et al. 2001). Crop residues from mature plants are generally depleted in nutrients compared to herbaceous plants. Residues from legumes are particularly rich in nitrogen, thanks to the symbiotic fixation capacity of these species. The meta-analysis carried out by Unkovich et al. (2010) shows that the fixation of N_2 can go up to 20–24 kg of N_2 t^{-1} MS for pasture legumes and 11–23 kg of N_2 t^{-1} MS for cultivated legumes, depending on the species.

Manure and various organic fertilizers, which play a major role in systems that integrate agriculture and livestock rearing, also have very variable compositions depending on the waste used and on the animal feed. With these factors being noted, the modes of storage of the organic matter also significantly influence its quality. Liquid residues (urines, purines) are often poorly collected, especially in small farms in developing countries. The storage conditions on the farm thus significantly affect nitrogen losses through volatilization and the loss of various nutrients through leaching (Rufino et al. 2006; Bernal et al. 2009). Tittonell et al. (2010) estimate that only 20–40% of NPK are preserved in manure across various storage practices in Kenya. Taking into account nutrient loss during the handling and storage of organic fertilizers, it is difficult to balance the removal of nutrients through harvesting by the use of manure alone (Fanjaniaina et al. 2021).

Organic fertilizers and residues from the harvest are important sources of nutrition for plants. However, in order to be available, the organic matter must first be mineralized by soil microorganisms and converted into inorganic forms. In favorable humidity conditions, organic residues and manure are rapidly mineralized. However, after this initial phase of mineralization (gross mineralization), the nutrients are either immobilized in other organic molecules and contribute to the storage of carbon (and nutrients) in soil organic matter, or they are made available for uptake in plants (net mineralization). Many studies on nitrogen showed that the C/N ratio in organic components is the main factor influencing the net release of nitrogen (Sradnick and Feller 2020): organic products like straw that are poor in nitrogen promote the immobilization of N, while those that are rich in N (like residues from legumes) are conducive to the release of nitrogen. Palm et al. (2001) estimated that the net mineralization of N occurs when the initial concentration of nitrogen in the organic product exceeds 2%. The biochemical composition of organic compounds may have a significant influence on the net mineralization. In agroforestry systems, Sauvadet et al. (2020) showed that fertility was enhanced by organic residues that were nutrient rich and could be easily decomposed by microorganisms due to their low tannin content. In multispecies systems (associated plants and agroforestry), the functional features of organic restitutions, which vary depending on their characteristics, impact soil functions, thereby improving the multifunctionality of agrosystems (Sauvadet et al. 2020). Overall, according to

Palm et al. (2001), only organic materials that are nitrogen rich ($N > 2.5\%$) and poor in recalcitrant compounds (lignin < 15%, phenols < 4%) are efficacious as fertilizers. When the organic materials are poor in N or richer in recalcitrant compounds, they must be combined with mineral fertilizers or other organic products of higher quality.

9.6.3. *Organic matter and acidification/alkalinization of soils*

The harvesting and removal of biomass from plots of land leads to acidification of the soil (Curtin and Wen 2004). As plants grow, they extract nutrients from the soil solution and incorporate them into their tissues. The assimilation of cations by plants is accompanied by a transfer of protons (H^+) from the plant to the surrounding soil solution. In contrast, the assimilation of anions is accompanied by a transfer of hydroxyl ions (OH^-), leading to the neutralization of the protons in the soil (van Breemen et al. 1983). The absorption of cations is generally greater than the absorption of anions (Noble et al. 1996), that is, there is a net excess absorption of cations, resulting in the storage of alkalis in plants as they grow and a concomitant acidification of the soil as the alkalis are removed from the soil solution. When the plant residues return to the soil after senescence, the alkalis in the plants return to the soil and neutralize the earlier acidification. On the other hand, if the alkalis are removed out of the land when the products are harvested, the acidification is not neutralized.

In sub-Saharan Africa, the removal of nutrients with the harvested products (grains, tubers, harvest residues used as fodder, etc.) represents around 50–70% of N, P and K losses, while erosion only accounts for about one-third of the losses (Stoorvogel et al. 1993). It must be noted that these observations are from agricultural systems with low fertilizer use, where there is not much acidification induced by the transformation and loss of nitrogen. In systems that are richer in nitrogen, the mineralization/nitrification of nitrogen, and its subsequent leaching, results in significant soil acidification and this often becomes the most significant mechanism in these systems (Miao et al. 2011).

The removal of nutrients through the biomass gradually results in a decrease in the levels of exchangeable cations and base saturation, an increase in the acidity of the exchangeable cations and a slight decrease in the soil pH. Therefore, the intensity of the acidification from biomass transfers is largely dependent on the agricultural practices in place. In the case of systems where the harvested biomass is largely grains or plant tubers, the depletion is often moderate. On the other hand, in the case of the harvest of fodder, there may be considerable alkali loss (Fanjaniaina et al. 2021).

9.6.4. *Organic material and phosphorus availability*

Phosphorus deficiency is generally the main factor limiting plant growth in tropical agrosystems (Raminoarison et al. 2020). Due to their removal with the harvests (see above discussion), the P reserves that are available to plants in the soil decrease. The sources for mining phosphorus, used to compensate for these losses, become rarer (Cordell et al. 2009).

It is therefore essential for P to be effectively cycled in organic matter, especially in developing countries (where ferralitic soils dominate), where mineral fertilizers are often too expensive for farmers. Organic material such as manure is an important source of phosphorus. The mineralization of organic phosphorus makes it possible to replenish the soil solution with orthophosphate ions, which can be used by plants. The phytoavailability of the phosphorus in some manures and other organic products is similar to that of soluble phosphate fertilizers (Oberson et al. 2001; Steffens et al. 2010). However, P levels in organic products are quite low: according to Palm et al. (2001), they vary from 0.5 to 5 g kg^{-1}; however, in traditional farming, where livestock effluent is often poorly managed, the levels are more often in the range of 0.5 g kg^{-1} (Fanjaniaina et al. 2021).

Apart from the direct supply of P, MO also plays an important role in modifying the sorption of P onto the soil components, thereby modifying the availability of P. As indicated earlier, the influx of MO has an alkalizing effect on the soil. The decreased P sorption, with an increase in the pH, results in an increase in the available P for a given total P content in the soil. An increase in the pH may result in decreased electrical potential on the surfaces, which increases electrostatic repulsion between the charged surface and the inorganic P, thereby decreasing the sorption of P. This process may be particularly significant in Ferralsols, which contain a large quantity of minerals whose surface charges vary with the pH, such as aluminum and iron oxides. Further, since organic anions (negatively charged) compete with orthophosphate ions for the same sorption sites, this may also contribute to increasing the availability of P.

The quality of the organic residues added to the soil also impacts phosphorus availability. Recent research (Sauvadet et al. 2021) has shown that the fertility of Ferralsols is highly dependent on the availability of phosphorus in the soil, which itself is dependent on the phosphorus levels in crop residues. In effect, the incorporation of P-rich crop residues results in increased phosphorus availability, increased pH and, finally, an increase in plant biomass production in pot trials. On the other hand, nitrogen-rich crop residues (e.g. residues from legumes) do not induce increased plant biomass production in pot trials due to the decrease in pH, which could affect phosphorus availability.

9.7. Conclusion

Ferralsols cover a large portion of tropical lands. These are soils that have undergone advanced weathering, resulting in very low residual concentrations of primary minerals and high levels of kaolinite and iron and aluminum oxides. Weathering processes and the secondary minerals that characterize these processes endow these soils with a number of common physicochemical properties that are quite unfavorable to soil fertility: low CEC and low levels of exchangeable alkali or alkaline-earth metal cations, which could induce significant deficiencies in plants; high acidity, associated with the exchangeable cations being saturated with aluminum, which could induce toxicities; a high phosphorus sorption capacity, resulting in very poor availability of P and nutritional deficiencies in P in plants. However, the generally stable structuration of these soils, along with their high porosity and good permeability, results in good physical fertility.

The majority of these soils are located in developing countries, where farmers face difficult socioeconomic conditions that often prevent them from remedying the unfavorable soil characteristics through the purchase of supplements and fertilizers. In these conditions, the judicious management of organic matter is the best means of enhancing soil fertility. An input of organic matter makes it possible to replenish the nutrients removed through crops and to limit certain deficiencies. It also makes it possible to modify the soil's physicochemical conditions, which results in lower soil acidity, an increased CEC and the retention of alkali or alkaline earth metal cations, a reduction in aluminum toxicity and increased phosphorus availability. This also results in increased biological activities, supporting the ecosystem services that enhance nutrient availability.

Biomass inputs can be increased by crop management. Either through rotation or association, the presence of plants that return large quantities of biomass to the soil is an important means of enriching the soil with organic matter. Legumes can play a specific role in enriching the soil in nitrogen through symbiotic fixation. However, the possible acidifying effect of these plants must be taken into account to prevent negative effects. In systems that integrate agriculture and livestock, biomass management at the level of individual farms can also be seen as a major lever for action. Effective biomass management through the various stages of handling and storing materials is essential to reduce nutrient losses.

Through the key role played by organic matter, the sustainable management of tropical soils is an integral challenge in the issues associated with conservation agriculture and the ecological intensification of agricultural production.

9.8. References

AFES (2008). *Référentiel pédologique*. Éditions Quæ, Versailles.

Allard, T., Gautheron, C., Bressan Riffel, S., Balan, E., Soares, B.F., Pinna-Jamme, R., Derycke, A., Morin, G., Bueno, G.T., do Nascimento, N. (2018). Combined dating of goethites and kaolinites from ferruginous duricrusts. Deciphering the Late Neogene erosion history of Central Amazonia. *Chemical Geology*, 479, 136–150.

Aubert, G. and Ségalen, P. (1966). Projet de classification des sols ferrallitiques. *Cahiers ORSTOM, Série Pédologie*, 4(4), 97–112.

Balan, E., Allard, T., Fritsch, E., Sélo, M., Falguères, C., Chabaux, F., Pierret, M.-C., Calas, G. (2005). Formation and evolution of lateritic profiles in the middle Amazon basin: Insights from radiation-induced defects in kaolinite. *Geochimica et Cosmochimica Acta*, 69, 2193–2204.

Balbino, L.C., Bruand, A., Brossard, M., Grimaldi, M., Hajnos, M., Guimaraes, M.F. (2002). Changes in porosity and microaggregation in clayey Ferralsols of the Brazilian Cerrado on clearing for pasture. *European Journal of Soil Science*, 53, 219–230.

Beauvais, A., Ruffet, G., Hénocque, O., Colin, F. (2008). Chemical and physical erosion rhythms of the West African Cenozoic morphogenesis: The 39 Ar- 40 Ar dating of supergene K-Mn oxides. *J. Geophys. Res.*, 113, F04007.

Becquer, T., Pétard, J., Duwig, C., Bourdon, E., Moreau, R., Herbillon, A.J. (2001). Mineralogical, chemical and charge properties of Geric Ferralsols from New Caledonia. *Geoderma*, 103, 291–306.

Becquer, T., Quantin, C., Rotte-Capet, S., Ghanbaja, J., Mustin, C., Herbillon, A.J. (2006). Sources of trace metals in Ferralsols in New Caledonia. *European Journal of Soil Science*, 57, 200–213.

Beillouin, D., Ben-Ari, T., Makowski, D. (2019). Evidence map of crop diversification strategies at the global scale. *Environ. Res. Lett.*, 14, 123001.

Bernal, M.P., Alburquerque, J.A., Moral, R. (2009). Composting of animal manures and chemical criteria for compost maturity assessment. A review. *Bioresource Technology*, 100, 5444–5453.

Blanco-Canqui, H., Shaver, T.M., Lindquist, J.L., Shapiro, C.A., Elmore, R.W., Francis, C.A., Hergert, G.W. (2015). Cover crops and ecosystem services: Insights from studies in temperate soils. *Agronomy Journal*, 107(6), 2449–2474.

Boudot, J.P., Becquer, T., Merlet, D., Rouiller, J. (1994). Aluminium toxicity in declining forests: A general overview with a seasonal assessment in the Vosges mountains (France). *Ann. Sci. For.*, 51, 27–51.

van Breemen, N., Mulder, J., Driscoll, C.T. (1983). Acidification and alkalinization of soils. *Plant and Soil*, 75, 283–308.

Burak D.L., van Oort F., Becquer T., Eoy E., Fontes M.P.F. (2013). Unravelling zinc and lead distributions in soils of the Brazilian central plateau: Insight from physical fractionation, optical microscopy and x-ray microfluorescence. *European Journal of Soil Science*, 64, 131–144.

Cordell, D., Drangert, J.O., White, S. (2009). The story of phosphorus: Global food security and food for thought. *Glob. Environ. Chang.*, 19, 292–305.

Cornell, R.M. and Schwertmann, U. (1996). *The Iron Oxides. Structure, Properties, Reactions, Occurrence and Uses.* VCH Publishers, Weinheim.

Cornu, S., Montagne, D., Vasconcelos, P.M. (2009). Dating constituent formation in soils to determine rates of soil processes: A review. *Geoderma*, 153, 293–303.

Curtin, D. and Wen, G. (2004). Plant cation–anion balance as affected by the ionic composition of the growing medium. *Plant and Soil*, 267, 109–115.

Fanjaniaina, M.L., Larvy Delarivière, J., Salgado, P., Tillard, E., Rabeharisoa, L., Becquer, T. (2021). Biomass harvesting leads to soil acidification: A study of mixed crop–livestock farming systems in Madagascar. *Crop & Pasture Science*, 72, 236–244.

Feder, F., Oliver, R., Rakotoarisoa, J., Muller, B., Scopel, E. (2020). Geochemical properties of variable charge soil explain the low nitrogen bioavailability. *Communications in Soil Science and Plant Analysis*, 51, 2022–2037.

Fritsch, E., Montes-Lauar, C.R., Boulet, R., Melfi, A.J., Balan, E., Magat, P. (2002). Lateritic and redoximorphic features in a faulted landscape near Manaus, Brazil. *European Journal of Soil Science*, 53, 203–217.

Gérard, F. (2016). Clay minerals, iron/aluminum oxides, and their contribution to phosphate sorption in soils – A myth revisited. *Geoderma*, 262, 213–226.

Haynes, R.J. and Mokolobate, M.S. (2001). Amelioration of Al toxicity and P deficiency in acid soils by additions of organic residues: A critical review of the phenomenon and the mechanisms involved. *Nutrient Cycling in Agroecosystems*, 59, 47–63.

Hinsinger, P. (2001). Bioavailability of soil inorganic P in the rhizosphere as affected by root induced chemical changes: A review. *Plant Soil*, 237, 173–195.

Hinsinger, P., Ndour, N.Y.B., Becquer, T., Chapuis-Lardy, L., Masse, D. (2017). Les enjeux liés au phosphore dans les sols tropicaux. In *Restauration de la productivité des sols tropicaux et méditerranéens : contribution à l'agroécologie*, Roose, E. (ed.). IRD, Marseille.

IUSS Working Group WRB (2015). World Reference Base for soil resources 2014: Update 2015 – International soil classification system for naming soils and creating legends for soil maps. Report, World Soil Resources Reports No. 106. FAO, Rome.

Larimer, A.L., Bever, J.D., Clay, K. (2010). The interactive effects of plant microbial symbionts: A review and meta-analysis. *Symbiosis*, 51, 139–148.

Lavelle, P., Bignell, D., Lepage, M., Wolters, V., Roger, P., Ineson, P., Heal, O.W., Dhillion, S. (1997). Soil function in a changing world: The role of invertebrate ecosystem engineers. *European Journal of Soil Biology*, 33, 159–193.

Lavelle, P., Spain, A., Fonte, S., Bedano, J.C., Blanchart, E., Galindo, V., Grimaldi, M., Jimenez, J.J., Velasquez, E., Zangerle, A. (2020). Soil aggregation, ecosystem engineers and the C cycle. *Acta Oecologica – International Journal of Ecology*, 105, 103561.

Lemaire, G., Franzluebbers, A., de Faccio Carvalho, P.C., Dedieu, B. (2014). Integrated crop-livestock systems: Strategies to achieve synergy between agricultural production and environmental quality. *Agric. Ecosyst. Environ.*, 190, 4–8.

Lepsch, I. (2011). *19 Liçoes de Pedologia.* Officina de Textos, Sao Paulo.

Lucas, Y., Luizao, F.J., Chauval, A., Rouiller, J., Nahon, D. (1993). The relation between biological activity of the rain-forest and mineral composition of soils. *Science*, 260, 521–523.

Martinez, P. and Souza, I.F. (2020). Genesis of pseudo-sand structure in Oxisols from Brazil – A review. *Geoderma Regional*, 22, e00292.

Meunier, A. (2005). *Clays.* Springer, Berlin, Heidelberg.

Miao, Y., Stewart, B.A., Zhang, F. (2011). Long-term experiments for sustainable nutrient management in China. A review. *Agronomy for Sustainable Development*, 31(2), 397–414.

de Moraes Sá, J.C., Cerri, C.C., Lal, R., Dick, W.A., de Cassia Piccolo, M., Feigl, B.E. (2009). Soil organic carbon and fertility interactions affected by a tillage chronosequence in a Brazilian Oxisol. *Soil and Tillage Research*, 104(1), 56–64.

Morel, C. (2002). Caractérisation de la phytodisponiblité du P du sol par la modélisation du transfert des ions phosphates entre le sol et la solution. Research Document. INPL-Ensaia, Nancy.

Nahon, D. (2003). Altérations dans la zone tropicale. Signification à travers les mécanismes anciens et/ou encore actuels. *Comptes Rendus Geoscience*, 335, 1109–1119.

Noble, A.D., Zenneck, I., Randall, P.J. (1996). Leaf litter ash alkalinity and neutralisation of soil acidity. *Plant and Soil*, 179, 293–302.

Oberson, A., Friesen, D.K., Rao, I.M., Bühler, S., Frossard, E. (2001). Phosphorus transformations in an Oxisol under contrasting land-use systems: The role of the soil microbial biomass. *Plant Soil*, 237, 197–210.

Palm, C.A., Gachengo, C.N., Delve, R.J., Cadisch, G., Giller, K.E. (2001). Organic inputs for soil fertility management in tropical agroecosystems: Application of an organic resource database. *Agriculture, Ecosystems & Environment*, 83(1–2), 27–42.

Parfitt, R.L. (1980). Chemical properties of variable charge soils. In *Soils with Variable Charge*, Theng, B.K.G. (ed.). New Zealand Society of Soil Science, Lower Hutt.

Penn, C.J. and Camberato, J.J. (2019). A critical review on soil chemical processes that control how soil pH affects phosphorus availability to plants. *Agriculture*, 9(6), 120.

Qafoku, N.P., Van Ranst, E., Noble, A., Baert, G. (2004). Variable charge soils: Their mineralogy, chemistry and management. *Advances in Agronomy*, 84, 159–215.

Raminoarison, M., Razafimbelo, T., Rakotoson, T., Becquer, T, Blanchart, E., Trap, J. (2020). Multiple-nutrient limitation of upland rainfed rice in Ferralsols: A greenhouse nutrient omission trial. *Journal of Plant Nutrition*, 43, 270–284.

Randriamanantsoa, L., Morel, C., Rabeharisoa, L., Douzet, J.M., Jansa, J., Frossard, E. (2013). Can the isotopic exchange kinetic method be used in soils with a very low water extractable phosphate content and a high sorbing capacity for phosphate ions? *Geoderma*, 200, 120–129.

Razakatiana, A.T.E., Trap, J., Baohanta, R.H., Raherimandimby, M., Le Roux, C., Duponnois, R., Ramanankierana, H., Becquer, T. (2020). Benefits of dual inoculation with arbuscular mycorrhizal fungi and rhizobia on Phaseolus vulgaris planted in a low-fertility tropical soil. *Pedobiologia*, 83, 150685.

Reatto, A., Bruand, A., Silva, E.M., Martins, E.S., Brossard, M. (2007). Hydraulic properties of the diagnostic horizon of Latosols of a regional toposequence across the Brazilian Central Plateau. *Geoderma*, 139, 51–59.

Rufino, M.C., Rowe, E.C., Delve, R.J., Giller, K.E. (2006) Nitrogen cycling efficiencies through resource-poor African crop–livestock systems. *Agriculture, Ecosystems & Environment*, 112, 261–282.

Santos, H.G., Jacomine, P.K.T., Anjos, L.H.C., Oliveira, V.A., Lumbreras, J.F., Coelho, M.R., Almeida, J.A., Araujo Filho, J.C., Oliveira, J.B., Cunha, T.J.F. (2018). *Sistema brasileiro de classificação de solos*. Embrapa, Brasília, DF.

Sauvadet, M., Saj, S., Freschet, G.T., Essobo, J.D., Enock, S., Becquer, T., Tixier, P., Harmand, J.M. (2020). Cocoa agroforest multifunctionality and soil fertility explained by shade tree litter traits. *Journal of Applied Ecology*, 57(3), 476–487.

Sauvadet, M., Trap, J., Damour, G., Plassard, C., Van den Meersche, K., Achard, R., Allinne, C., Autfray, P., Bertrand, I., Blanchart, E. et al. (2021). Agroecosystem diversification with legumes or non-legumes improves differently soil fertility according to soil type. *Sci. Total Environ.*, 795, 148934.

Schaefer, C.E.R. (2001). Brazilian latosols and their B horizon microstructure as long-term biotic constructs. *Australian Journal of Soil Research*, 39, 909–926.

Scopel, E., Triomphe, B., Affholder, F., Da Silva, F.A.M., Corbeels, M., Xavier, J.H.V., Lahmar, R., Recous, S., Bernoux, M., Blanchart, E. et al. (2013). Conservation agriculture cropping systems in temperate and tropical conditions, performances and impacts. A review. *Agron. Sustain. Dev.*, 33, 113–130.

Ségalen, P. (1994-1995). *Les sols ferrallitiques et leur répartition géographique. Tome 1 (1994). Introduction générale, les sols ferrallitiques: leur identification et leur environnement immédiat. Tome 2 (1995). Les sols ferrallitiques: les facteurs de formation et les sols ferrallitiques en Amérique. Tome 3 (1995). Les sols ferrallitiques en Afrique et en Extrême-Orient, Australie et Océanie: conclusions générales*. ORSTOM, Paris.

Shuster, D.L., Vasconcelos, P.M., Heim, J.A., Farley, K.A. (2005). Weathering geochronology by (U-Th)/He dating of goethite. *Geochimica et Cosmochimica Acta*, 69, 659–673.

Soil Survey Staff (1999). *Soil Taxonomy: A Basic System of Soil Classification for Making and Interpreting Soil Surveys*, 2nd edition. Natural Resources Conservation Service. U.S. Department of Agriculture Handbook 436, Washington, D.C.

Sparks, D.L. (2003). *Environmental Soil Chemistry*, 2nd edition. Academic Press, Oxford.

Sradnick, A. and Feller, C. (2020). A typological concept to predict the nitrogen release from organic fertilizers in farming systems. *Agronomy*, 10(9), 1448.

Steffens, D., Leppin, T., Luschin-Ebengreuth, N., Min Yang, Z., Schubert, S. (2010). Organic soil phosphorus considerably contributes to plant nutrition but is neglected by routine soil-testing methods. *J. Plant Nutr. Soil Sci.*, 173, 765–771.

Stoorvogel, J.J., Smaling, E.M.A., Windmeijer, P.N. (1993). Calculating soil nutrient balances in Africa at different scales. I. Supra-national scale. *Fertilizer Research*, 35, 227–235.

Tardy, Y. (1993). *Pétrologie des latérites et des sols tropicaux*. Masson, Paris.

Tittonell, P., Rufino, M.C., Janssen, B.H., Giller, K.E. (2010). Carbon and nutrient losses during manure storage under traditional and improved practices in smallholder crop-livestock systems: Evidence from Kenya. *Plant and Soil*, 328, 253–269.

Unkovich, M.J., Baldock, J., Peoples, M.B. (2010). Prospects and problems of simple linear models for estimating symbiotic N2 fixation by crop and pasture legumes. *Plant Soil*, 329, 75–89.

Vasconcelos, P.M. (1999). K-Ar and 40Ar/39Ar geochronology of weathering processes. *Annu. Rev. Earth Planet. Sci.*, 27, 183–229.

Vendrame, P.R.S., Brito, O.R., Martins, E.S., Quantin, C., Guimarães, M.F., Becquer, T. (2013). Acidity control in Latosols under long-term pastures in the Cerrado region, Brazil. *Soil Research*, 51, 253–261.

Zeng, M., de Vries, W., Bonten, L.T., Zhu, Q., Hao, T., Liu, X., Xu, M., Shi, X., Zhang, F, Shen, J. (2017). Model-based analysis of the long-term effects of fertilization management on cropland soil acidification. *Environmental Science & Technology*, 51(7), 3843–3851.

10

Urban Agricultural Soil

Sophie JOIMEL,[1] Tania DE ALMEIDA[2,3] and Baptiste GRARD[4]

[1]*UMR ECOSYS, AgroParisTech, INRAE, Université Paris Saclay, Palaiseau, France*

[2]*IMT-Nord Europe, Université de Lille, France*

[3]*JUNIA, ULR 4515 LGCgE, Université d'Artois, Arras, France*

[4]*Agroecology and Environment Research Unit, ISARA, Lyon, France*

10.1. Introduction

The "emerging city" of the late 20th century, characterized by its sprawl and peri-urbanization, saw the convergence of two processes: the urbanization of natural and agricultural environments and the ruralization/renaturation of the city (Chalas 1997). More than half the world's population lives in an urban environment today, and according to the United Nations Organisation, by 2050 this will grow to 75% of the world's population.[1] The urban environment is characterized by a high population and building density. Despite the many benefits they provide, green spaces are generally rare and not equally accessible to all.[2]

However, city-dwellers are increasingly demanding more "nature in the city" and green spaces that must be adapted to the constraints of the urban environment. In parallel, the city also considers the farmland and natural spaces on the periphery of the city as being available for urbanization. The city therefore continues to break away from the essential dialectic between humans and their environment (Sansot 1997) and ensures a certain urban dominance over the fate of areas dedicated to agriculture.

---

1 UNO (2018).

2 WHO (2016). Urban green spaces and health. WHO Regional Office for Europe, Copenhagen.

Agricultural Soil Science,
coordinated by Yves COQUET and Joël MICHELIN. © ISTE Ltd 2025.

Urban agriculture (Figure 10.1), which has seen a rise worldwide over the past few decades, especially in the Northern Hemisphere, makes it possible to respond to these expectations from city-dwellers. Nonetheless, the urban environment also poses certain challenges related to land and logistics, as well as the large diversity of urban soils, which are an essential support for agriculture but offer highly varied fertility and the risk of contamination by metals or organic pollutants.

Consequently, this chapter will explore the history of urban agriculture, the diversity of the urban soils involved and the challenges that cities face in terms of the services provided and their management.

Figure 10.1. *Photos of the microfarm run by Planète Lilas association, in the Lilas Departmental Park, Vitry-sur-Seine, May 2018. © Lea Wirz. For a color version of this figure, see www.iste.co.uk/coquet/agricultural.zip*

10.2. Redeploying urban agriculture

Urban agriculture is often seen as a modern form of agriculture. Nevertheless, as the phenomenon can be considered to date back to the sedentarization of humans (Choay 1972). Since then, urban agriculture has evolved and adapted over time based on current practices and the changes in land use within cities.[3] It has been defined in multiple ways, but all definitions agree on the importance of the links that exist between urban agriculture and the city and the diversity of its the forms and functions. Urban agriculture can therefore be defined as agriculture carried out within the city (intra-urban) or on the outskirts (peri-urban), producing, growing, transforming and distributing a variety of products, whether food or not, (re)using human and material resources, products and services that are found in and around the urban zone and supplying human and materials resources, products and services largely to this urban zone (Mougeot 2000).

In recent years, urban agriculture in Europe, and France in particular, has been characterized by increasing intra-urban development and, notably, by a large diversity of forms and functions (Aubry and Pourias 2013; Specht et al. 2014; Lohrberg et al. 2016). This growing interest in urban agriculture has resulted in many projects (cooperative or entrepreneurial), the creation of organizations that specialize in setting up production spaces in the city, and political support for the process. This is especially the case with Paris, where calls for participation in several recent projects (e.g. the "Parisculteurs" call for proposal[4]) have led to the emergence of agriculture within the city itself, supporting the participants in the project.

The forms of agriculture that have developed are shaped by the available space (e.g. rooftop gardening, brownfield and inside a building), the nature of the soil (e.g. natural, pseudo-natural and constructed), the farming system (e.g. on the ground or above ground, aquaponics and hydroponics) and on the type of production or again on the distribution channel. Urban agriculture can therefore be divided into five categories using a classification developed by AgroParisTech and Cerema[5,6]:

3 The concept of "city" here is synonymous with large city or metropolis.

4 Available at: www.parisculteurs.paris.

5 Available at: www.cerema.fr/fr/actualites/videos-pedagogiques-differentes-formes-agriculture-urbaine.

6 Centre d'études et d'expertise sur les risques, l'environnement, la mobilité et l'aménagement (Center for Studies and Expertise on Risk, Environment, Mobility and Planning).

– Interstitial productive spaces: citizen-led micro-gardening projects in public spaces (e.g. squares, parks, plots and sidewalks) or in private spaces (e.g. balconies, courtyards within residential or office buildings) with plants often cultivated in bins.

– Collective gardens: gardens ranging from tens to a few hundred m², worked on by a group of individuals (e.g. resident collectives and associations) which are organized to produce food products or other plants. These gardens may also be considered as community gardens, allotments (family gardens) or "insertion" gardens to facilitate social inclusion, thus leading to variable rules, site management practices and functions.

– Urban micro-farms: participative (i.e. involving volunteer) urban farms, measuring between a few hundred m² to several ha. They generally have a large portion of the work carried out on a voluntary basis, with all or part of the food produce being sold. They are developed for a diversity of functions from food production to social and environmental purposes.

– Specialized urban farms: agricultural companies whose main activity is above-ground farming (e.g. greenhouses on rooftops, vertical farming and indoor farms) using limited part of land (e.g. with farming systems such as hydroponics, aquaponics or crops substrates) and which also develop high-tech production processes. They are generally developed mainly to produce food.

– Peri-urban vegetable farms: farms on open land measuring 1.5–10 ha, whose main objective is often to produce food.

Figure 10.2. *Illustrations of the different forms of urban agriculture. (a) The vegetable garden in the canteen at Nantes, set up and farmed by Olivier Durand (May 2019) © Baptiste Grard. (b) The Montolivet allotments in Marseille: soil recreated from topsoil (2012) © Sophie Joimel. For a color version of this figure, see www.iste.co.uk/coquet/agricultural.zip*

Citizen demands for these different categories of urban agriculture change depending on urban context. However, historically and even today, in France and around the world, we see the predominant development of collective gardens (Figure 10.2), whether shared or allotments (Draper and Freedman 2010; Guitart et al. 2012; Poniży et al. 2021). They offer the unique combination of economic, social and environmental functions (Bell 2016), as well as leisure activities (Consalès 2004), while also providing a wide range of ecosystem services to users and citizens (Duchemin et al. 2009) (e.g. vegetable production) over small areas. In 2018, there were over 1,000 collective garden sites in the Ile-de-France region (Paris and its surrounding area). Recently, farming on or inside buildings rapidly developed at the global level in a diversity of contexts (Spech et al. 2014). Farming on buildings developed in response to the paucity of space in urban areas. A recent study on rooftop agriculture around the world (Appolloni et al. 2021) showed that the majority of existing projects involve vegetable green roofs that use substrates or a constructed soil (84% of the 185 roofs studied), and the setting up of greenhouses on rooftops remains quite limited at the moment. Additionally, these roofs primarily serve socio-educational purposes, related to the quality of urban life. This example is a good illustration of the challenge faced by any form of urban agriculture: access to, or developing soil of sufficient quality to provide the ecosystem services expected from these ecosystems, such as food production and water retention or again a habitat for biodiversity.

10.3. Urban soils: the basis of urban agriculture

10.3.1. *The large diversity in urban soils*

There are many different definitions for urban soils depending on whether the approach used is oriented toward spatialization or toward the origin of the soil. In the first case (an approach used especially by actors working on land-use planning), soil is said to be "urban soil" when it is located within an urban region.

The second approach, favored by pedologists and agronomists, requires the quantification of the degree of "anthropic disturbance" (Lehman and Stahrr 2007), that is, how much the soil has been transformed or constructed by human activities.

Most urban soils have properties that are heavily influenced by human activities due to their usage and the intense disturbance (Bechet et al. 2009) and are thus called "anthropized" soils.

Within these anthropized soils, depending on the degree of anthropization, we can identify a wide variety of urban soils, as can be seen in the study by Morel et al. (2014):

– Sealed soils: soils that are significantly remolded by adding anthropogenic material in the soil profile, such as brick or concrete. These are notably used as support in infrastructure projects. The soil functions are affected by the fact that the soil is made impermeable, and by the nature of the materials used in the soil.

– Constructed soils: remolded soils of very heterogeneous origins (e.g. brownfield, rooftop vegetable garden and a garden) that can be constructed from a variety of materials. Constructed soils, which can sometimes consist entirely of urban residues, are also called Anthroposols or Technosols, depending on the pedological reference used (Baize and Girard 2009; IUSS Working Group WRB 2015).

– "Pseudo-natural" soils: soils that have not been disturbed very much through human activities. These are notably used in parks or gardens. The first horizon may be modified by intensive practices such as organic matter enrichment.

Figure 10.3. *Diversity of urban soils (pseudo-natural vegetated soil, constructed vegetated soil, sealed soil) © H. Huot, B. Grard (adapted from Joimel et al. 2021). For a color version of this figure, see www.iste.co.uk/coquet/agricultural.zip*

The history of soil use in urban areas indicates that the majority of pedological cover in these spaces is made up of Technosols or Anthroposols. Combined with the different functions carried out here, this diversity of soils creates a spatial heterogeneity that is seen both at the horizontal level and vertical level (Morel et al. 2005; Bechet et al. 2009).

10.3.2. *Varied physicochemical properties*

A soil's ability to provide ecosystem goods and services is determined by its quality (Morel et al. 2014), which is defined based on its physical, chemical and biological properties. The common effects of anthropization on a soil's physicochemical parameters are observable.

10.3.2.1. *Soils with neutral or alkaline pH*

Anthropized soils often have a high pH (Lorenz and Kandeler 2005; Morel et al. 2005). In pseudo-natural vegetated soils, as in gardens, basic pH values (between 7 and 8.5) have been observed in various French cities (Joimel et al. 2016).

10.3.2.2. *Soils rich in organic matter*

Urban soils are often considered to be poor in nutrients. However, high organic matter content has been seen in allotment gardens (up to 10%, as against 1–4% in agricultural soils) with high phosphorus levels (sometimes excessive) (Joimel et al. 2016), thus adding to the prior observations on urban soils, especially in parks (Takahashi et al. 2008; Säumel et al. 2012; Antisari et al. 2015).

In the case of constructed Technosols, they are "created" from a variety of materials that are generally derived from organic products – especially potting soil – or residues (e.g. compost from household waste or green waste). This was observed in many urban rooftop agriculture projects that were selected as Pariculteurs projects, or as an approach used by certain company of the sector[7].

Those parent materials contain between 200 and 500 g $Corg.kg^{-1}$ (Grard et al. 2020). This results in particularly high organic matter content on the developed Technosol (around 35% on average, Joimel et al. 2021).

7 Available at: www.culturesenville.fr.

Inset: focus on the JASSUR project
Period: 2013–2016 Subject of the study: collective gardens on a country-wide scale in France. Objectives of the project: to study the quality of soils and the services provided by collective gardens. Project coordinators: Schwartz (INRAE-UL) and Consales (CNRS-UM Aix-Marseille). Funding from ANR. Website: https://anr.fr/Projet-ANR-12-VBDU-0011. Selection of articles: – Joimel et al. (2016, 2017, 2018, 2020, 2021, 2022a); – Bechet et al. (2018); – Consalès et al. (2016); – Zhong et al. (2022); – See: https://theconversation.com/a-la-rencontre-des-petites-betes-des-toitures-vegetalisees-107452.

10.3.2.3. *Multi-contamination by pollutants (metallic and organic)*

Several studies have documented the metallic contamination of urban soils (Morel et al. 2005; Cheng et al. 2015; Joimel et al. 2016), more specifically, allotment gardens (Morel and Schwartz 1999; Joimel et al. 2016, 2021; Burghardt and Schneider 2018). Concentrations of metals in agroecosystems are either inherited from the parent materials of the soil, or introduced through human activities (He et al. 2005; Joimel et al. 2021). In some cases, a large quantity of metals may result from both sources, which leads to abnormally high concentrations in the soils (Jean-Soro et al. 2015). Cultivation practices that aim to enhance soil fertility are also a source of pollutants. The production of food-based biomass may induce metallic contamination through the use of fertilizers and pesticides containing trace metals, either from additives and impurities (Belon et al. 2012), or from amendments such as sewage sludge (Cambier et al. 2009). Metal contamination of urban agriculture can also be linked to urban activities such as industrial emissions and road traffic (Bechet et al. 2009, El Hamiani et al. 2010) (Figure 10.4), whether current or historical (Douay et al. 2007).

Earlier studies carried out in several countries such as the United States or Hungary have shown a large range of concentrations of metals in garden soils (Szolnoki et al. 2013; Mitchell et al. 2014; Cheng et al. 2015). Notably, Cd, Cu, Pb and Zn levels exceeded geochemical background values in French urban vegetable gardens (Joimel et al. 2016, 2021). It must also be noted that trace metal elements concentration in agricultural soils are usually lower than in urban soils (Joimel et al. 2016) (Figure 10.5).

Figure 10.4. *Urban agriculture sites are often located in direct proximity to one or more potentially polluting transport infrastructure projects (e.g. railways and highways). This allotment garden in Marseille is located on a former factory, between a railroad and a highway (2012) © Sophie Joimel. For a color version of this figure, see www.iste.co.uk/coquet/agricultural.zip*

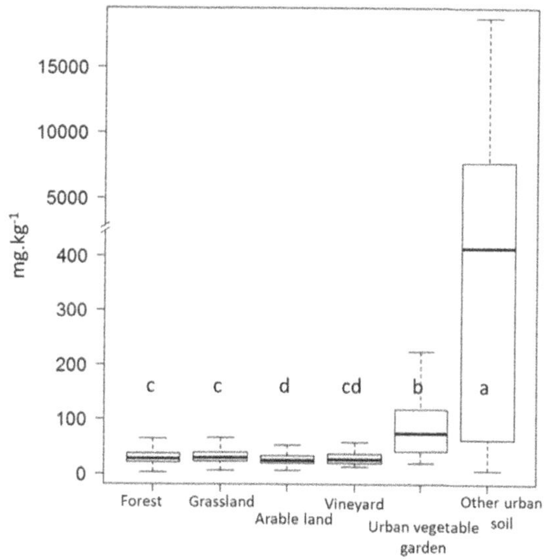

Figure 10.5. *Urban soils, especially those used for agriculture, have higher trace metal elements content than other soils, due to contamination linked to anthropic activities (e.g. practices and urban environment). For example, lead content in French soils classified based on various uses: forest (n = 582), grassland (n = 623), arable land (n = 820), vineyards and orchards (n = 48), urban vegetable garden (n = 104), and other urban soils (n = 221). The letters indicate the significant differences between uses (adapted from Joimel et al. 2016)*

As with metals, organic molecules can also be found in soils (e.g. polycyclic aromatic hydrocarbons (PAHs), pesticides, solvents and dioxins) depending on the environment in which the site is located and its past use. Atmospheric deposition, the use of inputs (e.g. organic wastes and pesticides) or the use of urban substrates by gardeners can influence the contaminant levels in the surface horizon, where the roots systems of crops are concentrated. However, there is still very little data on the presence of organic contaminants in urban agricultural soils. As far as the users of these gardens are concerned, the perceived health risks of organic contaminants are smaller than those associated with metal contaminants (Kim et al. 2014).

These different studies on (in)organic contaminants were chiefly carried out in pseudo-natural vegetated soils. There has been very little work on constructed soils used for food production. Early studies (Grard et al. 2020; Joimel et al. 2021), however, are encouraging as they do not highlight contamination in soils used for vegetable green roofs. However, these are very young soils and as with in-situ soils, the nature of the products used for constructing these soils may pose a health risk if they contain metals or organic pollutants.

10.3.2.4. *Very large variability*

It is also important to note that urban soils have highly varied physicochemical properties (Joimel et al. 2016). Thus, extreme physical and chemical properties can be seen in different urban soils: for example, a pH of 3, or extreme metal levels, or, on the contrary, an absence of contamination (Joimel et al. 2016). Even when used for identical purposes, high variability is seen in soil properties. For example, organic matter content may vary highly in family garden, ranging from 1% to 10% (Joimel 2015; Joimel et al. 2021). Organic carbon content has also been seen to vary from 60 to 170 g. kg^{-1} (Research project SEMOIRS[8]) in the case of urban micro-farms. This variability can be explained by the diversity of locations (intra-urban or on the periphery), the origin of the parent materials (in situ soils on former vegetable gardens, excavations or again displaced humus soil) and its use (collective garden or professional farm).

Urban soils are also young soils, especially in the case of constructed Technosols, and they undergo frequent changes of usage. Finally, urban soils are often characterized by spatial heterogeneity due to anthropic elements, also called technogenic elements, and the fraction of these constituents increases toward the center of cities (El Khalil et al. 2008; Nehls et al. 2013).

---

8 Available at: https://ecosys.versailles-saclay.hub.inrae.fr/recherche/projets/projets-termines/semoirs.

10.3.3. *Interesting biological properties*

The Global Assessment on Biodiversity and Ecosystem Services (IPBES 2019) estimates that around a million animal and plant species are threatened with extinction in the coming decades. Furthermore, soils shelter around a quarter of our planet's biodiversity (Decaëns et al. 2006). Soil organisms play an essential role, especially with respect to soil fertility, water quality and carbon storage. Given the physicochemical properties of urban soils (e.g. contamination, extreme pH values or compaction), we may expect that these soils would have low biodiversity, with negative effects on organisms of all sizes (e.g. earthworms, coleoptera and microorganisms). On the contrary however, non-sealing urban soils seem to offer a refuge to biodiversity in the city. Thus, 1.6 times more species of springtails (Figure 10.6) (known to be soil quality indicators) are seen in urban vegetable gardens than in agricultural lands, and they are eight times more abundant (Joimel et al. 2017). Similar or even more pronounced results have been observed in vegetable green roofs (Joimel et al. 2018) for other groups of organisms as well (Joimel et al. 2021b). Furthermore, the presence of rare species, often specific to the urban environment (e.g. in springtails: *Coecobrya tenebricosa* or *Folsomia similis*) has also been demonstrated in the case of other arthropods (Madre et al. 2013), reptiles and birds.

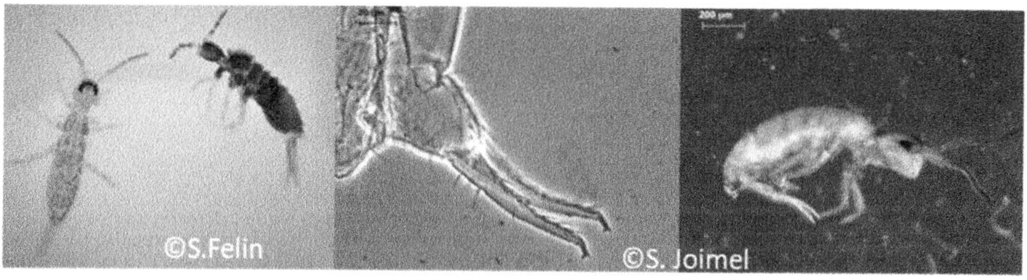

Figure 10.6. *Springtails (soil mesofauna, i.e. smaller than 2 mm) play a role in breaking down organic matter in soil. For a color version of this figure, see www.iste.co.uk/coquet/agricultural.zip*

These organisms must nonetheless adapt to the constraints of urban soils and, consequently, springtails have specific characteristics related to their life on the surface, rather than in the depths of the soil (Joimel et al. 2018).

Constructed soils used for vegetable green roofs therefore have a heterogeneous mix of common species of springtails, as well as rarer species, in extreme abundance. These are brought in through the parent materials with which the soils are made (e.g. compost) or through the wind blowing from surrounding green spaces (Joimel et al. 2018).

Questions do arise related to the functionality of these particular communities in urban environments (Joimel et al. 2021a) as well as the risk of biotic homogenization, where the same communities are found across cities independent of the climate (Joimel et al. 2021c). However, while this risk of homogenization has been demonstrated for earthworms in parks (Vergnes et al. 2017) it does not currently seem relevant to other groups (e.g. microorganisms or springtails) in urban environments (Joimel et al. 2021b, 2021c) (SEMOIRS project).

There are still many gaps in understanding colonization, biotic interactions and habitat relationships between urban soil organisms (Joimel et al. 2022b). It can be seen that a very small percentage of studies are carried out in an urban environment focusing on soil biodiversity (< 3%).

Inset: focus on the TROL project
Period: 2017–2019 Subject of study: green roofs (around 20 in Paris and its surrounding area). Objectives of the project: to study the biodiversity and quality of soils on rooftop vegetable gardens and extensive rooftop farming. Project coordinators: Joimel and Chenu (INRAE-AgroParisTech). Funding from Labex BASC. See: https://www6.inrae.fr/basc_eng/Research/Emergence-projects/TROL. Selection of articles: – Joimel et al. (2018); – Joimel et al. (2021); – See: https://theconversation.com/a-la-rencontre-des-petites-betes-des-toitures-vegetalisees-107452.

10.4. Managing ecosystem services provided by urban soils

10.4.1. *Ecosystem services – a definition*

Since some forms of urban agriculture are multi-purpose in nature, there are many expectations surrounding the services that will be provided (Aerts et al. 2016). Popularized by the Millennium Ecosystem Assessment in 2005 (MEA 2005), the concept of "ecosystem service" and the use of this concept by a part of the scientific community and political decision-makers results from a desire to better understand and draw attention on the impact that humans have on ecosystems, as well as the benefits these ecosystems provide in terms of services (see Chapter 7). Growing in popularity over the last 20 years, the term was introduced in 1970 by the Massachusetts Institute of Technology[9] and aimed to draw greater attention to the environmental impacts of human activities. Following this report, the involvement of scientists and political decision-makers led to the creation of the Millennium Ecosystem Assessment, which made it possible to widely disseminate this concept and create a common classification of ecosystem services. This project was followed by different conventions and classifications, such as the Common International Classification for Ecosystem Services (CICES) in 2009.

If we refer to the definition given in the MEA, ecosystem services are "the benefits that humans derive from nature". These benefits connect an ecosystem's biophysical components with human well-being, placed at the center of a socioeconomic system. Depending on the classifications used, it is possible to identify three or four categories of services:

i) supply services, which may correspond to the production of food, nonfood biomass, or the supply of fresh water;

ii) regulation services, such as water storage, pollution mitigation and climate change;

iii) cultural services which encompass education, landscapes and tourism;

iv) in certain cases, a fourth category may be defined, called support services, which notably covers services provided by biodiversity, and may be seen as essential for the realization of all other services.

The operationalization of this concept, that is, being able to qualify and quantity the services provided or even using this study to plan development, has been the subject of various debate within the scientific community.

Those debates have permitted the emergence of definition of indicators to use, their relationship to the biophysical properties of the soil, as well as how to interpret these indicators (Boerema et al. 2017). This concept has also drawn much criticism for its anthropocentric nature, which could lead to an economic assessment of something

9 The report was entitled "Study of Critical Environmental Problems".

that has nonmonetary value, or could lead to an evaluation that is contrary to the objectives of biodiversity conservation (Schröter et al. 2014).

The nature and type of soil being considered, especially through their physical, chemical and biological properties, strongly influence the nature and quantity of the ecosystem services provided. Morel et al. (2014) demonstrate the link between soil type and the ecosystem services provided by categorizing urban soils into pseudo-natural soils, engineered soils (i.e. constructed Technosols), soil resulting from deposits and seal soils. The soil type therefore influences the services that urban sites provides.

10.4.2. *Soils that serve many purposes*

10.4.2.1. *Food production in an urban environment*

Urban agriculture projects are inserted into channels for the distribution and consumption of various food products: from personal consumption to more or less short circuit marketing. These projects, whether they are on open ground or in constructed soil, offer various agronomic practices that could impact both the quality of the soils as well as the quality of the services provided.

Studies that examine the food production from urban agriculture in cities have multiplied in recent years, as seen in the review of existing literature by Weidner et al. (2019). This synthesis shows that urban agriculture can provide between 1% and 246% of a city's food requirements, depending on the crop system, available spaces, the forms of urban agriculture and the type of food products considered. However, these studies make very little use of data gathered from urban agriculture cropping system. In fact, prospective studies that look at what urban agriculture could represent in terms of food production are very often not based on data from urban farms or gardens, but on databases on other contexts.

However, a rare study of 11 family and shared gardens in Montreal (Canada) and Paris (France) (Pourias et al. 2015) was able to show a large variability in production levels between gardeners, with an average ranging from 1 to 2 $kg.m^{-2}$. This variability may be linked to the motivations of the gardeners, but also to climate and soil quality. Researchers working on the SEMOIRS project also observed a high variability in yield, with similar levels of production 1.4 ± 1.1 $kg.m^{-2}$ (Grard et al. 2022), related to soil characteristics which varied widely between the six sites studied, especially in terms of organic matter content, nutrients and density.

Inset: focus on the SEMOIRS project
Period: 2018–2020
Subject of study: urban micro-farms (seven in Paris and its surrounding area).
Objectives of the project: to study ecosystem services provided by urban micro-farms and their soils.
Project coordinators: Chenu and Grard (INRAE-AgroParisTech).
Funding from ADEME.
See: https://www6.versailles-grignon.inrae.fr/ecosys/Recherche/Projets/Projets-termines/SEMOIRS.
Selection of articles: – Grard et al. (2021); – See: https://theconversation.com/bd-pourquoi-mettre-des-fermes-dans-les-villes-142107.

Furthermore, the variability in the yield from urban agriculture can also be explained by proportion of land dedicated to food production. Various studies show that whether at the level of a garden or an urban farm, only a portion of the space, and therefore the soil, is dedicated to food production. In urban farms or gardens, the area dedicated to food production is estimated to be around 50%, according to Chang and Morel (2018) (project SEMOIRS) going up to 76% (Pourias 2014).

The high variability seen in the food yield can also be explained by the variability in practices such as irrigation or tillage. An important practice among these that must be noted is the significant use of organic waste, whether to improve urban agricultural soil, or for constructing soils. This organic waste is mainly derived from the city, especially from the recycling of waste, such as green waste compost, residues from mushroom cultivation, or wood waste (Guitart et al. 2012; Edmondson et al. 2020; Grard et al. 2020).

These contributions can go up to 30 m^3 of green waste compost over two years, on a farm with usable agricultural area of 2,000 m^2, or 700 m^3 of materials brought in to construct the soil in a vegetable green roof measuring around 900 m^2 (SEMOIRS project).

Some studies have demonstrated the role of organic matter inputs in urban market gardening. In a study carried out over five years, studying different Technosols made up only of urban residues on rooftops, researchers working on the T4P project showed that it was possible to attain high levels of food production, with a five-year average production ranging from 8.3 to 15.4 $kg.m^{-2}.year^{-1}$ (Grard et al. 2020).

Inset: focus on the T4P (Experimental productive rooftop, pilot project) project

Period: 2012 to present

Subject of the study: experimental vegetable green roof (around 900 m^2).

Objectives of the project: studying Technosols formed only from urban residues on rooftops and analyzing the ecosystem services provided.

Project coordinators: Aubry, Grard, Chenu and Joimel (INRAE-AgroParisTech).

Funding from Chaire LabEnvironnement, as well as different projects that used the experimental site (see projects SEMOIRS and TROL in particular).

See: http://www2.agroparistech.fr/T4P-un-Projet-de-recherche-innovant-pour-des-Toits-Parisiens-Productifs.html.

Selection of articles:

– See: https://theconversation.com/bd-sur-le-toit-des-legumes-et-de-la-science-107386;

– Grard et al. (2015, 2017, 2020).

10.4.2.2. *Water regulation service*

Managing water flows in an urban environment has become a major challenge today both in terms of planning as well as functioning. Water management systems in many cities often get saturated with multiple spells of intense rain and the increased sealing of soils. Given this challenge, many solutions for the retention of water at the roof level have been supported by a number of public policies, such as the Paris municipality, through its ParisPluie[10] project. Through this project, urban gardens and farms among other green infrastructure can play a role in allowing for the retention or even harvesting of part of the rainwater.

In the case of Technosols made from urban residues, a study carried out as part of the T4P project demonstrated a significant potential for rainwater harvesting, with between 74% and 81% of rainwater being retained (Grard et al. 2022). Other studies have also demonstrated this capacity for water retention in farms or gardens on open land (Richards et al. 2015; Whittinghill et al. 2015, 2016). However, the specific features of certain urban soils, including constructed soils, and especially soils with high organic matter and nutrient content, may affect the quality of percolation water through the release of nutrients or carbon (Richards et al. 2015; Whittinghill et al. 2015, 2016; Grard et al. 2018).

Furthermore, irrigation, a key issue in vegetable cultivation including in an urban environment, is often promoted as having a positive impact on urban agriculture projects, which can use grey water, reclaimed water and/or rainwater, thus making use of waste or a resource in the urban environment. However, the few studies carried out on this subject currently show that the majority of these projects use

10 Available at: https://www.paris.fr/pages/le-plan-parispluie-5618.

potable water (Csortan et al. 2020; Whittinghill and Sarr 2021; Grard et al. 2022), chiefly because of financial constraints (e.g. expensive installations), logistical challenges (e.g. lack of space and technical constraints) and health challenges (e.g. fear of pathogens). The irrigation management practices (e.g. automatic or manual, using a sprinkler or drip system, allotted time) can also impact the moisture of the in situ soils and, a fortiori, their retention capacities. A study of a Technosol from vegetable green roof project across 6,000 m^2 belonging to the New York company Brooklyn Grange showed that this acted as a net source of water. Thus, over the duration of the study, 11% of rainwater volume was discharged from the roof. The authors connected this observation with the irrigation practice: soil maintained at very high moisture levels, including during rainfall, without actual utilization of the soil's water retention capacity.

This example demonstrates the importance of studying how urban agriculture projects use water, thereby permitting installations and operations that will provide ecosystem services. Using the appropriate tools and materials are also a key factor in adapting practices.

10.4.2.3. *Biogeochemical cycles and carbon storage*

Certain urban soils, rich in organic matter, represent a non-negligible stock of carbon. The very few studies carried out on this topic have assessed carbon stocks in urban soils (Cambou et al. 2018) as well as for rooftop vegetable gardens (Whittinghill et al. 2014). Carbon stocks in non-sealed urban soils in New York and Paris are non-negligible in quantity: 44%–110% greater than carbon stocks in agricultural soils in a 0–30 cm layer, measuring around 10–11 kgC.m^{-2} (Cambou et al. 2018). These results are similar to those obtained by the SEMOIRS project, showing a high variability of stocks ranging from 80 to 406 kgC.ha^{-1} across seven urban micro-farms in open land and on rooftops (0–30 cm). On average, the soils used in urban micro-farms have carbon stocks that are four to six times greater than those found in vineyard or orchard soils (Grard et al. 2022). However, the contribution of urban agriculture to carbon sequestration in a region's soil is still limited, given the current area of soil used for agriculture as compared to other uses. Furthermore, soils used for other purposes have additional C storage due to the greater depth of these soils, while many urban agriculture spaces have shallow soils (30 cm).

10.5. Toward decision-making tools

In the past few years, many decision-making tools have been developed for urban agriculture in order to meet to the needs of agriculture project leaders, the different actors in urban planning (e.g. urban planners and landscape planners) or community technical services. From a simple guide (e.g. the guide published by the Regional Health Department in France on collective gardening) to planning tools, some of these decision-making tools are based on soil quality and are sometimes specific to urban agriculture[11]. For example, the DESTISOL'AU tool[12], which, when finalized, will provide urban planners, with recommendations, especially in upstream stages of a project, with respect to the potential use of urban soils for agri-urban projects.

Other decision-making tools take into consideration urban agriculture and the associated soil at a city-planning level (UQUALISOL)[13] by orienting urbanization towards areas where the soils are better suited to agriculture.

10.6. Conclusion

Although they are often overlooked, urban agricultural soils are highly variable in quality. These soils undergo biological, physical and chemical modification due to human activities, both past and present, that occur within the space of urban agriculture (e.g. irrigation and input of organic matter), as well as activities outside this space, related to their location within the urban matrix (e.g. road traffic).

Urban agronomic practices that could modify the functions carried out by urban agriculture involve significant challenges in management, especially in the context of the development of urban agriculture.

The significant development of constructed soils within urban agriculture is also a challenge. This is an emerging topic of research and we need more knowledge on its functioning and its impact on environment and health.

While decision-making tools are already available for planning urban agriculture projects (DESTISOL'AU), it is nonetheless important to develop new tools that can be directly accessed by project leaders in order to allow them to better manage these soils and the services provided.

---

11 Nature4Cities, available at: https://www.cerema.fr/fr/actualites/plateforme-aide-decision-mettre-oeuvre-solutions-fondees.

12 Available at: https://www.inrae.fr/actualites/Destisol%27AU.

13 Available at: http://multimedia.ademe.fr/catalogues/CTecosystemes/fiches/methode23p8586.pdf.

Urban agricultural soils form an environment that can host high biodiversity, produce food-based biomass and contribute to the regulation of the climate, waste and water.

Urban agricultural soils, whether on open land or in built-up areas, appear to be an essential component in the functioning of a city to enable it to cope with global changes.

10.7. References

Aerts, R., Dewaelheyns, V., Achten, W.M.J. (2016). Potential ecosystem services of urban agriculture: A review. *PeerJ Preprints*, 4, e2286v1.

Antisari, L.V., Orsini, F., Marchetti, L., Vianello, G., Gianquinto, G. (2015). Heavy metal accumulation in vegetables grown in urban gardens. *Agronomy for Sustainable Development*, 35(3), 1139–1147.

Appolloni, E., Orsini, F., Specht, K., Thomaier, S., Sanye-Mengual, E., Pennisi, G., Gianquinto, G. (2021). The global rise of urban rooftop agriculture: A review of worldwide cases. *Journal of Cleaner Production*, 296, 126556.

Aubry, C. and Pourias, J. (2013). L'agriculture urbaine fait déjà partie du "métabolisme urbain". In *Natures urbaines : l'agriculture au cœur métropoles ? Nature et agriculture pour la ville. Les nouveaux désirs des citadins s'imposent.*, Torre, A. (ed.) Déméter, Paris.

Baize, D. and Girard, M.-C. (2009). *Référentiel pédologique 2008*. Éditions Quæ, Versailles.

Balvanera, P., Siddique, I., Dee, L., Paquette, A., Isbell, F., Gonzalez, A., Byrnes, J., O'Connor, M.I., Hungate, B.A., Griffin, J.N. (2014). Linking biodiversity and ecosystem services: Current uncertainties and the necessary next steps. *BioScience*, 64, 49–57. doi: 10.1093/biosci/bit003.

Bechet, B., Carré, F., Florentin, L., Leyval, C., Montanarella, L., Morel, J.L., Raimbault, G., Rodrigues, F., Rossignol, J., Schwartz, C. (2009). Caractérisation et fonctionnement des sols urbains. In *Sous les pavés, la terre*, Cheverry, C. and Gascuel, C. (eds). Le Seuil, Paris.

Bechet, B., Joimel, S., Jean-Soro, L., Hursthouse, A., Agboola, A., Leitão, T.E., Costa, H., do Rosário Cameira, M., Le Guern, C., Schwartz, C. (2018). Spatial variability of trace elements in allotment gardens of four European cities: Assessments at city, garden, and plot scale. *Journal of Soils and Sediments*, 18(2), 391–406.

Bell, S., Fox-Kämper, R., Keshavarz, N., Benson, M., Caputo, S., Noori, S., Voigt, A. (eds). (2016). *Urban Allotment Gardens in Europe*. Routledge, London.

Belon, E., Boisson, M., Deportes, I.Z., Eglin, T.K., Feix, I., Bispo, A.O., Galsomies, L., Leblond, S., Guellier, C.R. (2012). An inventory of trace elements inputs to French agricultural soils. *Science of the Total Environment*, 439, 87–95.

Boerema, A., Rebelo, A.J., Bodi, M.B., Esler, K.J., Meire, P. (2017). Are ecosystem services adequately quantified? *Journal of Applied Ecology*, 54(2), 358–370.

Burghardt, W. and Schneider, T. (2018). Bulk density and content, density and stock of carbon, nitrogen and heavy metals in vegetable patches and lawns of allotments gardens in the northwestern Ruhr area, Germany. *Journal of Soils and Sediments*, 18(2), 407–417.

Cambier, P., Schvartz, C., van Oort, F. (2009). *Contaminations métalliques des agrosystèmes et écosystèmes péri-industriels*. Éditions Quæ, Versailles.

Cambou, A., Shaw, R.K., Huot, H., Vidal-Beaudet, L., Hunault, G., Cannavo, P., Nold, F., Schwartz, C. (2018). Estimation of soil organic carbon stocks of two cities, New York City and Paris. *Science of the Total Environment*, 644, 452–464.

Chalas, Y. (1997). Territoires contemporains et représentations : des vieux paradigmes urbanistiques aux nouvelles figures de la ville. *Revue de géographie alpine*, 85(4), 11–36.

Chang, M. and Morel, K. (2018). Reconciling economic viability and socio-ecological aspirations in London urban microfarms. *Agronomy for Sustainable Development*, 38(1). doi: 10.1007/s13593-018-0487-5.

Cheng, Z., Paltseva, A., Li, I., Morin, T., Huot, H., Egendorf, S., Su, Z., Yolanda, R., Singh, K., Lee, L. (2015). Trace metal contamination in New York City garden soils. *Soil Science*, 180(4/5), 167–174.

Choay, F. (1972). Sémiologie et urbanisme. *Le sens de la ville*, 11–30.

Consalès, J.N. (2004). Les jardins familiaux à Marseille, Gênes et Barcelone : laboratoires territoriaux de l'agriculture urbaine dans l'Arc Méditerranéen. PhD Thesis, Université Aix-Marseille 1, Marseille.

Consalès, J.N., Joimel, S., Cordier, F., Jareno, C., Chenot, E. (2016). De l'argument à l'action : la biodiversité au service des jardins familiaux. *Projets de paysage*, 13. doi: 10.4000/paysage.9603.

Csortan, G., Ward, J., Roetman, P. (2020). Productivity, resource efficiency and financial savings: An investigation of the current capabilities and potential of South Australian home food gardens. *PLoS One*, 15(4), e0230232.

Decaëns, T., Jiménez, J.J., Gioia, C., Measey, G.J., Lavelle, P. (2006). The values of soil animals for conservation biology. *European Journal of Soil Biology*, 42(suppl.1), S23–S38. doi: 10.1016/j.ejsobi.2006.07.001.

Douay, F., Roussel, H., Fourrier, H., Heyman, C., Chateau, G. (2007). Investigation of heavy metal concentrations on urban soils, dust and vegetables nearby a former smelter site in Mortagne du Nord, Northern France. *Journal of Soils and Sediments*, 7(3), 143–146.

Draper, C. and Freedman, D. (2010). Review and analysis of the benefits, purposes, and motivations associated with community gardening in the United States. *Journal of Community Practice*, 18(4), 458–492.

Duchemin, E., Wegmuller, F., Legault, A.-M. (2009). Urban agriculture: Multi-dimensional tools for social development in poor neighbourhoods. *Field Actions Science Reports: The Journal of Field Actions*, 1, 43–52.

Edmondson, J.L., Childs, D.Z., Dobson, M.C., Gaston, K.J., Warren, P.H., Leake, J.R. (2020). Feeding a city – Leicester as a case study of the importance of allotments for horticultural production in the UK. *Science of The Total Environment* 705, 135930. doi: 10.1016/j.scitotenv.2019.135930.

El Hamiani, O., El Khalil, H., Lounate, K., Sirguey, C., Hafidi, M., Bitton, G., Schwartz, C., Boularbah, A. (2010). Toxicity assessment of garden soils in the vicinity of mining areas in Southern Morocco. *Journal of Hazardous Materials*, 177(1–3), 755–761.

El Khalil, H., El Hamiani, O., Bitton, G., Ouazzani, N. Boularbah, A. (2008). Heavy metal contamination from mining sites in South Morocco: Monitoring metal content and toxicity of soil runoff and groundwater. *Environmental Monitoring and Assessment*, 136(1), 147–160.

Grard, B.J.-P., Bel, N., Marchal, N., Madre, N., Castell, J.F., Cambier, P., Houot, S., Manoucheri, N., Besancon, S., Michel, J.C. et al. (2015). Recycling urban waste as possible use for rooftop vegetable garden. *Future of Food: J. Food Agric. Soc.*, 3, 21–34.

Grard, B.J.-P., Chenu, C., Manouchehri, N., Houot, S., Frascaria-Lacoste, N., Aubry, C. (2017). Rooftop farming on urban waste provides many ecosystem services. *Agron. Sustain. Dev.*, 38, 2. doi: 10.1007/s13593-017-0474-2.

Grard, B.J.-P., Chenu, C., Manouchehri, N., Houot, S., Frascaria-Lacoste, N., Aubry, C. (2018). Rooftop farming on urban waste provides many ecosystem services. *Agronomy for Sustainable Development*, 38(1), 1–12.

Grard, B.J.-P., Manouchehri, N., Aubry, C., Frascaria-Lacoste, N., Chenu, C. (2020). Potential of technosols created with urban by-products for rooftop edible production. *International Journal of Environmental Research and Public Health*, 17(9), 3210.

Grard, B.J.-P., Joimel, S., Vieublé-Gonod, L., Giacche, G., Aubry, C., Consales, J.H., Séré, G., Manouchehri, N., Haudin, C.S., Auclerc, A. et al. (2021). Evaluation des services écosystémiques fournis par les sols de micro-fermes urbaines : méthodologie et retours d'expériences. *Étude et gestion des sols*, 28, 31–47.

Grard, B.J.-P., Chenu, C., Boulanger-Joimel, S., Séré, G., Giacchè, G., Aubry, C., Manouchehri, N., Lagneau, A., Gonod, L.V. Consalès, J.-N. (2022). Evaluation des services écosystémiques rendus par les micro-fermes urbaines et leurs sols – bilan du projet de recherche semoirs [2018–2021]. Research Report, INRAE, Agroparistech, Université de Lorraine, Aix-Marseille Université, Institut d'aménagement et d'urbanisme de la région d'Ile-de-France (IAURIF).

Guitart, D., Pickering, C., Byrne, J. (2012). Past results and future directions in urban community gardens research. *Urban Forestry & Urban Greening*, 11(4), 364–373.

He, Z.L., Yang, X.E., Stoffella, P.J. (2005). Trace elements in agroecosystems and impacts on the environment. *Journal of Trace Elements in Medicine and Biology*, 19, 125–140. doi: 10.1016/j.jtemb.2005.02.010.

IUSS Working Group WRB (2014). World reference base for soil resources. International soil classification system for naming soils and creating legends for soil maps: Updated 2015. Report, Food and Agriculture Organization of the United Nations, Rome.

Jean-Soro, L., Le Guern, C., Bechet, B., Lebeau, T., Ringeard, M.-F. (2015). Origin of trace elements in an urban garden in Nantes, France. *Journal of Soils and Sediments*, 15(8), 1802–1812.

Joimel, S. (2015). Biodiversité et caractéristiques physico-chimiques des sols de jardins associatifs urbains français. PhD Thesis, Université de Lorraine, Lorraine.

Joimel, S., Cortet, J., Jolivet, C.C., Saby, N.P.A., Chenot, E.-D., Branchu, P., Consalès, J.-N., Lefort, C., Morel, J.-L., Schwartz, C. (2016). Physico-chemical characteristics of topsoil for contrasted forest, agricultural, urban and industrial land uses in France. *Science of the Total Environment*, 545, 40–47.

Joimel, S., Schwartz, C., Hedde, M., Kiyota, S., Krogh, P.H., Nahmani, J., Pérès, G., Vergnes, A., Cortet, J. (2017). Urban and industrial land uses have a higher soil biological quality than expected from physicochemical quality. *Science of the Total Environment*, 584, 614–621.

Joimel, S., Grard, B., Auclerc, A., Hedde, M., Le Doaré, N., Salmon, S., Chenu, C. (2018). Are Collembola "flying" onto green roofs? *Ecological Engineering*, 111, 117–124.

Joimel, S., Schwartz, C., Bonfanti, J., Hedde, M., Krogh, P.H., Peres, G., Pernin, C., Rakoto, A., Salmon, S., Santorufo, L. et al. (2021a). Functional and taxonomic diversity of collembola as complementary tools to assess land use effects on soils biodiversity. *Frontiers in Ecology and Evolution*, 9, 630919. doi: 10.3389/fevo.2021.630919.

Joimel, S., Cortet, J., Consalès, J.N., Branchu, P., Haudin, C.-S., Morel, J.L., Schwartz, C. (2021b). Contribution of chemical inputs on the trace elements concentrations of surface soils in urban allotment gardens. *Journal of Soils and Sediments*, 21(1), 328–337.

Joimel S., Grard B., Vieublé-Gonod L., Chenu C. (2021c). Fonctionnement écologique des villes : et si on pensait aux sols ? *Metropolitiques* [Online]. Available at: https://metropolitiques.eu/Le-fonctionnement-ecologique-des-villes-et-si-on-pensait-auxsols.html.

Joimel, S., Grard, B., Chenu, C., Cheval, P., Mondy, S., Lelièvre, M., Auclerc, A., Vieublé-Gonod, L. (2022a). One green roof type, one Technosol, one ecological community. *Ecological Engineering*, 175, 106475. doi: 10.1016/j.ecoleng.2021.106475.

Joimel, S., Jules, A., Vieublé-Gonod, L. (2022b). Collembola dispersion, selection, and biological interactions in urban ecosystems: A review. *Environmental Chemistry Letters*, 20(3), 2123–2133. doi: 10.1007/s10311-022-01406-z.

Kim, B.F., Poulsen, M.N., Margulies, J.D., Dix, K.L., Palmer, A.M., Nachman, K.E. (2014). Urban community gardeners' knowledge and perceptions of soil contaminant risks. *PLoS One*, 9(2), e87913.

Lehmann, A. and Stahr, K. (2007). Nature and significance of anthropogenic urban soils. *Journal of Soils and Sediments*, 7(4), 247–260.

Lohrberg, F., Lička, L., Scazzosi, L., Timpe, A. (2016). *Urban Agriculture Europe*. Jovis, Berlin.

Lorenz, K. and Kandeler, E. (2005). Biochemical characterization of urban soil profiles from Stuttgart, Germany. *Soil Biology and Biochemistry*, 37(7), 1373–1385.

Madre, F., Vergnes, A., Machon, N., Clergeau, P. (2013). A comparison of 3 types of green roof as habitats for arthropods. *Ecological Engineering*, 57, 109–117.

Millennium Ecosystem Assessment (2005). *Ecosystems and Human Well-being*, volume 5. Island Press, Washington, D.C.

Mitchell, R.G., Spliethoff, H.M., Ribaudo, L.N., Lopp, D.M., Shayler, H.A., Marquez-Bravo, L.G., Lambert, V.T., Ferenz, G.S., Russell-Anelli, J.M., Stone, E.B. (2014). Lead (Pb) and other metals in New York City community garden soils: Factors influencing contaminant distributions. *Environmental Pollution*, 187, 162–169.

Morel, J.L. and Schwartz, C. (1999). Qualité et gestion des sols des jardins familiaux. *Comptes rendus de l'académie d'agriculture de France*, 85(2), 107–123.

Morel, J.L., Schwartz, C., Florentin, L., de Kimpe, C. (2005). *Urban Soils*. Elsevier, Amsterdam, and Academic Press, Cambridge.

Morel, J.L., Chenu, C., Lorenz, K. (2014). Ecosystem services provided by soils of urban, industrial, traffic, mining, and military areas (SUITMAs). *Journal of Soils and Sediments*, 15(8), 1659–1666.

Mougeot, L.J.A. (2000). Urban agriculture: Definition, presence, potentials and risks. *Growing Cities, Growing Food: Urban Agriculture on the Policy Agenda*, 1, 42.

Nehls, T., Rokia, S., Mekiffer, B., Schwartz, C., Wessolek, G. (2013). Contribution of bricks to urban soil properties. *Journal of Soils and Sediments*, 13(3), 575–584.

Poniży, L., Latkowska, M.J., Breuste, J., Hursthouse, A., Joimel, S., Külvik, M., Leitão, T.E., Mizgajski, A., Voigt, A., Kacprzak, E. (2021). The rich diversity of urban allotment gardens in Europe: Contemporary trends in the context of historical, socio-economic and legal conditions. *Sustainability*, 13(19), 11076.

Pourias, J. (2014). Production alimentaire et pratiques culturales en agriculture urbaine. Analyse agronomique de la fonction alimentaire des jardins associatifs urbains à Paris et Montréal. PhD Thesis, AgroParisTech and Université du Québec, Montreal.

Pourias, J., Duchemin, E., Aubry, C. (2015). Products from urban collective gardens: Food for thought or for consumption? Insights from Paris and Montreal. *Journal of Agriculture, Food Systems, and Community Development*, 5(2), 175–199.

Richards, P.J., Farrell, C., Tom, M., Williams, N.S.G., Fletcher, T.D. (2015). Vegetable raingardens can produce food and reduce stormwater runoff. *Urban Forestry & Urban Greening*, 14(3), 646–654.

Sansot, P. and Dufrenne, M. (2004). *Poétique de la ville*. Payot & Rivages, Paris.

Säumel, I., Kotsyuk, I., Hölscher, M., Lenkereit, C., Weber, F., Kowarik, I. (2012). How healthy is urban horticulture in high traffic areas? Trace metal concentrations in vegetable crops from plantings within inner city neighbourhoods in Berlin, Germany. *Environmental Pollution*, 165, 124–132.

Schröter, M., Van der Zanden, E.H., van Oudenhoven, A.P.E., Remme, R.P., Serna-Chavez, H.M., De Groot, R.S., Opdam, P. (2014). Ecosystem services as a contested concept: A synthesis of critique and counter-arguments. *Conservation Letters*, 7(6), 514–523.

Specht, K., Siebert, R., Hartmann, I., Freisinger, U.B., Sawicka, M., Werner, A., Thomaier, S., Henckel, D., Walk, H., Dierich, A. (2014). Urban agriculture of the future: An overview of sustainability aspects of food production in and on buildings. *Agriculture and Human Values*, 31(1), 33–51.

Szolnoki, Z.S., Farsang, A., Puskás, I. (2013). Cumulative impacts of human activities on urban garden soils: Origin and accumulation of metals. *Environmental Pollution*, 177, 106–115.

Takahashi, T., Amano, Y., Kuchimura, K., Kobayashi, T. (2008). Carbon content of soil in urban parks in Tokyo, Japan. *Landscape and Ecological Engineering*, 4(2), 139–142.

Vergnes, A., Blouin, M., Muratet, A., Lerch, T.Z., Mendez-Millan, M., Rouelle-Castrec, M., Dubs, F. (2017). Initial conditions during Technosol implementation shape earthworms and ants diversity. *Landscape and Urban Planning*, 159, 32–41. doi: 10.1016/j.landurbplan.2016.10.002.

Weidner, T., Yang, A., Hamm, M.W. (2019). Consolidating the current knowledge on urban agriculture in productive urban food systems: Learnings, gaps and outlook. *Journal of Cleaner Production*, 209, 1637–1655.

Whittinghill, L.J. and Sarr, S. (2021). Practices and barriers to sustainable urban agriculture: A case study of Louisville, Kentucky. *Urban Science*, 5(4), 92.

Whittinghill, L.J., Rowe, D.B., Schutzki, R., Cregg, B.M. (2014). Quantifying carbon sequestration of various green roof and ornamental landscape systems. *Landscape and Urban Planning*, 123, 41–48.

Whittinghill, L.J., Rowe, D.B., Andresen, J.A., Cregg, B.M. (2015). Comparison of stormwater runoff from sedum, native prairie, and vegetable producing green roofs. *Urban Ecosystems*, 18(1), 13–29.

Whittinghill, L.J., Hsueh, D., Culligan, P., Plunz, R. (2016). Stormwater performance of a full scale rooftop farm: Runoff water quality. *Ecological Engineering*, 91, 195–206.

Zhong, X., Joimel, S., Schwartz, C., Sterckeman, T. (2022). Assessing the future trends of soil trace metal contents in French urban gardens. *Environ. Sci. Pollut. Res.*, 29, 3900–3917. doi: 10.1007/s11356-021-15679-4.

List of Authors

Lionel ALLETTO
UMR AGIR
INRAE
Université de Toulouse
Castanet-Tolosan
France

Thierry BECQUER
UMR Eco&Sols
Institut Agro Montpellier
CIRAD
INRAE
IRD
Université de Montpellier
France

Yves COQUET
UMR ECOSYS
AgroParisTech
INRAE
Université Paris Saclay
Palaiseau
France

Isabelle COUSIN
Info&Sol
INRAE
Orléans
France

Tania DE ALMEIDA
IMT-Nord Europe
Université de Lille
and
JUNIA
ULR 4515 LGCgE
Université d'Artois
Arras
France

Jean-Marc GILLIOT
UMR ECOSYS
AgroParisTech
INRAE
Université Paris Saclay
Palaiseau
France

Laure Vieublé GONOD
UMR ECOSYS
AgroParisTech
INRAE
Université Paris Saclay
Palaiseau
France

Baptiste GRARD
Agroecology and Environment
Research Unit
ISARA
Lyon
France

Sophie JOIMEL
UMR ECOSYS
AgroParisTech
INRAE
Université Paris Saclay
Palaiseau
France

Philippe MARTIN
UMR SADAPT
AgroParisTech
INRAE
Université Paris Saclay
Palaiseau
France

Joël MICHELIN
UMR ECOSYS
AgroParisTech
INRAE
Université Paris Saclay
Palaiseau
France

David MONTAGNE
UMR ECOSYS
AgroParisTech
INRAE
Université Paris-Saclay
Palaiseau
France

Audrey NIBOYET
UMR iEES Paris
Sorbonne Université
Université Paris Cité
UPEC
CNRS
INRAE
IRD
Paris
and
AgroParisTech
Palaiseau
France

Cécile QUANTIN
UMR GEOPS, CNRS
Université Paris-Saclay
Orsay
France

Louis DE REDON
SESG
AgroParisTech
INRAE
Université Paris Saclay
Palaiseau
and
Institut de Recherche Juridique de la
Sorbonne (IRJS)
Paris
France

Ottone SCAMMACCA
UMR ECOSYS
AgroParisTech
INRAE
Université Paris-Saclay
Palaiseau
and
UMR Prodig
AgroParisTech
IRD, CNRS
Université Paris 1 Panthéon-Sorbonne
Aubervilliers
France

Emmanuelle VAUDOUR
UMR ECOSYS
AgroParisTech
INRAE
Université Paris Saclay
Palaiseau
France

Christian WALTER
UMR SAS
INRAE
Institut Agro Rennes-Angers
France

Index

A, B

C

D, E

F

G, H, I

K, L, M

N, O